Texts in Theoretical Computer Science
An EATCS Series

Springer
Berlin
Heidelberg
New York
Hong Kong
London
Milan
Paris
Tokyo

Simona Ronchi Della Rocca
Luca Paolini

The Parametric Lambda Calculus

A Metamodel for Computation

Springer

Authors

Prof. Simona Ronchi Della Rocca
Università di Torino
Dipartimento di Informatica
corso Svizzera 185
10149 Torino, Italy
ronchi@di.unito.it
www.di.unito.it/~ronchi

Dr. Luca Paolini
Università di Torino
Dipartimento di Informatica
corso Svizzera 185
10149 Torino, Italy
paolini@di.unito.it
www.di.unito.it/~paolini

Series Editors

Prof. Dr. Wilfried Brauer
Institut für Informatik der TUM
Boltzmannstr. 3, 85748 Garching, Germany
Brauer@informatik.tu-muenchen.de

Prof. Dr. Grzegorz Rozenberg
Leiden Institute of Advanced Computer Science
University of Leiden
Niels Bohrweg 1, 2333 CA Leiden, The Netherlands
rozenber@liacs.nl

Prof. Dr. Arto Salomaa
Turku Centre for Computer Science
Lemminkäisenkatu 14 A, 20520 Turku, Finland
asalomaa@utu.fi

Library of Congress Cataloging-in-Publication Data

Ronchi Della Rocca, S. (Simona)
The parametric lambda calculus : A metamodel for computation / Simona Ronchi Della Rocca, Luca Paolini.
p. cm. – (Texts in theoretical computer science)
Includes bibliographical references and index.

1. Lambda calculus. I. Paolini, Luca, 1970– II. Title. III. Series.
QA9.5.R66 2004 511.3'5–dc22 2003069100

ACM Computing Classification (1998): F.4, F.3, I.2.3, D.2

ISBN 978-3-642-05746-5

Springer-Verlag is a part of Springer Science+Business Media

springeronline.com

© Springer-Verlag Berlin Heidelberg 2010
Printed in Germany

The use of general descriptive names, trademarks, etc. in this publication does not imply, even in the absence of a specific statement, that such names are exempt from the relevant protective laws and therefore free for general use.

Cover Design: KünkelLopka, Heidelberg

Printed on acid-free paper 45/3142/GF - 5 4 3 2 1 0

To Corrado Böhm, from which Simona and –
by transitivity – Luca learned the pleasure
of research and the interest in λ-calculus

Preface

The λ-calculus was invented by Church in the 1930s with the purpose of supplying a logical foundation for logic and mathematics [25]. Its use by Kleene as a coding for computable functions makes it the first programming language, in an abstract sense, exactly as the Turing machine can be considered the first computer machine [57]. The λ-calculus has quite a simple syntax (with just three formation rules for terms) and a simple operational semantics (with just one operation, substitution), and so it is a very basic setting for studying computation properties.

The first contact between λ-calculus and real programming languages was in the years 1956-1960, when McCarthy developed the LISP programming language, inspired from λ-calculus, which is the first "functional" programming language, i.e., where functions are first-class citizens [66]. But the use of λ-calculus as an abstract paradigm for programming languages started later as the work of three important scientists: Strachey, Landin and Böhm. Strachey used the λ-notation as a descriptive tool to represent functional features in programming when he posed the basis for a formal semantics of programming languages [92]. Landin formalized the idea that the semantics of a programming language can be given by translating it into a simpler language that is easier to understand. He identified such a target language in λ-calculus and experimented with this idea by giving a complete translation of ALGOL60 into λ-calculus [64]. Moreover, he declared in [65] that a programming language is nothing more than λ-calculus plus some "syntactic sugar". Böhm was the first to use λ-calculus as an effective programming language, defining, with W. Gross, the CUCH language, which is a mixture of λ-calculus and the Curry combinators language, and showing how to represent in it the most common data structures [19].

But, until the end of the 1960s, λ-calculus suffered from the lack of a formal semantics. In fact, while it was possible to codify in it all the computable functions, the meaning of a generic λ-term not related to this coding was unclear. The attempt to interpret λ-terms as set-theoretic functions failed, since it would have been necessary to interpret it into a set D isomorphic to the set of functions from D to D, which is impossible since the two spaces always have different cardinality. Scott [88, 89] solved the problem by interpreting λ-calculus in a lattice isomorphic to the space of its continuous functions, thus giving it a clear mathematical interpretation. So the technique of interpretation by translation, first developed by Landin, became a standard tool to study the denotational semantics of programming languages; almost all textbooks in denotational semantics follow this approach [91, 98].

But there was a gap between λ-calculus and the real functional programming languages. The majority of real functional languages have a "call-by-value" parameter passing policy, i.e., parameters are evaluated before being passed to a function, while the reduction rule of λ-calculus reflects a "call-by-name" policy, i.e., a policy where parameters are passed without being evaluated. In the folklore there was the idea that a call-by-value behaviour could be mimicked in λ-calculus just by defining a suitable reduction strategy. Plotkin proved that this intuition was wrong and that λ-calculus is intrinsically call-by-name [78]. So, in order to describe the call-by-value evaluation, he proposed a different calculus, which has the same syntax as λ-calculus, but a different reduction rule.

The aim of this book is to introduce both the call-by-name and the call-by-value λ-calculi and to study their syntactical and semantical properties, on which their status of paradigmatic programming languages is based. In order to study them in a uniform way we present a new calculus, the $\lambda\Delta$-calculus, whose reduction rule is parametric with respect to a subset Δ of terms (called the set of input values) that enjoy some suitable conditions. Different choices of Δ allow us to define different languages, in particular the two λ-calculus variants we are speaking about. The most interesting feature of $\lambda\Delta$-calculus is that it is possible to prove important properties (like confluence) for a large class of languages in just one step. We think that $\lambda\Delta$-calculus can be seen as the foundation of functional programming.

Organization of the Book

The book is divided into four parts, each one composed of different chapters. The first part is devoted to the study of the syntax of $\lambda\Delta$-calculus. Some syntactical properties, like confluence and standardization, can be studied for the whole Δ class. Other properties, like solvability and separability, cannot be treated in a uniform way, and they are therefore introduced separately for different instances of Δ.

In the second part the operational semantics of $\lambda\Delta$-calculus is studied. The notion of operational semantics can be given in a parametric way, by supplying not only a set of input values but also a set of output values Θ, enjoying some very natural properties. A universal reduction machine is defined, parametric into both Δ and Θ, enjoying a sort of correctness property in the sense that, if a term can be reduced to an output value, then the machine stops, returning a term operationally equivalent to it. Then four particular reduction machines are presented, three for the call-by-name λ-calculus and one for the call-by-value λ-calculus, thereby presenting four operational behaviours that are particularly interesting for modeling programming languages. Moreover, the notion of extensionality is revised, giving a new parametric definition that depends on the operational semantics we want to consider.

The third part is devoted to denotational semantics. The general notion of a model of $\lambda\Delta$-calculus is defined, and then the more restrictive and useful notion of a filter model, based on intersection types, is given. Then four particular filter models are presented, each one correct with respect to one of the operational semantics studied in the previous part. For two of them completeness is also proved. The other two models are incomplete: we prove that there are no filter models enjoying the completeness property with respect to given operational semantics, and we build two complete models by using a technique based on intersection types. Moreover, the relation between the filter models and Scott's models is given.

The fourth part deals with the computational power of $\lambda\Delta$-calculus. It is well known that λ-calculus is Turing complete, in both its call-by-name and call-by-value variants, i.e. it has the power of the computable functions. Here we prove something more, namely that each one of the reduction machines we present in the third part of this book can be used for computing all the computable functions.

Use of the Book

This book is dedicated to researchers, and it can be used as a textbook for master's or PhD courses in Foundations of Computer Science. Moreover, we wish to advise the reader that its aim is not to cover all possible topics concerning λ-calculus, but just those syntactical and semantics properties which can be used as tools for the foundation of programming languages. The reader interested in studying λ-calculus in itself can use the classical textbook by Barendregt [9], or other more descriptive ones such as [51] or [60]. The reader interested in a typed approach can read Mitchell's text [69] for an introduction, in which two chapters are dedicated to simply typed λ-calculus and its model, and the book of Hindley for a complete development of the topic [49].

Acknowledgement. Both authors would like to thank all the people of the "lambda-group" at the Dipartimento di Informatica of the Università di Torino for their support and collaboration. Moreover they are grateful to Roger Hindley and Elaine Pimentel for pointing out some inaccuracies. Luca Paolini thanks Pino Rosolini for the useful and interesting discussions about the topics of this book. Simona Ronchi Della Rocca did the final revision of the book during a sabbatical period. Some friends offered her hospitality and a stimulating scientific environment: Betti Venneri, Gigi Liquori, Rocco De Nicola, Pierre Lescanne and Philippe De Groote. To all of them she wants to express her gratitude. Last but not least, both the authors thank the publisher Ingeborg Mayer, whose patient assistance made possible the publication of this book.

Torino, May 2004

Simona Ronchi Della Rocca
Luca Paolini

Contents

Part I. Syntax

1. The Parametric λ-Calculus 3
1.1 The Language of λ-Terms 3
1.2 The $\lambda\Delta$-Calculus 6
1.2.1 Proof of Confluence and Standardization Theorems 14
1.3 Δ-Theories 21

2. The Call-by-Name λ-Calculus 25
2.1 The Syntax of $\lambda\Lambda$-Calculus 25
2.1.1 Proof of Λ-Solvability Theorem 27
2.1.2 Proof of Böhm's Theorem 28

3. The Call-by-Value λ-Calculus 35
3.1 The Syntax of the $\lambda\Gamma$-Calculus 35
3.1.1 $\Xi\ell$-Confluence and $\Xi\ell$-Standardization 41
3.1.2 Proof of Potential Γ-Valuability and Γ-Solvability Theorems 43
3.1.3 Proof of Γ-Separability Theorem 49
3.2 Potentially Γ-Valuable Terms and Λ-Reduction 58

4. Further Reading 61

Part II. Operational Semantics

5. Parametric Operational Semantics 65
5.1 The Universal Δ-Reduction Machine 70

6. Call-by-Name Operational Semantics 73
6.1 **H**-Operational Semantics 73
6.2 **N**-Operational Semantics 77
6.3 **L**-Operational Semantics 81
6.3.1 An Example 85

7. Call-by-Value Operational Semantics 89
7.1 **V**-Operational Semantics 89
7.1.1 An Example 93

8. Operational Extensionality 95
8.1 Operational Semantics and Extensionality 95
8.1.1 Head-Discriminability 99

9. Further Reading 101

Part III. Denotational Semantics

10. $\lambda\Delta$-Models 105
10.1 Filter $\lambda\Delta$-Models 108

11. Call-by-Name Denotational Semantics 119
11.1 The Model $\mathcal{H}$ 119
11.1.1 The $\leq_\infty$-Intersection Relation 129
11.1.2 Proof of the $\mathcal{H}$-Approximation Theorem 132
11.1.3 Proof of Semiseparability, $\mathcal{H}$-Discriminability and $\mathcal{H}$-Characterization Theorems 136
11.2 The Model $\mathcal{N}$ 144
11.2.1 The $\leq_{\bowtie}$-Intersection Relation 151
11.2.2 Proof of $\mathcal{N}$-Approximation Theorem 154
11.2.3 Proof of $\mathcal{N}$-Discriminability and $\mathcal{N}$-Characterization Theorems 157
11.3 The Model $\mathcal{L}$ 162
11.3.1 Proof of $\mathcal{L}$-Approximation Theorem 168
11.3.2 Proof of Theorems 11.3.15 and 11.3.16 170
11.4 A Fully Abstract Model for the **L**-Operational Semantics 172
11.5 Crossing Models 178
11.5.1 The Model $\mathcal{H}$ 178
11.5.2 The Model $\mathcal{N}$ 179
11.5.3 The Model $\mathcal{L}$ 179

12. Call-by-Value Denotational Semantics 181
12.1 The Model $\mathcal{V}$ 181
12.1.1 The $\leq_{\surd}$-Intersection Relation 190
12.1.2 Proof of Theorem 12.1.6 192
12.1.3 Proof of the $\mathcal{V}$-Approximation Theorem 195
12.1.4 Proof of Theorems 12.1.24 and 12.1.25 198
12.2 A Fully Abstract Model for the **V**-Operational Semantics 201

13. Filter $\lambda\Delta$-Models and Domains ... 207
13.1 Domains ... 207
13.1.1 $\mathcal{H}$ as Domain ... 214
13.1.2 $\mathcal{N}$ as Domain ... 216
13.1.3 $\mathcal{L}$ as Domain ... 217
13.1.4 $\mathcal{V}$ as Domain ... 218
13.1.5 Another Domain ... 219

14. Further Reading ... 221

Part IV. Computational Power

15. Preliminaries ... 225
15.1 Kleene's Recursive Functions ... 225
15.2 Representing Data Structures ... 227

16. Representing Functions ... 233
16.1 Call-by-Name Computational Completeness ... 233
16.2 Call-by-Value Computational Completeness ... 237
16.3 Historical Remarks ... 239

Bibliography ... 241

Index ... 247

Part I

Syntax

1. The Parametric λ-Calculus

A calculus is a language equipped with some reduction rules. All the calculi we consider in this book share the same language, which is the language of λ-calculus, while they differ each other in their reduction rules. In order to treat them in an uniform way we define a parametric calculus, the $\lambda\Delta$-calculus, which gives rise to different calculi by different instantiations of the parameter Δ. In Part I we study the syntactical properties of the $\lambda\Delta$-calculus, and in particular those of its two most important instances, the call-by-name and the call-by-value λ-calculi. The $\lambda\Delta$-calculus has been introduced first in [85] and further studied in [74]. We use the terminology of [9].

1.1 The Language of λ-Terms

Definition 1.1.1 (The language Λ).
Let Var *be a countable set of variables. The set Λ of λ-terms is a set of words on the alphabet* $\mathrm{Var} \cup \{\ (\ ,\ )\ ,\ .\ ,\ \lambda\ \}$ *inductively defined as follows:*

- $x \in \mathrm{Var}$ *implies* $x \in \Lambda$,
- $M \in \Lambda$ *and* $x \in \mathrm{Var}$ *implies* $(\lambda x.M) \in \Lambda$ (abstraction),
- $M \in \Lambda$ *and* $N \in \Lambda$ *implies* $(MN) \in \Lambda$ (application).

λ-terms will be ranged over by Latin capital letters. Sets of λ-terms will be denoted by Greek capital letters.

Sometimes, we will refer to λ-terms simply as terms. The symbol $\equiv$ will denote syntactical identity of terms. We will use the following abbreviations in order to avoid an excessive number of parenthesis: $\lambda x_1 ... x_n.M$ will stand for $(\lambda x_1.(...(\lambda x_n.M)...))$, and $MN_1N_2...N_n$ will stand for $(...((MN_1)N_2)...N_n)$. Moreover, $\vec{M}$ will denote a sequence of terms $M_1, \ldots, M_n$ for some $n \geq 0$, and $\lambda\vec{x}.M$ and $\vec{M}\vec{N}$ will denote respectively $\lambda x_1 \ldots x_n.M$ and $M_1 \ldots M_m N_1 \ldots N_n$ for some $n, m \geq 0$. The length of the sequence $\vec{N}$ is denoted by $\|\vec{N}\|$. By abusing the notation, $N \in \vec{N}$ denotes that the term N occurs in the sequence $\vec{N}$.

Example 1.1.2. $\lambda x.xx$, $\lambda x.x(\lambda z.zy)$, $\lambda y.(\lambda x.x)(\lambda uv.u)$ are examples of λ-terms. Some λ-terms have standard names for historical reasons. The names

that will be extensively used in this book are:
$I \equiv \lambda x.x$, $K \equiv \lambda xy.x$, $O \equiv \lambda xy.y$, $D \equiv \lambda x.xx$, $E \equiv \lambda xy.xy$.

Definition 1.1.3 (Subterms).
A term N is a subterm *of M if and only if one of the following conditions arises:*

- $M \equiv N$,
- $M \equiv \lambda x.M'$ *and N is a subterm of M',*
- $M \equiv PQ$ *and N is a subterm either of P or of Q.*

A term N occurs in a term M if and only if N is a subterm of M.

Example 1.1.4. The set of subterms of the term $\lambda x.x(\lambda z.zy)$ is

$$\{\lambda x.x(\lambda z.zy), x(\lambda z.zy), \lambda z.zy, zy, x, z, y\}.$$

The symbol "λ" plays the role of binder for variables, as formalized in the next definition.

Definition 1.1.5 (Free variables).

(i) *The set of* free variables *of a term M, denoted by* $\mathrm{FV}(M)$, *is inductively defined as follows:*
- $M \equiv x$ *implies* $\mathrm{FV}(M) = \{x\}$,
- $M \equiv \lambda x.M'$ *implies* $\mathrm{FV}(M) = \mathrm{FV}(M') - \{x\}$,
- $M \equiv PQ$ *implies* $\mathrm{FV}(M) = \mathrm{FV}(P) \cup \mathrm{FV}(Q)$.

A variable is bound *in M when it is not free in M.*

(ii) *A term M is* closed *if and only if* $\mathrm{FV}(M) = \emptyset$. *A term is* open *if it is not closed. For every subset of terms $\Theta \subseteq \Lambda$, we will denote with Θ^0 the restriction of Θ to closed terms.*

Example 1.1.6. $\mathrm{FV}(\lambda z.(\lambda x.x(\lambda z.zy))(\lambda xyz.yz)) = \{y\}$, $\mathrm{FV}(\lambda z.x(\lambda x.xy)) = \{x, y\}$, and $\mathrm{FV}((\lambda yx.x)y) = \{y\}$.

The replacement of a free variable by a term is the basic syntactical operation on Λ on which the definition of reduction rules will be based. But the replacement must respect the status of the variables: e.g., x can be replaced by $M \equiv \lambda y.zy$ in $\lambda u.xu$, so obtaining the term $\lambda u.(\lambda y.zy)u$, while the same replacement cannot take place in the term $\lambda z.xz$, since in the obtained term $\lambda z.(\lambda y.zy)z$ the free occurrence of z in M would become bound. The notion is formalized in the next definition.

Definition 1.1.7. *The statement "M* is free for *x* in *N" is defined by induction on N as follows:*

- *M is free for x in x;*
- *M is free for x in y;*
- *If M is free for x both in P and Q then M is free for x in PQ;*

- *If M is free for x in N and $x \not\equiv y$ and $y \notin \mathrm{FV}(M)$ then M is free for x in $\lambda y.N$.*

Example 1.1.8. $\lambda xy.xz$ is free for x and y in $(\lambda u.x)(\lambda u.xz)$ but is not free for u in both $\lambda xz.u$ and $\lambda zu.u$.

Let M be free for x in N; so $N[M/x]$ denotes the *simultaneous replacement* of all free occurrences of x in N by M. Clearly,

$$\mathrm{FV}(N[M/x]) = \begin{cases} \mathrm{FV}(N) & \text{if } x \notin \mathrm{FV}(N), \\ (\mathrm{FV}(N) - \{x\}) \cup \mathrm{FV}(M) & \text{otherwise.} \end{cases}$$

For example, $(\lambda x.u(xy))[xy/u]$ is not defined because xy is not free for u in $\lambda x.u(xy)$, while $(\lambda x.u(xu))[u(\lambda z.z)/u] \equiv \lambda x.u(\lambda z.z)(xu(\lambda z.z))$.

Let $\|\vec{N}\| = \|\vec{x}\|$; both $\vec{M}[N_1/x_1, ..., N_n/x_n]$ and $\vec{M}[\vec{N}/\vec{x}]$ are abbreviations for the simultaneous replacement of x_i by N_i in every M_j $(0 \leq i \leq \|\vec{x}\| = n$, $0 \leq j \leq \|\vec{M}\|)$.

In the standard mathematical notation, the name of a bound variable is meaningless; for example, $\sum_{1 \leq i \leq n} i$ and $\sum_{1 \leq j \leq n} j$ both denote the sum of the first n natural numbers. Also in the language Λ, it is natural to consider the terms modulo names of bound variables. The renaming is formalized in the next definition.

Definition 1.1.9 (α-Reduction).

(i) $\lambda x.M \to_\alpha \lambda y.M[y/x]$ *if y is free for x in M and $y \notin \mathrm{FV}(M)$.*
(ii) $=_\alpha$ *is the reflexive, symmetric, transitive and contextual closure of* $\to_\alpha$.

Example 1.1.10. $\lambda x.x =_\alpha \lambda y.y =_\alpha \lambda z.z$, $\lambda xy.x =_\alpha \lambda xz.x$ and $\lambda xy.x =_\alpha \lambda yx.y$. On the other hand, $\lambda x.y \neq_\alpha \lambda x.x$ and $\lambda x.yx \neq_\alpha \lambda y.yy$.

In the entire book, we will consider terms modulo $=_\alpha$.

Thus we can also safely extend the notation $N[M/x]$ to the case where M is not free for x in N. In this case $N[M/x]$ denotes the result of replacing x by M in a term $N' =_\alpha N$ such that M is free for x in N'. Clearly such an N' always exists and the notation is well posed. So $(\lambda x.u(xy))[xy/u]$ is α-equivalent to the term $\lambda z.xy(zy)$.

An alternative way of denoting a simultaneous replacement is by explicitly using the notion of substitution. A *substitution* is a function from variables to terms. If $\mathbf{s}$ is a substitution and $\mathrm{FV}(M) = \{x_1, ..., x_n\}$, $\mathbf{s}(M)$ denotes $M[\mathbf{s}(x_1)/x_1, ..., \mathbf{s}(x_n)/x_n]$.

An important syntactical tool that will be extensively used in the following chapters is the notion of *context*. Informally, a context is a term that can contain some occurrences of a *hole* (denoted by the constant [.]) that can be filled by a term.

Definition 1.1.11 (Context).
Let Var *be a countable set of variables, and* $[.]$ *be a constant* (the hole).

(i) *The set* Λ_C *of* contexts *is a set of words on* $\mathrm{Var} \cup \{ \ (\ , \ ) \ , \ . \ , \ \lambda \ , \ [.] \ \}$ *inductively defined as follows:*
- $[.] \in \Lambda_C$;
- $x \in \mathrm{Var}$ *implies* $x \in \Lambda_C$;
- $C[.] \in \Lambda_C$ *and* $x \in \mathrm{Var}$ *implies* $(\lambda x.C[.]) \in \Lambda_C$;
- $C_1[.] \in \Lambda_C$ *and* $C_2[.] \in \Lambda_C$ *implies* $(C_1[.]C_2[.]) \in \Lambda_C$.

Contexts will be denoted by $C[.], C'[.], C_1[.], \ldots$.

(ii) *A context of the shape:* $(\lambda\vec{x}.[.])\vec{P}$ *is a* head context.

(iii) *Let* $C[.]$ *be a context and* M *be a term. Then* $C[M]$ *denotes the term obtained by replacing by* M *every occurrence of* $[.]$ *in* $C[.]$.

We will use the same abbreviation notation for contexts that we used for terms.

Note that filling a hole in a context is *not* a substitution; in fact, free variables in M can become bound in $C[M]$. For example, filling the hole of $\lambda x.[.]$ with the free variable x results in the term $\lambda x.x$.

1.2 The λΔ-Calculus

We will present some λ-calculi, all based on the language Λ, defined in the previous section, each one characterized by different reduction rules.

The λΔ-calculus is the language Λ equipped with a set $\Delta \subseteq \Lambda$ of input values, satisfying some closure conditions. Informally, input values represent partially evaluated terms that can be passed as parameters. Call-by-name and call-by-value parameter passing can be seen as the two most radical choices; parameters are not evaluated in the former policy, while in the latter they are evaluated until an output result is reached.

Most of the known variants of λ-calculus can be obtained from this parametric calculus by instantiating Δ in a suitable way. The set Δ of input values and the reduction $\rightarrow_\Delta$ induced by it are defined in Definition 1.2.1.

Definition 1.2.1. *Let* $\Delta \subseteq \Lambda$.

(i) *The* Δ-reduction *(*$\rightarrow_\Delta$*) is the contextual closure of the following rule:*

$$(\lambda x.M)N \rightarrow M[N/x] \text{ if and only if } N \in \Delta.$$

$(\lambda x.M)N$ *is called a* Δ-redex *(or simply redex), and* $M[N/x]$ *is called its* Δ-contractum *(or simply contractum).*

(ii) $\rightarrow^*_\Delta$ *and* $=_\Delta$ *are respectively the reflexive and transitive closure of* $\rightarrow_\Delta$ *and the symmetric, reflexive and transitive closure of* $\rightarrow_\Delta$.

(iii) *A set* $\Delta \subseteq \Lambda$ *is said* set of input values *when the following conditions are satisfied:*

- $\mathrm{Var} \subseteq \Delta$ (Var-closure);
- $P, Q \in \Delta$ *implies* $P[Q/x] \in \Delta$, *for each* $x \in \mathrm{Var}$ (substitution closure);
- $M \in \Delta$ *and* $M \rightarrow_\Delta N$ *imply* $N \in \Delta$ (reduction closure).

(iv) *A term* is in Δ-normal form (Δ-*nf*) *if it has not* Δ-*redexes and it* has a Δ-normal form, *or it is* Δ-normalizing *if it reduces to a* Δ-*normal form; the set of* Δ-*nf is denoted by* Δ-*NF.*

(v) *A term is* Δ-strongly normalizing *if it is* Δ-*normalizing, and moreover there is not an infinite* Δ-*reduction sequence starting from it.*

The closure conditions on the set of input values need some comment. Since, as already said, input values represent partially evaluated terms, it is natural to ask that this partial evaluation is preserved by reduction, which is the rule on which the evaluation process is based. The substitution closure comes naturally from the fact that variables always belong to the set of input values.

In this book the symbol Δ will denote a generic set of input values. We will omit the prefix Δ in cases where it is clear from the context.

Example 1.2.2. Let I, K, O, D be the terms defined in the Example 1.1.2, and let M, N be input values. Then $IM \rightarrow_\Delta M$, so I has the behaviour of the identity function, $KMN \rightarrow^*_\Delta M$, $OMN \rightarrow^*_\Delta N$, $DM \rightarrow_\Delta MM$. If $D \in \Delta$ then $DD \rightarrow^*_\Delta DD$.

Now some possible sets of input values will be defined.

Definition 1.2.3. (i) $\Gamma = \mathrm{Var} \cup \{\lambda x.M \mid M \in \Lambda\}$.

(ii) Λ_I *is the language obtained from the grammar generating* Λ, *given in Definition 1.1.1, by modifying the formation rule for abstraction in the following way:*

$$(\lambda x.M) \in \Lambda_I \text{ if and only if } M \in \Lambda \text{ and } x \in \mathrm{Var} \text{ and } x \text{ occurs in } M.$$

The next property shows that there exists some set of input values, although not all sets of terms are sets of input values.

Property 1.2.4. (i) Λ is a set of input values.
(ii) Γ is a set of input values.
(iii) Λ_I is a set of input values.
(iv) Λ-NF is not a set of input values.
(v) $\mathrm{Var} \cup \Lambda\text{-NF}^0$ is a set of input values.
(vi) $\Upsilon = \mathrm{Var} \cup \{\lambda x.P \mid x \in \mathrm{FV}(P)\}$ is not a set of input values.

Proof. The first case is obvious. In cases 2, 3, and 5, it is easy to check that the closure properties of Definition 1.2.1 are satisfied. Λ-NF is not closed under substitution. It is easy to see that Υ is closed under substitution, but it is not closed under reduction. In fact, $\lambda x.KIx \in \Upsilon$, while $\lambda x.KIx \to_{\Upsilon} \lambda x.I \notin \Upsilon$. □

The choice $\Delta = \Lambda$ gives rise to the classical call-by-name λ-calculus [25], while $\Delta = \Gamma$ gives rise to a *pure* version (i.e. without constants) of the call-by-value λ-calculus, first defined by Plotkin [78].

The fact that $\mathrm{Var} \cup \Lambda\text{-NF}^0$ is a correct set of input values was first noticed in [39].

It is easy to check that every term M has the following shape:

$$\lambda x_1 ... x_n.\zeta M_1 ... M_m \qquad (n, m \geq 0),$$

where $M_i \in \Lambda$ are the *arguments* of M ($1 \leq i \leq m$) and ζ is the *head* of M. Here ζ is either a variable (*head variable*) or an application of the shape $(\lambda z.P)Q$, which can be either a redex (*head redex*) or not (*head block*), depending on the fact that Q belongs or not to the set Δ.

The natural interpretation of an abstraction term $\lambda x.M$ is a function whose formal parameter is x. The interpretation of an application $(\lambda x.M)N$, when $N \in \Delta$, is the application of the function $\lambda x.M$ to the actual parameter N, and so the Δ-reduction rule models the replacement of the formal parameter x by the actual parameter N in the body M of the function. Thus the Δ-normal form of a term, if it exists, can be seen as the final result of a computation.

The following fundamental theorem implies that this interpretation is correct, i.e. if the computation process stops, then the result is unique.

Theorem 1.2.5 (Confluence). [26, 74]
Let $M \to_{\Delta}^{*} N_1$ and $M \to_{\Delta}^{*} N_2$. There is Q such that both $N_1 \to_{\Delta}^{*} Q$ and $N_2 \to_{\Delta}^{*} Q$.

Proof. The proof is in Sect. 1.2.1. □

Corollary 1.2.6. *The Δ-normal form of a term, if it exists, is unique.*

Proof. Assume by absurdum that a term M has two different normal forms M_1 and M_2. Then, by the confluence theorem, there is a term N such that both M_1 and M_2 Δ-reduce to N, against the hypothesis that both are normal forms. □

It is natural to ask if the closure conditions on input values, given in Definition 1.2.1, are necessary in order to assure the confluence of the calculus. It can be observed that they are not strictly necessary, but a weaker version of them is needed.

Let $P \in \Delta$ be such that, for every $Q \not\equiv P$ such that $P \rightarrow^*_\Delta Q$, $Q \notin \Delta$. Thus $(\lambda x.M)P$ reduces both to $M[P/x]$ and to $(\lambda x.M)Q$, which do not have a common reduct, since the last term will be never a redex. Thus the weaker version of *reduction closure* that is necessary is the following: $M \in \Delta$ and $M \rightarrow^*_\Delta N$ imply that there is $P \in \Delta$ such that $N \rightarrow^*_\Delta P$.

On the other hand, let $N, P \in \Delta$ but for all Q such that $N[P/x] \rightarrow^*_\Delta Q$, $Q \notin \Delta$. Thus $(\lambda x.(\lambda y.M)N)P$ reduces both to $(\lambda y.M[P/x])N[P/x]$ and to $(M[N/y])[P/x]$, which do not have a common reduct. Thus the weaker version of the *substitution closure* that is necessary is the following: $P, Q \in \Delta$ implies there is $R \in \Delta$ such that $P[Q/x] \rightarrow^*_\Delta R$.

Assume $M \rightarrow^*_\Delta N$, and assume that there is more than one Δ-reduction sequence from M to N. The standardization theorem says that, in case the set of input values enjoys a particular property, there is a "standard" reduction sequence from M to N, reducing the redexes in a given order.

Let us introduce formally the notion of standard reduction sequence.

Definition 1.2.7. (i) *A symbol λ in a term M is* active *if and only if it is the first symbol of a Δ-redex of M.*

(ii) *The* Δ-sequentialization $(M)^\circ$ *of a term M is a function from Λ to Λ defined as follows:*
- $(xM_1...M_m)^\circ = x(M_1)^\circ...(M_m)^\circ$;
- $((\lambda x.P)QM_1...M_m)^\circ = (\lambda x.P)^\circ(Q)^\circ(M_1)^\circ...(M_m)^\circ$, *if* $Q \in \Delta$;
- $((\lambda x.P)QM_1...M_m)^\circ = (Q)^\circ(\lambda x.P)^\circ(M_1)^\circ...(M_m)^\circ$, *if* $Q \notin \Delta$;
- $(\lambda x.P)^\circ = \lambda x.(P)^\circ$.

(iii) *The* degree *of a redex R in M is the numbers of λ's that both are active in M and occur on the left of $(R)^\circ$ in $(M)^\circ$.*

(iv) *The* principal redex *of M, if it exists, is the redex of M with minimum degree. The* principal reduction $M \rightarrow^p_\Delta N$ *denotes that N is obtained from M by reducing the principal redex of M. Moreover, $\rightarrow^{*p}_\Delta$ is the reflexive and transitive closure of $\rightarrow^p_\Delta$.*

(v) *A sequence $M \equiv P_0 \rightarrow_\Delta P_1 \rightarrow_\Delta \ldots \rightarrow_\Delta P_n \rightarrow_\Delta N$ is* standard *if and only if the degree of the redex contracted in P_i is less than or equal to the degree of the redex contracted in P_{i+1}, for every $i < n$.*

We denote by $M \rightarrow^\circ_\Delta N$ a standard reduction sequence from M to N.

It is important to notice that the degree of a redex can change during the reduction; in particular, the redex of minimum degree always has degree zero. Moreover, note that the reduction sequences of length 0 and 1 are always standard. It is easy to check that, for every M, the Λ-sequentialization is $(M)^\circ \equiv M$; thus in this case the redex of degree 0 is always the leftmost one.

Example 1.2.8. (i) Let $\Delta = \Lambda$, and let $M \equiv (\lambda x.x(KI))(II)$. Thus M has degree 0, KI has degree 1 and II has degree 2 (in the term M). The following reduction sequence is standard:
$(\lambda x.x(KI))(II) \rightarrow_\Lambda (II)(KI) \rightarrow_\Lambda I(KI) \rightarrow_\Lambda I(\lambda y.I)$.

(ii) Let M be as before, and let $\Delta = \Gamma$. Thus II has degree 0, and KI has degree 1. Note that now M is no more a redex. The following reduction sequence is standard:
$(\lambda x.x(KI))(II) \to_\Gamma (\lambda x.x(KI))I \to_\Gamma I(KI) \to_\Gamma I(\lambda y.I) \to_\Gamma \lambda y.I.$

(iii) Let M be as before, and let $\Delta = \mathrm{Var} \cup \Lambda\text{-NF}^0$. Thus KI has degree 0 and II has degree 1. Also in this case M is not a redex. The following reduction sequence is standard:
$(\lambda x.x(KI))(II) \to_\Delta (\lambda x.x(KI))I \to_\Delta (\lambda x.x(\lambda y.I))I.$

The notion of a standard set of input values, which is given in Definition 1.2.9 is key for having the standardization property.

Definition 1.2.9 (Standard input values).
A set Δ of input values is standard *if and only if $M \notin \Delta$ and $M \to_\Delta^* N$ by reducing at every step a not principal redex imply $N \notin \Delta$.*

Now the standardization property can be stated.

Theorem 1.2.10 (Standardization). [74]
Let Δ be standard. $M \to_\Delta^* N$ implies there is a standard reduction sequence from M to N.

Proof. The proof is in Sect. 1.2.1. □

The next property shows that some sets of input values are standard, while some are not standard.

Property 1.2.11. (i) Λ and Γ are standard.
(ii) $\mathrm{Var} \cup \Delta\text{-NF}^0$ is standard, for every Δ.
(iii) Λ_I is not standard.

Proof. (i) Λ is trivially standard. Let us consider Γ; we will prove that, if $M \notin \Gamma$, and $M \to_\Gamma N$ through a not principal reduction, then $N \notin \Gamma$. $M \notin \Gamma$ implies that M has one of the following shapes:
1. $yM_1...M_m$ $(m > 1)$.
2. $(\lambda x.M_1)M_2...M_m$ $(m \geq 2)$ and either $(\lambda x.M_1)M_2$ is a redex or it is a head block.

Case 1 is trivial, since M can never be reduced to a term in Γ.
In case 2, if $M_2 \in \Gamma$ then the principal redex is $(\lambda x.M_1)M_2$, while if $M_2 \notin \Gamma$ then if $M_2 \notin \Gamma$-NF the principal redex is in M_2; if $M_2 \in \Gamma$-NF then the principal redex is in some M_j $(j \leq 3)$. So the reduction of a not principal redex cannot produce a term belonging to Γ.

(ii) $\mathrm{Var} \cup \Delta\text{-NF}^0$ is standard since not principal reductions preserve the presence of the redex of minimum degree.

(iii) Consider the term, $M \equiv \lambda x.x(DD)((\lambda z.I)I)$. Clearly $M \notin \Lambda_I$ and the principal redex of M is DD. So $M \to_{\Lambda_I} \lambda x.x(DD)I \in \Lambda_I$ and in this reduction the reduced redex is not principal, while for every sequence of $\to_{\Lambda_I}^{*p}$ reductions; $M \to_{\Lambda_I}^{*p} M \notin \Lambda_I$. □

It is easy to see that the substitution closure on input values, given in Definition 1.2.1, is necessary in order to assure the standardization property.

In fact, let $M, N \in \Delta$ and $M[N/x] \notin \Delta$. The following non-standard reduction sequence $(\lambda x.IM)N \rightarrow_{\Delta} (\lambda x.M)N \rightarrow_{\Delta} M[N/x]$ does not have a standard counterpart, in fact $I(M[N/x]) \not\rightarrow_{\Delta} M[N/x]$.

Theorem 1.2.12. *The condition that Δ is standard is necessary and sufficient for the $\lambda\Delta$-calculus enjoys the standardization property.*

Proof. The sufficiency of the condition is a consequence of the Standardization Theorem. To prove its necessity, assume Δ is not standard; we can find a term $M \notin \Delta$ such that $M \rightarrow^*_{\Delta} N \in \Delta$, without reducing the principal redex. Hence $IM \rightarrow_{\Delta} IN \rightarrow_{\Delta} N$, by reducing first a redex of degree different from 0 and then a redex of degree 0. Clearly, there is no way of commuting the order of reductions. □

An important consequence of the standardization property is the fact that the reduction sequence reducing, at every step, the principal redex is normalizing, as shown in Corollary 1.2.13.

Corollary 1.2.13. *Let Δ be standard.*
If $M \rightarrow^*_{\Delta} N$ and N is a normal form then $M \rightarrow^{*p}_{\Delta} N$.

Proof. By Corollary 1.2.6 and by the definition of the standard set of input values. □

Example 1.2.14. (i) Let $\Delta = \Lambda$. The term $KI(DD)$ has Λ-normal form I. In fact, the principal Λ-reduction sequence is $KI(DD) \rightarrow_{\Lambda} (\lambda y.I)(DD) \rightarrow_{\Lambda} I$, while the Λ-reduction sequence choosing at every step the rightmost Λ-redex never stops. Notice that, if we choose $\Delta = \Gamma$, $KI(DD)$ has not Γ-normal form.

(ii) The term $II(II(II))$ is Λ-strongly normalizing and Γ-strongly normalizing, while $KI(DD)$ is neither Λ-strongly normalizing nor Γ-strongly normalizing.

(iii) Let $\Delta = \mathrm{Var} \cup \Lambda\text{-NF}^0$. Thus $I(K(xx))$ is the Δ-normal form of term $I(II)(K(xx))$.

Remark 1.2.15. The first notion of standardization was given, for the $\lambda\Lambda$-calculus, by Curry and Feys [34, 35]. With respect to their notion, if $M \rightarrow^*_{\Lambda} N$ then there is a standard reduction sequence from M to N, but this reduction sequence is not necessarily unique. For instance, $\lambda x.x(II)(II) \rightarrow_{\Lambda} \lambda x.xI(II) \rightarrow_{\Lambda} \lambda x.xII$ and $\lambda x.x(II)(II) \rightarrow_{\Lambda} \lambda x.x(II)I \rightarrow_{\Lambda} \lambda x.xII$ are both standard reduction sequences. Klop [58] introduced a notion of strong standardization, according to which, if $M \rightarrow^*_{\Lambda} N$, then there is a unique strongly standard reduction sequence from M to N, and he designed an algorithm for transforming a reduction sequence into a strongly standard one. According to his notion, in the example before only the first reduction sequence is standard. Our definition, when restricted to the $\lambda\Lambda$-calculus, is quite similar to the strong standardization. In fact, according to our definition, the standard reduction sequence is unique, but in some degenerated case: e.g. for $\Delta = \Lambda$, there are infinite reduction sequences from $x(DD)$ to $x(DD)$, each one performing a different number of Λ-reductions.

Plotkin [78] extended the notion of standardization to the $\lambda\Gamma$-calculus. His notion of standardization is not strong using Klop's terminology. Our definition, when restricted to $\lambda\Gamma$-calculus, is similar to a strong version of Plotkin's standardization. The advantage of our notion of standardization is the validity of Corollary 1.2.13, i.e. the fact that the principal reduction is Δ-normalizing.

A notion that will play an important role in what follows is that one of solvability.

Definition 1.2.16. (i) *An head context* $(\lambda\vec{x}[.])\vec{P}$ *is* Δ-valuable *if and only if each* $P \in \vec{P}$ *is such that* $P \in \Delta$.

(ii) *A term* M *is* Δ-solvable *if and only if there is a* Δ*-valuable* head context $C[.] \equiv (\lambda\vec{x}.[.])\vec{N}$ *such that:*

$$C[M] =_\Delta I.$$

(iii) *A term is* Δ-unsolvable *if and only if it is not* Δ*-solvable.*

Note that $(\lambda\vec{x}.[.])\vec{N} =_\Delta I$ means $(\lambda\vec{x}.[.])\vec{N} \rightarrow^*_\Delta I$, since I is in Δ-nf, for every Δ.

We will abbreviate Δ-solvable and Δ-unsolvable respectively as solvable and unsolvable, when the meaning is clear from the context. Informally speaking, a solvable term is a term that is in some sense *computationally meaningful.* In fact, let $M \in \Lambda^0$ be solvable, and let P be an input value; we can always find a sequence $\vec{N}$ of terms such that $M\vec{N}$ reduces to P: just take the sequence $\vec{Q}$ such that $M\vec{Q} =_\Delta I$, which exists since M is solvable, and pose $\vec{N} \equiv \vec{Q}P$. So a closed solvable term can mimic the behaviour of any term, if applied to suitable arguments.

It would be interesting to syntactically characterize the solvable terms. Unfortunately, there is not a general characterization for the $\lambda\Delta$-calculus, so we will study this problem for some particular instances of Δ.

Example 1.2.17. (i) Consider the two sets of input values Λ and Γ. In both calculi, the term I is solvable, while DD is unsolvable. $\lambda x.x(DD)$ is an example of a term that is Λ-solvable and Γ-unsolvable. In fact, $(\lambda x.x(DD))O \rightarrow^*_\Lambda I$, while there is no term P such that $P(DD) \rightarrow^*_\Gamma I$, since $DD \notin \Gamma$ and $DD \rightarrow^*_\Gamma DD$.

(ii) Let Φ be the set of input values $\mathrm{Var} \cup \Lambda\text{-NF}^0$. Then $I(\lambda x.I(xx)) \in \Phi\text{-NF}$ is a Φ-unsolvable term.

In order to understand the behaviour of unsolvable terms, it is important to stress some of their closure properties.

Property 1.2.18. (i) The unsolvability is preserved by substitution of variables by input values.
(ii) The unsolvability is preserved by Δ-valuable head contexts.

Proof. Let M be unsolvable.

(i) By contraposition let us assume $M[P/z]$ to be solvable for some input values P. Then there is a Δ-valuable head context $C[.] \equiv (\lambda\vec{x}.[.])\vec{Q}$ such that $C[M[P/z]] \rightarrow^*_\Delta I$.
Without loss of generality, we can assume $\|\vec{Q}\| > \|\vec{x}\|$. Indeed, in the case $\|\vec{Q}\| \leq \|\vec{x}\|$, we can choose a closed solvable term N such that there is $\vec{R}$ such that $N\vec{R} \rightarrow^*_\Delta I$ and $\|\vec{R}\| = \|\vec{x}\| - \|\vec{Q}\|$, and then consider the Δ-valuable context $C[.]N\vec{R}$. So let $\vec{Q} \equiv \vec{Q_1}\vec{Q_2}$, where $\|\vec{Q_1}\| = \|\vec{x}\|$.
$(\lambda\vec{x}.M[P/z])\vec{Q_1}\vec{Q_2} \rightarrow^*_\Delta I$ implies $(\lambda\vec{x}.(\lambda z.M)P)\vec{Q_1}\vec{Q_2} \rightarrow^*_\Delta I$ (since $P \in \Delta$). This in turn implies $(\lambda z.(\lambda\vec{x}.M)\vec{Q_1})(P[\vec{Q_1}/\vec{x}])\vec{Q_2} \rightarrow^*_\Delta I$ and $(\lambda z\vec{x}.M)(P[\vec{Q_1}/\vec{x}])\vec{Q_1}\vec{Q_2} \rightarrow^*_\Delta I$, because by α-equivalence we can assume $z \notin \mathrm{FV}(Q_1)$ and $z \notin \vec{x}$. But $P[\vec{Q_1}/\vec{x}] \in \Delta$ (since input values are closed under substitution) which means that the Δ-valuable head context $C'[.] \equiv (\lambda z\vec{x}.[.])(P[\vec{Q_1}/\vec{x}])\vec{Q_1}\vec{Q_2}$ is such that $C'[M] \rightarrow^*_\Delta I$.

(ii) By contraposition let us assume $C'[M]$ to be solvable for some Δ-valuable head context $C'[.] \equiv (\lambda\vec{z}.[.])\vec{P}$. Then there is a Δ-valuable head context $C[.] \equiv (\lambda\vec{x}.[.])\vec{Q}$, such that $C[C'[M]] \rightarrow^*_\Delta I$. If $\vec{z} \equiv \vec{z_0}\vec{z_1}$ and $\|\vec{P}\| = \|\vec{z_0}\|$ then $C[C'[M]] \rightarrow^*_\Delta C[\lambda\vec{z_1}.M[\vec{P}/\vec{z_0}]] \rightarrow^*_\Delta I$, thus $M[\vec{P}/\vec{z_0}]$ is solvable, and by the previous part of this property M is also solvable. Otherwise $\vec{P} \equiv \vec{P_0}\vec{P_1}$, $\|\vec{P_1}\| > 1$ and $\|\vec{P_0}\| = \|\vec{z}\|$. Thus

$$C[C'[M]] \rightarrow^*_\Delta C[M[\vec{P_0}/\vec{z}]\vec{P_1}] \equiv (\lambda\vec{x}.M[\vec{P_0}/\vec{z}]\vec{P_1})\vec{Q} \rightarrow^*_\Delta I.$$

Without loss of generality we can assume $\|\vec{Q}\| > \|\vec{x}\|$, $\vec{Q} \equiv \vec{Q_0}\vec{Q_1}$ and $\|\vec{Q_0}\| = \|\vec{x}\|$. So

$$(\lambda\vec{x}.M[\vec{P_0}/\vec{z}]\vec{P_1})\vec{Q} \rightarrow^*_\Delta (M[\vec{P_0}/\vec{z}]\vec{P_1})[\vec{Q_0}/\vec{x}]\vec{Q_1} \equiv$$
$$(M[\vec{P_0}/\vec{z}][\vec{Q_0}/\vec{x}])(\vec{P_1}[\vec{Q_0}/\vec{x}])\vec{Q_1} \rightarrow^*_\Delta I,$$

which implies $(M[\vec{P_0}/\vec{z}][\vec{Q_0}/\vec{x}])$ solvable. Again the proof follows from part (i) of this property. □

We will see that in all the calculi we will study in the following, the property of solvability is not preserved by either substitution or by head contexts. As an example in the $\lambda\Lambda$-calculus xD is Λ-solvable, but $xD[D/x]$ is not Λ-solvable.

1.2.1 Proof of Confluence and Standardization Theorems

Both the proofs are based on the notion of parallel reduction.

Definition 1.2.19. *Let Δ be a set of input values.*

(i) *The* deterministic parallel reduction $\hookrightarrow_\Delta$ *is inductively defined as follows:*
   1. $x \hookrightarrow_\Delta x$;
   2. $M \hookrightarrow_\Delta N$ *implies* $\lambda x.M \hookrightarrow_\Delta \lambda x.N$;
   3. $M \hookrightarrow_\Delta M'$,$N \hookrightarrow_\Delta N'$ *and* $N \in \Delta$ *imply* $(\lambda x.M)N \hookrightarrow_\Delta M'[N'/x]$;
   4. $M \hookrightarrow_\Delta M'$,$N \hookrightarrow_\Delta N'$ *and* $N \notin \Delta$ *imply* $MN \hookrightarrow_\Delta M'N'$.

(ii) *The* nondeterministic parallel reduction $\Rightarrow_\Delta$ *is inductively defined as follows:*
   1. $x \Rightarrow_\Delta x$;
   2. $M \Rightarrow_\Delta N$ *implies* $\lambda x.M \Rightarrow_\Delta \lambda x.N$;
   3. $M \Rightarrow_\Delta M'$,$N \Rightarrow_\Delta N'$ *and* $N \in \Delta$ *imply* $(\lambda x.M)N \Rightarrow_\Delta M'[N'/x]$;
   4. $M \Rightarrow_\Delta M'$,$N \Rightarrow_\Delta N'$ *imply* $MN \Rightarrow_\Delta M'N'$.

Roughly speaking, the deterministic parallel reduction reduces in one step all the redexes present in a term, while the nondeterministic one reduces a subset of them.

Example 1.2.20. Let $M \equiv I(II)$. If $\Delta \equiv \Lambda$ then $M \hookrightarrow_\Delta I$, while $M \Rightarrow_\Delta M$, $M \Rightarrow_\Delta II$ and $M \Rightarrow_\Delta I$. If $\Delta \equiv \Gamma$ then $M \hookrightarrow_\Delta II$ while $M \Rightarrow_\Delta M$ and $M \Rightarrow_\Delta II$.

The following lemma shows the relation between the $\Rightarrow_\Delta$ and $\rightarrow_\Delta$ reductions.

Lemma 1.2.21. *Let Δ be a set of input values.*

(i) $M \rightarrow_\Delta N$ *implies* $M \Rightarrow_\Delta N$.
(ii) $M \Rightarrow_\Delta N$ *implies* $M \rightarrow^*_\Delta N$.
(iii) $\rightarrow^*_\Delta$ *is the transitive closure of* $\Rightarrow_\Delta$.

Proof. Easy. □

$\Rightarrow_\Delta$ enjoys a useful substitution property.

Lemma 1.2.22. *Let $M \Rightarrow_\Delta M'$ and $N \Rightarrow_\Delta N'$.*
If $N \in \Delta$ then $M[N/x] \Rightarrow_\Delta M'[N'/x]$.

Proof. By induction on M. Let us prove just the most difficult case, i.e. the term M is a Δ-redex. Let $M \equiv (\lambda z.P)Q$, $Q \in \Delta$, $P \Rightarrow_\Delta P'$, $Q \Rightarrow_\Delta Q'$ and $M' \equiv P'[Q'/z]$. By induction $P[N/x] \Rightarrow_\Delta P'[N'/x]$ and $Q[N/x] \Rightarrow_\Delta Q'[N'/x]$, where $Q'[N'/x] \in \Delta$ for the closure conditions on Δ. Thus

$$((\lambda z.P)Q)[N/x] \equiv (\lambda z.P[N/x])Q[N/x] \Rightarrow_\Delta$$

$$P'[N'/x][Q'[N'/x]/z] \equiv (P'[Q'/z])[N'/x]$$

by point 3 of the definition of $\Rightarrow_\Delta$. □

The next property, whose proof is obvious, states that, for every term M, there is a unique term N such that $M \hookrightarrow_\Delta N$.

Property 1.2.23. $M \hookrightarrow_\Delta P$ and $M \hookrightarrow_\Delta Q$ implies $P \equiv Q$.

Proof. Trivial. □

Let $[M]_\Delta$ be the term such $M \hookrightarrow_\Delta [M]_\Delta$. In the literature $[M]_\Delta$ is called the *complete development* of M (see [93]). The following lemma holds.

Lemma 1.2.24. *$M \Rightarrow_\Delta N$ implies $N \Rightarrow_\Delta [M]_\Delta$.*

Proof. By induction on M.

- If $M \equiv x$, then $N \equiv x$ and $[M]_\Delta \equiv x$.
- If $M \equiv \lambda x.P$ then $N \equiv \lambda x.Q$, for some Q such that $P \Rightarrow_\Delta Q$. By induction $Q \Rightarrow_\Delta [P]_\Delta$, and so $N \Rightarrow_\Delta \lambda x.[P]_\Delta \equiv [M]_\Delta$.
- If $M \equiv P_1P_2$ and it is not a Δ-redex, then $N \equiv Q_1Q_2$ for some Q_1 and Q_2 such that $P_1 \Rightarrow_\Delta Q_1$ and $P_2 \Rightarrow_\Delta Q_2$. So, by induction, $Q_1 \Rightarrow_\Delta [P_1]_\Delta$ and $Q_2 \Rightarrow_\Delta [P_2]_\Delta$, which implies $N \Rightarrow_\Delta [P_1]_\Delta[P_2]_\Delta \equiv [M]_\Delta$.
- If $M \equiv (\lambda x.P_1)P_2$ is a redex (i.e. $P_2 \in \Delta$) then either $N \equiv (\lambda x.Q_1)Q_2$ or $N \equiv Q_1[Q_2/x]$, for some Q_i such that $P_i \Rightarrow_\Delta Q_i$ $(1 \leq i \leq 2)$. By induction, $Q_i \Rightarrow_\Delta [P_i]_\Delta$ $(1 \leq i \leq 2)$. Note that $[P_2]_\Delta \in \Delta$ by Lemma 1.2.21.(ii). In both cases, $N \Rightarrow_\Delta [P_1]_\Delta[[P_2]_\Delta/x] \equiv [M]_\Delta$, in the former case simply by induction, and in the latter both by induction and by Lemma 1.2.22. □

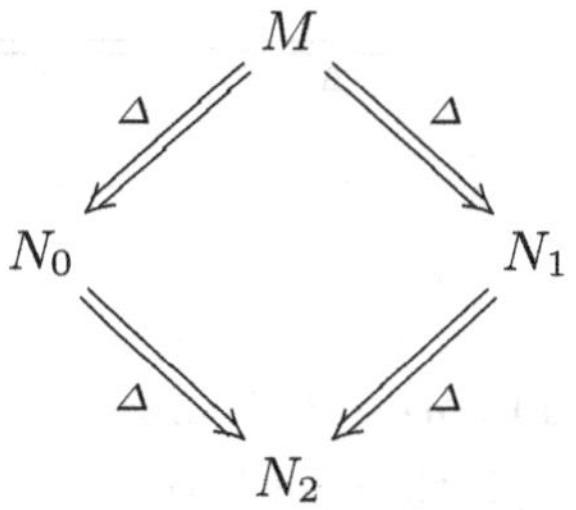

Fig. 1.1. Diamond property.

The proof of confluence follows the Takahashi pattern [93], which is a simplification of the original proof made by Taït and Martin Löf for classical $\lambda\Lambda$-calculus. It is based on the property that a reduction that is the transitive closure of another one enjoying the Diamond Property is confluent.

Lemma 1.2.25 (Diamond property of $\Rightarrow_\Delta$).
If $M \Rightarrow_\Delta N_0$ and $M \Rightarrow_\Delta N_1$ then there is N_2 such that both $N_0 \Rightarrow_\Delta N_2$ and $N_1 \Rightarrow_\Delta N_2$.

Proof. By Lemma 1.2.24, $M \Rightarrow_\Delta N$ implies $N \Rightarrow_\Delta [M]_\Delta$. So, if $M \Rightarrow_\Delta M_1$ and $M \Rightarrow_\Delta M_2$, then both $M_1 \Rightarrow_\Delta [M]_\Delta$ and $M_2 \Rightarrow_\Delta [M]_\Delta$, as shown in Fig. 1.1 (pag. 15). □

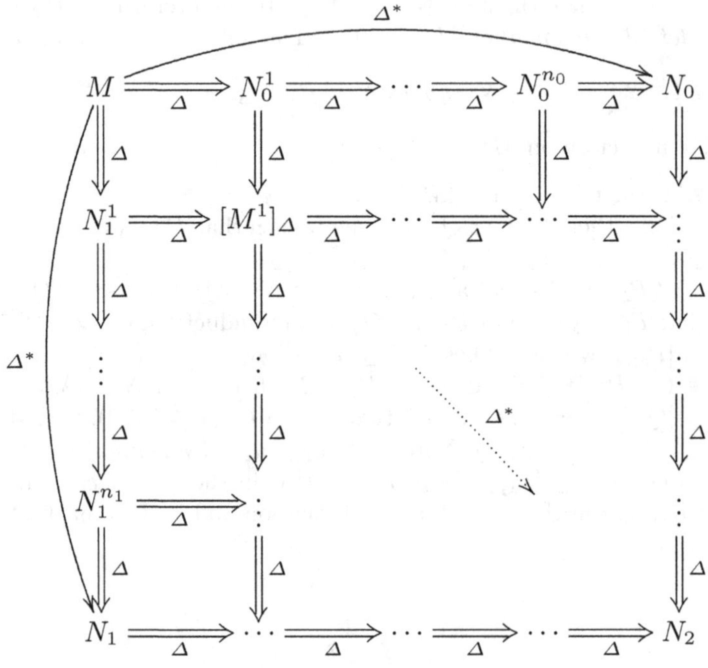

Fig. 1.2. Diamond closure.

▼ **Proof of Confluence Theorem** (Theorem 1.2.5 pag. 8).

By Property 1.2.21.(iii), $\to_\Delta^*$ is the transitive closure of $\Rightarrow_\Delta$. This means that there are $N_0^1, ..., N_0^{n_0}, N_1^1, ..., N_1^{n_1}$ $(n_0, n_1 \geq 1)$ such that $M \Rightarrow_\Delta N_0^1 ... \Rightarrow_\Delta N_0^{n_0} \Rightarrow_\Delta N_0$ and $M \Rightarrow_\Delta N_1^1 ... \Rightarrow_\Delta N_m^{n_1} \Rightarrow_\Delta N_1$. Then the proof follows by repeatedly applying the diamond property of $\Rightarrow_\Delta$ (diamond closure), as shown in Fig. 1.2. ■

The rest of this subsection is devoted to the proof of the standardization theorem. First, we need to establish some technical results.

Let $M \Rightarrow_\Delta^\circ N$ denote "$M \to_\Delta^\circ N$ and $M \Rightarrow_\Delta N$".

The following lemma, at the point (ii), shows that a nondeterministic parallel reduction can always be transformed into a standard reduction sequence.

Lemma 1.2.26. *Let $\vec{P}, \vec{Q}$ be two sequences of terms such that $\|\vec{P}\| = \|\vec{Q}\|$; moreover, let $P_i \in \Delta$ and $P_i \Rightarrow^{\circ}_{\Delta} Q_i$ for all $i \leq \|\vec{P}\|$.*

(i) *If $M \Rightarrow^{\circ}_{\Delta} N$ then $M[\vec{P}/\vec{x}] \Rightarrow^{\circ}_{\Delta} N[\vec{Q}/\vec{x}]$.*
(ii) *If $M \Rightarrow_{\Delta} N$ then $M \Rightarrow^{\circ}_{\Delta} N$.*

Proof. Parts (i) and (ii) can be proved by mutual induction on M.

(i) By Lemma 1.2.22, $M[\vec{P}/\vec{x}] \Rightarrow_{\Delta} N[\vec{Q}/\vec{x}]$, hence it suffices to show that $M[\vec{P}/\vec{x}] \rightarrow^{\circ}_{\Delta} N[\vec{Q}/\vec{x}]$.
Let $M \equiv \lambda y_1 ... y_h.\zeta M_1 ... M_m$ $(h, m \in \mathbb{N})$, where either ζ is a variable or $\zeta \equiv (\lambda z.T)U$.
If $h > 0$, then the proof follows by induction.
Let $h = 0$, thus $N \equiv \xi N_1 ... N_m$ such that $\zeta \Rightarrow^{\circ}_{\Delta} \xi$ and $M_i \Rightarrow^{\circ}_{\Delta} N_i$; furthermore, let $M'_i \equiv M_i[\vec{P}/\vec{x}]$ and $N'_i \equiv N_i[\vec{Q}/\vec{x}]$ $(1 \leq i \leq m)$.
The proof is organized according to the possible shapes of ζ.

1. Let ζ be a variable. If $m = 0$ then the proof is trivial, so let $m > 0$. There are two cases to be considered.
   1.1. $\zeta \notin \vec{x}$, so $\xi[\vec{Q}/\vec{x}] \equiv \zeta$. By induction $M_i[\vec{P}/\vec{x}] \rightarrow^{\circ}_{\Delta} N_i[\vec{Q}/\vec{x}]$ and the standard reduction sequence is

$$\zeta M'_1 ... M'_m \rightarrow^{\circ}_{\Delta} \zeta N'_1 M'_2 ... M'_m \rightarrow^{\circ}_{\Delta} \; ..... \; \rightarrow^{\circ}_{\Delta} \zeta N'_1 ... N'_m.$$

   1.2. $\zeta \equiv x_j \in \vec{x}$ $(1 \leq j \leq l)$, so $\xi[\vec{Q}/\vec{x}] \equiv Q_j$. But $P_j \Rightarrow^{\circ}_{\Delta} Q_j$ means that there is a standard sequence $P_j \equiv S_0 \rightarrow_{\Delta} \; ..... \; \rightarrow_{\Delta} S_n \equiv Q_j$ $(n \in \mathbb{N})$. Two cases can arise.
   1.2.1. $\forall i \leq n$, $S_i \not\equiv \lambda z.S'$. Then the following reduction sequence

$$\sigma : S_0 M'_1 ... M'_m \rightarrow_{\Delta} \; ..... \; \rightarrow_{\Delta} S_n M'_1 ... M'_m$$

   is standard. Since by induction $M_i[\vec{P}/\vec{x}] \rightarrow^{\circ}_{\Delta} N_i[\vec{Q}/\vec{x}]$, there is a standard reduction sequence

$$\tau : S_n M'_1 ... M'_m \rightarrow^{\circ}_{\Delta} S_n N'_1 M'_2 ... M'_m \rightarrow^{\circ}_{\Delta} \; ..... \; \rightarrow^{\circ}_{\Delta} S_n N'_1 ... N'_m.$$

   Note that $S_0 M'_1 ... M'_m \equiv M[\vec{P}/\vec{x}]$ and $S_n N'_1 ... N'_m \equiv N[\vec{Q}/\vec{x}]$, so σ followed by τ is the desired standard reduction sequence.
   1.2.2. There is a minimum $k \leq n$ such that $S_k \equiv \lambda z.S'$.
   By induction on (ii), $M_1 \Rightarrow^{\circ}_{\Delta} N_1$. Therefore, by induction $M_1[\vec{P}/\vec{x}] \Rightarrow^{\circ}_{\Delta} N_1[\vec{Q}/\vec{x}]$, where $M_1[\vec{P}/\vec{x}] \rightarrow^{\circ}_{\Delta} N_1[\vec{Q}/\vec{x}]$ is $M_1[\vec{P}/\vec{x}] \equiv R_0 \rightarrow_{\Delta} \; ..... \; \rightarrow_{\Delta} R_p \equiv N_1[\vec{Q}/\vec{x}]$ $(p \in \mathbb{N})$. There are two subcases:

1.2.2.1. $\forall i \leq p$, $R_i \notin \Delta$. Then the following reduction sequence:

$$\begin{aligned} \sigma' : M[\vec{P}/\vec{x}] \equiv S_0 R_0 M_2' ... M_m' \to_\Delta ..... \to_\Delta S_k R_0 M_2' ... M_m' \\ \to_\Delta ..... \to_\Delta S_k R_p M_2' ... M_m' \\ \to_\Delta S_{k+1} R_p M_2' ... M_m' \to_\Delta ..... \to_\Delta S_n R_p M_2' ... M_m' \end{aligned}$$

is also standard. Moreover, since $M_i[\vec{P}/\vec{x}] \to_\Delta^\circ N_i[\vec{P}/\vec{x}]$, the following reduction sequence:

$$\begin{aligned} \tau' : S_n R_p M_2' ... M_m' \to_\Delta^\circ \\ S_n R_p N_2' M_3' ... M_m' \to_\Delta^\circ ..... \to_\Delta^\circ S_n R_p N_2' ... N_m' \end{aligned}$$

is also standard. Clearly σ' followed by τ' is the desired standard reduction sequence.

1.2.2.2. There is a minimum $q \leq p$ such that $R_q \in \Delta$. So

$$\begin{aligned} \sigma'' : \quad M[\vec{P}/\vec{x}] \equiv S_0 R_0 M_2' ... M_m' \to_\Delta ..... \to_\Delta S_k R_0 M_2' ... M_m' \\ \to_\Delta ..... \to_\Delta S_k R_q M_2' ... M_m' \to_\Delta S_{k+1} R_q M_2' ... M_m' \\ \to_\Delta ..... \to_\Delta S_n R_q M_2' ... M_m' \to_\Delta ..... \to_\Delta S_n R_p M_2' ... M_m' \end{aligned}$$

is a standard reduction sequence. The desired standard reduction sequence is σ'' followed by τ'.

2. Let $\zeta \equiv (\lambda z.T)U$. Thus $N \equiv (\lambda z.\bar{T})\bar{U} N_1 ... N_m$ or $N \equiv \bar{T}[\bar{U}/z] N_1 ... N_m$, where $T \Rightarrow_\Delta \bar{T}$, $U \Rightarrow_\Delta \bar{U}$ and $M_i \Rightarrow_\Delta N_i$ $(1 \leq i \leq m)$.
   By induction, $U' \equiv U[\vec{P}/\vec{x}] \Rightarrow_\Delta^\circ \bar{U}[\vec{Q}/\vec{x}] \equiv U''$, $T' \equiv T[\vec{P}/\vec{x}] \Rightarrow_\Delta^\circ \bar{T}[\vec{Q}/\vec{x}] \equiv T''$ and $M_i' \equiv M_i[\vec{P}/\vec{x}] \Rightarrow_\Delta^\circ N_i[\vec{Q}/\vec{x}] \equiv N_i'$ $(1 \leq i \leq m)$.
   Let $U' \equiv R_0 \to_\Delta ... \to_\Delta R_p \equiv U''$ $(p \in \mathbb{N})$ be the standard sequence $U' \to_\Delta^\circ U''$. Without loss of generality let us assume $z \notin \vec{x}$.
   2.1. Let $N \equiv (\lambda z.\bar{T})\bar{U} N_1 ... N_m$. There are two cases.
      2.1.1. $\forall i \leq p$ $R_i \notin \Delta$. Then the standard reduction sequence $M[\vec{P}/\vec{x}] \to_\Delta^\circ N[\vec{Q}/\vec{x}]$ is

$$\begin{aligned} (\lambda z.T') R_0 M_1' ... M_m' \to_\Delta ..... \to_\Delta (\lambda z.T') R_p M_1' ... M_m' \\ \to_\Delta^\circ (\lambda z.T'') R_p M_1' ... M_m' \to_\Delta^\circ (\lambda z.T'') R_p N_1' M_2' ... M_m' \\ \to_\Delta^\circ ..... \to_\Delta^\circ (\lambda z.T'') R_p N_1' ... N_m'. \end{aligned}$$

      2.1.2. There is a minimum $q \leq p$ such that $R_q \in \Delta$. Thus the desired standard reduction sequence is:

$$\begin{aligned} (\lambda z.T') R_0 M_1' ... M_m' \to_\Delta ..... \to_\Delta (\lambda z.T') R_q M_1' ... M_m' \\ \to_\Delta^\circ (\lambda z.T'') R_q M_1' ... M_m' \to_\Delta ..... \to_\Delta (\lambda z.T'') R_p M_1' ... M_m' \\ \to_\Delta^\circ (\lambda z.T'') R_p N_1' M_2' ... M_m' \to_\Delta^\circ ..... \to_\Delta^\circ (\lambda z.T'') R_p N_1' ... N_m'. \end{aligned}$$

2.2. Let $N \equiv \bar{T}[\bar{U}/z]N_1...N_m$. So, there is a minimum $q \leq p$ such that $R_q \in \Delta$; let μ be the standard reduction sequence:

$$M[\vec{P}/\vec{x}] \equiv (\lambda z.T')R_0M'_1...M'_m \rightarrow_\Delta ..... \rightarrow_\Delta (\lambda z.T')R_qM'_1...M'_m \rightarrow_\Delta T'[R_q/z]M'_1...M'_m.$$

$T \Rightarrow^\circ_\Delta \bar{T}$, by induction on (ii). Furthermore, since $R_q \Rightarrow^\circ_\Delta U''$, it follows by induction that $T[\vec{P}/\vec{x}][R_q/z] \Rightarrow^\circ_\Delta \bar{T}[\vec{Q}/\vec{x}][U''/z]$.
Let $T[\vec{P}/\vec{x}][R_q/z] \equiv T_0 \rightarrow_\Delta ..... \rightarrow_\Delta T_t \equiv \bar{T}[\vec{Q}/\vec{x}][U''/z]$ be the corresponding standard reduction sequence. Two subcases can arise:

2.2.1. $\forall i \leq t$, $T_i \not\equiv \lambda z.S'$. The desired standard reduction sequence is μ followed by:

$$T'[R_p/z]M'_1...M'_m \equiv T[\vec{P}/\vec{x}][R_p/z]M'_1...M'_m \rightarrow_\Delta T_1M'_1...M'_m \rightarrow_\Delta ..... \rightarrow_\Delta T_tM'_1...M'_m \rightarrow^\circ_\Delta \; ..... \rightarrow^\circ_\Delta T_tN'_1...N'_m \equiv [\vec{Q}/\vec{x}]$$

2.2.2. Let $k \leq t$ be the minimum index such that $T_k \equiv \lambda y.T'_k$. The construction of the standard reduction sequence depends on the fact that M_2 may or may not become an input value, but, in every case, it can be easily built as in the previous cases.

(ii) The cases $M \equiv x$ and $M \equiv \lambda z.M'$ are easy.

1. Let $M \equiv PQ \Rightarrow_\Delta P'Q' \equiv N$, $P \Rightarrow_\Delta P'$ and $Q \Rightarrow_\Delta Q'$.
By induction, there are standard sequences $P \equiv P_0 \rightarrow_\Delta ... \rightarrow_\Delta P_p \equiv P'$ and $Q \equiv Q_0 \rightarrow_\Delta ... \rightarrow_\Delta Q_q \equiv Q'$.
If $\forall i \leq p \;\; P_i \not\equiv \lambda z.P'_i$, then $M \rightarrow^\circ_\Delta N$ is $P_0Q_0 \rightarrow^\circ_\Delta P_pQ_0 \rightarrow^\circ_\Delta P_pQ_q$.
Otherwise, let k be the minimum index such that $P_k \equiv \lambda z.P'_k$.
   - If $\forall j \leq q \;\; Q_j \notin \Delta$, then $M \rightarrow^\circ_\Delta N$ is

$$P_0Q_0 \rightarrow_\Delta ..... \rightarrow_\Delta P_kQ_0 \rightarrow^\circ_\Delta P_kQ_q \rightarrow_\Delta P_{k+1}Q_q \rightarrow_\Delta ..... \rightarrow_\Delta P_pQ_q.$$

   - If there is a minimum h such that $Q_h \in \Delta$, the standard sequence is $P_0Q_0 \rightarrow^\circ_\Delta P_kQ_0 \rightarrow^\circ_\Delta P_kQ_h \rightarrow_\Delta P_{k+1}Q_h \rightarrow^\circ_\Delta P_pQ_h \rightarrow^\circ_\Delta P_pQ_q$.
2. Let $M \equiv (\lambda x.P)Q \Rightarrow_\Delta P'[Q'/x] \equiv N$ where $P \Rightarrow_\Delta P'$, $Q \Rightarrow_\Delta Q'$ and $Q \in \Delta$. Hence $P \Rightarrow^\circ_\Delta P'$ and $Q \Rightarrow^\circ_\Delta Q'$ follow by induction, so $P[Q/x] \Rightarrow^\circ_\Delta P'[Q'/x]$, by induction on (i). Thus, the desired standard reduction sequence is $(\lambda x.P)Q \rightarrow_\Delta P[Q/x] \rightarrow^\circ_\Delta P'[Q'/x]$. □

In order to prove the standardization theorem some auxiliary definitions are necessary.

Definition 1.2.27. *Let $M, N \in \Lambda$.*

(i) *$M \rightarrow^i_\Delta N$ denotes that N is obtained from M by reducing a redex that is not the principal redex.*
(ii) *$M \Rightarrow^i_\Delta N$ denotes $M \Rightarrow_\Delta N$ and $M \rightarrow^{*i}_\Delta N$.*

According to this new terminology, a set of input values is standard, in the sense of Definition 1.2.9 (pag. 10), if and only if $M \notin \Delta$ and $M \to_{\Delta}^{*i} N$ imply $N \notin \Delta$.

Lemma 1.2.28. *$M \Rightarrow_{\Delta} N$ implies there is P such that $M \to_{\Delta}^{*p} P \Rightarrow_{\Delta}^{i} N$.*

Proof. Trivial, by Lemma 1.2.26.(ii).
Notice that it can be $M \equiv P$, by definition of $\to_{\Delta}^{*p}$. □

Example 1.2.29. Let $M \equiv (\lambda xy.I(\lambda z.IK(II)))I \Rightarrow_{\Gamma} \lambda yz.IKI$.
Therefore $M \to_{\Gamma}^{p} \lambda y.I(\lambda z.IK(II)) \to_{\Gamma}^{p} \lambda yz.IK(II) \Rightarrow_{\Gamma}^{i} \lambda yz.IKI$ and clearly $\lambda yz.IK(II) \in \Gamma$.

Note that if Δ is standard and R is the principal redex of M and $M \to_{\Delta}^{*i} N$, then R is the principal redex of N.

Lemma 1.2.30. *Let Δ be standard.*
$M \Rightarrow_{\Delta}^{i} P \to_{\Delta}^{p} N$ implies $M \to_{\Delta}^{*p} Q \Rightarrow_{\Delta}^{i} N$, for some Q.

Proof. By induction on M. If either $M \equiv \lambda x.M'$, or the head of M is a variable, then the proof follows by induction. Otherwise, let $M \equiv (\lambda y.M_0)M_1...M_m$; thus it must be $P \equiv (\lambda y.P_0)P_1...P_m$. Note that $M \Rightarrow_{\Delta}^{i} P$ implies $M_i \Rightarrow_{\Delta} P_i$ ($1 \leq i \leq m$). Now there are two cases, according to whether $P_1 \in \Delta$ or not.

- Let $P_1 \in \Delta$; it follows that P_1 is the argument of the principal redex of P, thus $N \equiv P_0[P_1/y]P_2...P_m$.
  Let $M_1 \in \Delta$. Then we can build the following reduction sequence:
  $M \equiv (\lambda y.M_0)M_1...M_m \to_{\Delta}^{p} M_0[M_1/y]...M_m \Rightarrow_{\Delta} P_0[P_1/y]P_2...P_m$, which can be transformed into a standard one by Lemma 1.2.28.
  Let $M_1 \notin \Delta$ and $P_1 \in \Delta$; since the set Δ is standard, $M_1 \Rightarrow_{\Delta} P_1 \in \Delta$ if and only if $M_1 \to_{\Delta}^{*p} P_1' \Rightarrow_{\Delta}^{i} P_1$, where $P_1' \in \Delta$. But this would imply that in the reduction $M \Rightarrow_{\Delta}^{i} P$ the principal redex of M_1 has been reduced; but by definition the principal redex of M_1 coincides with the principal redex of M, against the hypothesis that $M \Rightarrow_{\Delta}^{i} P$. So this case is not possible.
- Let $P_1 \notin \Delta$. Then there is $j \geq 0$ such that the principal redex of P_j is the principal redex of P. Let $j \geq 2$; so $\forall k \leq j$ P_k is a normal form. So $N \equiv (\lambda y.P_0)P_1...P_j'..P_m$, where $P_j \to_{\Delta}^{p} P_j'$. From the hypothesis that $M \Rightarrow_{\Delta}^{i} P$, it follows that $M_i \equiv P_i$ ($0 \leq i \leq j-1$), and $M_i \Rightarrow_{\Delta} P_i$ ($j < i \leq m$). Then by induction there is P_j^* such that $M_j \to_{\Delta}^{*p} P_j^* \Rightarrow_{\Delta}^{i} P_j'$, and we can build the following reduction sequence:

$$(\lambda y.M_0)M_1...M_m \to_{\Delta}^{*p} (\lambda y.M_0)M_1...P_j^* P_{j+1}...P_m \Rightarrow_{\Delta} (\lambda y.M_0)M_1...P_j'...P_m$$

  which can be transformed into a standard one by Lemma 1.2.28.
  The case $j < 2$ is similar. □

This Lemma has a key corollary.

Corollary 1.2.31. *Let Δ be standard.*
If $M \to_\Delta^* N$ then $M \to_\Delta^{*p} Q \underbrace{\Rightarrow_\Delta^i \ldots \Rightarrow_\Delta^i}_{k} N$, for some Q and some k.

Proof. Note that if $P \to_\Delta P'$ then $P \Rightarrow_\Delta P'$. So $M \to_\Delta^* N$ implies $M \Rightarrow_\Delta N_1 \Rightarrow_\Delta \ldots \Rightarrow_\Delta N_n \Rightarrow_\Delta N$. So, by repeatedly applying Lemma 1.2.28 and Lemma 1.2.30 we reach the proof. □

Now we are able to prove the standardization theorem.

▼ **Proof of Standardization Theorem** (Theorem 1.2.10 pag. 10).
The proof is given by induction on N. From Corollary 1.2.31, $M \to_\Delta^* N$ implies $M \to_\Delta^{*p} Q \to_\Delta^{*i} N$ for some Q. Obviously, the reduction sequence $\sigma : M \to_\Delta^{*p} Q$ is standard by definition of $\to_\Delta^p$. Note that, by definition of $\to_\Delta^{*i}$, $Q \to_\Delta^{*i} N$ implies that Q and N have the same structure, i.e. $Q \equiv \lambda x_1 \ldots x_n.\zeta Q_1 \ldots Q_n$ and $N \equiv \lambda x_1 \ldots x_n.\zeta' N_1 \ldots N_n$, where $Q_i \to_\Delta^* N_i$ $(i \leq n)$ and either ζ and ζ' are the same variable, or $\zeta \equiv (\lambda x.R)S$, $\zeta' \equiv (\lambda x.R')S'$, $R \to_\Delta^* R'$ and $S \to_\Delta^* S'$.
The case when ζ is a variable follows by induction. Otherwise, by induction there are standard reduction sequences $\sigma_i : Q_i \to_\Delta^\circ N_i$ $(1 \leq i \leq n)$, $\tau_R : R \to_\Delta^\circ R'$ and $\tau_S : S \to_\Delta^\circ S'$. Let $S \equiv S_0 \to_\Delta \ldots\ldots \to_\Delta S_k \equiv S'$ $(k \in \mathbb{N})$.
If $\forall i \leq k$ $S_i \notin \Delta$ then the desired standard reduction sequence is σ followed by $\tau_S, \tau_R, \sigma_1, \ldots, \sigma_n$.
Otherwise, there is $S_h \in \Delta$ $(h \leq k)$. In this case, let $\tau_S^0 : S_0 \to_\Delta \ldots\ldots \to_\Delta S_h$ and $\tau_S^1 : S_{h+1} \to_\Delta \ldots\ldots \to_\Delta S_k$; the desired standard reduction sequence is σ followed by $\tau_S^0, \tau_R, \tau_S^1, \sigma_1, \ldots, \sigma_n$. ■

1.3 Δ-Theories

In order to model computation, Δ-equality is too weak. As an example, let Δ be either Λ or Γ. If we want to model the termination property, both the terms DD and $(\lambda x.xxx)(\lambda x.xxx)$ represent programs that run forever, while the two terms are $\neq_\Delta$ each other. Indeed $DD \to_\Delta DD$ and $(\lambda x.xxx)(\lambda x.xxx) \to_\Delta (\lambda x.xxx)(\lambda x.xxx)(\lambda x.xxx)$. So it would be natural to consider them equal in this particular setting. But if we want to take into account not only termination but also the size of terms, they need to be different; in fact, the first one reduces to itself while the second increases its size during the reduction. As we will see in the following, for all instances of Δ we will consider, all interesting interpretations of the calculus also equate terms that are not $=_\Delta$.

Let us introduce the notion of Δ-theory.

Definition 1.3.1. (i) *$\mathcal{T} \subseteq \Lambda \times \Lambda$ is a* congruence *whenever:*
- *$(M, M) \in \mathcal{T}$ for each $M \in \Lambda$,*

- *$(M,N) \in \mathcal{T}$ implies $(N,M) \in \mathcal{T}$,*
- *$(M,P) \in \mathcal{T}$ and $(P,N) \in \mathcal{T}$ imply $(M,N) \in \mathcal{T}$,*
- *$(M,N) \in \mathcal{T}$ implies $(C[M], C[N]) \in \mathcal{T}$, for all contexts $C[.]$.*

(ii) *$\mathcal{T} \subseteq \Lambda \times \Lambda$ is a* Δ-theory *if and only if it is a congruence and $M =_\Delta N$ implies $(M,N) \in \mathcal{T}$.*

We will denote $(M,N) \in \mathcal{T}$ also by $M =_\mathcal{T} N$.

Clearly a Δ-theory equating all terms would be completely uninteresting. So we will ask for consistency.

Definition 1.3.2. (i) *A Δ-theory $\mathcal{T}$ is* consistent *if and only if there are $M,N \in \Lambda$ such that $M \neq_\mathcal{T} N$. Otherwise $\mathcal{T}$ is inconsistent.*
(ii) *A Δ-theory $\mathcal{T}$ is* input consistent *if and only if there are $M,N \in \Delta$ such that $M \neq_\mathcal{T} N$. Otherwise $\mathcal{T}$ is input inconsistent.*
(iii) *A Δ-theory $\mathcal{T}$ is* maximal *if and only if it has no consistent extension, i.e. for all $M,N \in \Lambda$ such that $M \neq_\mathcal{T} N$, any Δ-theory $\mathcal{T}'$ containing $\mathcal{T}$ and such that $M =_{\mathcal{T}'} N$ is inconsistent.*

Property 1.3.3. Let $\mathcal{T}$ be a Δ-theory.
If $\mathcal{T}$ is input consistent then it is consistent.

Proof. Obvious. □

In the last section of this book, we will see that in order to use a $\lambda\Delta$-calculus for computing, we need to work inside theories that are both consistent and input consistent.

Δ-theories can be classified according to their behaviour with respect to the Δ-solvable terms.

Definition 1.3.4. (i) *A Δ-theory is* sensible *if it equates all Δ-unsolvable terms.*
(ii) *A Δ-theory is* semisensible *if it never equates a Δ-solvable term and a Δ-unsolvable term.*

Another important notion for Δ-theories is that of separability. In fact, this property help us to understand what equalities cannot be induced by a theory.

Definition 1.3.5. *Let Δ be a set of input values.*
Two terms M,N are Δ-separable *if and only if there is a context $C[.]$ such that $C[M] =_\Delta x$ and $C[N] =_\Delta y$ for two different variables x and y.*

Property 1.3.6. Let M,N be Δ-separable.
If $\mathcal{T}$ is a Δ-theory such that $M =_\mathcal{T} N$ then $\mathcal{T}$ is input inconsistent.

Proof. Let $C[.]$ be the context separating M and N, i.e. $C[M] =_\Delta x$ and $C[N] =_\Delta y$ for two different variables x and y. Since $=_\mathcal{T}$ is a congruence, $M =_\mathcal{T} N$ implies $C[M] =_\mathcal{T} C[N]$, and so, since $\mathcal{T}$ is closed under $=_\Delta$, $x =_\mathcal{T} y$. But this implies $\lambda xy.x =_\mathcal{T} \lambda xy.y$, i.e. $K =_\mathcal{T} O$. But, since $=_\mathcal{T}$ is a congruence, this implies $KMN =_\mathcal{T} OMN$ for all terms M, N. In particular, if $M, N \in \Delta$ then $M =_\mathcal{T} N$ by Δ-reduction. □

A theory is *fully extensional* if all terms in it (not only abstractions) have a functional behaviour. So, in a fully extensional theory, the equality between terms must be extensional (in the usual sense), i.e., it must satisfy the property:

$$\text{(EXT)} \qquad Mx = Nx \Rightarrow M = N \qquad\qquad x \notin \mathrm{FV}(M) \cup \mathrm{FV}(N).$$

Clearly $=_\Delta$ does not satisfy (EXT). In fact, (EXT) holds for $=_\Delta$ only if it is restricted to terms that reduce to an abstraction: indeed, $xy =_\Delta (\lambda z.xz)y$, but $x \neq_\Delta \lambda z.xz$.

The least extensional extension of $=_\Delta$ is induced by the η-reduction rule, defined as follows:

Definition 1.3.7 (η-Reduction).

(i) *The η-reduction ($\rightarrow_\eta$) is the contextual closure of the following rule:* $\lambda x.Mx \rightarrow_\eta M$ *if and only if* $x \notin \mathrm{FV}(M)$*;* $\lambda x.Mx$ *is a η-redex and M is its contractum;*
(ii) $M \rightarrow_{\Delta\eta} N$ *if N is obtained from M by reducing either a Δ or a η redex in M;*
(iii) $\rightarrow^*_{\Delta\eta}$ *and* $=_{\Delta\eta}$ *are respectively the reflexive and transitive closure of* $\rightarrow_{\Delta\eta}$ *and the symmetric, reflexive and transitive closure of* $\rightarrow_{\Delta\eta}$*.*

The next theorem shows an interesting result for η-reduction.

Theorem 1.3.8. $=_{\Delta\eta}$ *is the least extensional extension of* $=_\Delta$*.*

Proof. It is immediate to check that $=_{\Delta\eta}$ is extensional. In fact, for $x \notin \mathrm{FV}(M)$, $Mx =_{\Delta\eta} Nx$ implies $\lambda x.Mx =_{\Delta\eta} \lambda x.Nx$ (since $=_{\Delta\eta}$ is a congruence), and this implies, $M =_{\Delta\eta} N$ by $=_\eta$.
On the other hand, let $\mathcal{T}$ be a fully extensional Δ-theory, i.e. $Mx =_\mathcal{T} Nx$ implies $M =_\mathcal{T} N$. For $x \notin \mathrm{FV}(M)$, $(\lambda x.Mx)x =_\mathcal{T} Mx$, since $(\lambda x.Mx)x \rightarrow_\Delta Mx$, and thus by (EXT), $\lambda x.Mx =_\mathcal{T} M$. So $\mathcal{T}$ is closed under $=_\eta$. □

In the literature, full extensionality is called simply extensionality. We use this name to stress the fact that it is also possible to define weaker notions of extensionality. We will develop this topic in Sect. 8.1.

2. The Call-by-Name λ-Calculus

A parameter passing policy is said to be *call-by-name* if the parameters need not be evaluated in order to be supplied to the function. In our setting, this means that all terms can be considered as input values. So, in order to mimic this policy with the parametric $\lambda\Delta$-calculus, it is sufficient to define $\Delta = \Lambda$. Then all terms are input values, and every application of the shape $(\lambda x.M)N$ is a redex. The $\lambda\Lambda$-calculus coincides with the standard λ-calculus, defined by Church [25], and the reduction $\rightarrow_\Lambda$ is the well known β-reduction.

2.1 The Syntax of $\lambda\Lambda$-Calculus

By the definition of $\rightarrow_\Lambda$, in the $\lambda\Lambda$-calculus the head of a term is either a variable or a redex. If the head of M is a variable then M *is in Λ-head normal form* (Λ-hnf), namely M is of the shape $\lambda x_1...x_n.zM_1...M_m$ $(n, m \in \mathbb{N})$. M *has a Λ-head normal form* if it reduces to a term in Λ-hnf. Λ-HNF denotes the set of all Λ-head normal forms.

It is easy to see that M *is in Λ-normal form* (Λ-nf) if and only if both its head is a variable and its arguments are in Λ-normal form too. So the set of terms having Λ-hnf strictly includes the set of terms having Λ-nf. Consider, for example, the term $\lambda x.x(DD)$; it is in Λ-hnf, but it does not have Λ-nf. An example of a term having neither Λ-hnf nor Λ-nf is DD.

A term *is in Λ-lazy head normal form* (Λ-lhnf) if and only it is either an abstraction or a head normal form. A term *has a Λ-lazy head normal form* if and only if it reduces to a lazy head normal form. Λ-LHNF denotes the set of all Λ-lazy head normal forms. Clearly $\lambda x.DD$ is a Λ-lhnf, but it has neither Λ-hnf nor Λ-nf. In the literature, a Λ-lazy head normal form is called weak-head normal form. We changed this terminology to stress the fact that to reach a Λ-lazy head normal form it is not necessary to reduce the Λ-redexes that do not occur under the scope of a λ-abstraction. Both Λ-head normal forms and Λ-lazy head normal forms are important classes of terms from the computational point of view.

The general definition of Δ-solvability is given in Definition 1.2.16. In the $\lambda\Lambda$-calculus, solvable terms have a very nice syntactical characterization.

Theorem 2.1.1 (Λ-Solvability).
A term is Λ-solvable if and only if it has a Λ-head normal form.

Proof. The proof is in Sect. 2.1.1. □

Let us notice that the Λ-head normal form of a term is not unique. Consider $\lambda x.(\lambda uv.u)x(DD)(II)$. It reduces both to $\lambda x.x(II)$ and to $\lambda x.xI$, which are both Λ-head normal forms. But it is easy to show that all the Λ-hnf's obtained by Λ-reduction from the same term share some structural properties.

First we need to introduce some naming. If $M \equiv \lambda x_1 \dots x_n.zM_1 \dots M_m$, then n is the Λ-*order* of M and m is its Λ-*degree.*

Property 2.1.2. Let M be Λ-solvable. Then there are unique $n, m \in \mathbb{N}$ such that $M \to^*_\Lambda N$ and N in Λ-hnf, imply that the Λ-order and the Λ-degree of N are respectively n and m.

Proof. By contraposition, let M have two Λ-head normal forms with different Λ-order and Λ-degree, i.e. $M \to^*_\Lambda P_1 \equiv \lambda x_1 ... x_n.xM_1...M_m$ and $M \to^*_\Lambda P_2 \equiv \lambda x_1...x_p.xN_1...N_q$, where $n \neq p$ and/or $m \neq q$. By the confluence theorem, it must be a term Q such that both $P_1 \to^*_\Lambda Q$ and $P_2 \to^*_\Lambda Q$. But this is impossible, since the only redexes can occur in M_i or in N_j ($1 \leq i \leq m$, $1 \leq j \leq q$), and their reduction cannot change any of n, m, p, q. □

The notion of Λ-order of a term can be easily extended to terms not in Λ-head normal form.

Definition 2.1.3. *A term M has* Λ-order n *if and only if n is the largest i such that $M =_\Lambda \lambda x_1 \dots x_i.N$. If such an n does not exist M has Λ-order ∞.*

Example 2.1.4. DD and $xM_1 \dots M_m$ $(m \geq 0)$ have Λ-order 0; while both $\lambda x_1 \dots x_n.DD$ and $\lambda x_1 \dots x_n.z$ have Λ-order n and $(\lambda xy.xx)(\lambda xy.xx)$ has Λ-order ∞, since $\forall k \in \mathbb{N}$ $(\lambda xy.xx)(\lambda xy.xx) \to^*_\Lambda \lambda x_0...x_k.(\lambda xy.xx)(\lambda xy.xx)$.

A particularly interesting Λ-theory is the theory $\Lambda\eta$. The Λ-normal forms play an important role in this theory, as shown in the next theorem.

Theorem 2.1.5 (Böhm's theorem). [15]
Let $M, N \in \Lambda$-NF. If $M \neq_{\Lambda\eta} N$ then M and N are Λ-separable.

Proof. The proof is in Sect. 2.1.2. □

Böhm's theorem has an interesting semantical consequence, namely that two Λ-nf's that are $\neq_{\Lambda\eta}$ cannot be equated in any consistent or input consistent Λ-theory (note that, for the $\lambda\Lambda$-calculus, consistency and input consistency coincide).

Corollary 2.1.6. *Let M, N be two Λ-normal forms and let $M \neq_{\Lambda\eta} N$. For every Λ-theory $\mathcal{T}$, if $M =_\mathcal{T} N$ then $\mathcal{T}$ is (input) inconsistent.*

Proof. The proof is identical to the proof of Property 1.3.6, just putting $\Delta = \Lambda$. □

Given a Λ-theory, there is an easy way of proving its full extensionality, as shown in the following property.

Property 2.1.7. Let $E \equiv \lambda xy.xy$, and let $\mathcal{T}$ be a Λ-theory.
$I =_{\mathcal{T}} E$ if and only if $\mathcal{T}$ is fully extensional.

Proof. ($\Rightarrow$) $I =_{\mathcal{T}} E$ implies $IM =_{\mathcal{T}} EM$, which implies, by Λ-reduction $M =_{\mathcal{T}} \lambda x.Mx$, where $x \notin \mathrm{FV}(M)$. The proof follows, by Theorem 1.3.8.
($\Leftarrow$) By Theorem 1.3.8 $x =_{\mathcal{T}} \lambda y.xy$ where $y \notin \mathrm{FV}(M)$; so $\lambda x.x =_{\mathcal{T}} \lambda xy.xy$, since $=_{\mathcal{T}}$ is a congruence. □

We will prove in Chap. 16 that the $\lambda\Lambda$-calculus can be considered as a programming language, in the sense that it is possible to define some evaluation machines performing the Λ-reduction, and the $\lambda\Lambda$-calculus equipped by each of this machines, has the computational power of all the partial computable functions. The key property on which this result is based is the fact that every term in the $\lambda\Lambda$-calculus has a fixed point.

Theorem 2.1.8 (Call-by-name fixed point).
Every term $M \in \Lambda$ has a fixed point, i.e. for every term M there is a term N such that $MN =_\Lambda N$.

Proof. Let $Y \equiv \lambda x.(\lambda y.x(yy))(\lambda y.x(yy))$. It is readily checked that, for every M, $YM =_\Lambda M(YM)$. Hence YM is a fixed point of M. □

The term Y in the proof of the previous theorem is called a *call-by-name fixed point operator* since, when applied to a term M, it produces one of its fixed points.

2.1.1 Proof of Λ-Solvability Theorem

First we need to prove a property.

Property 2.1.9. (i) The lack of Λ-hnf is preserved by substitution, i.e. if M does not have Λ-hnf then $M[N/y]$ does not have Λ-hnf either, for all $x \in \mathrm{Var}$ and $N \in \Lambda$.
(ii) The lack of Λ-hnf is preserved by head contexts, i.e. if M does not have Λ-hnf then $(\lambda\vec{x}.M)\vec{N}$ does not have Λ-hnf either, for all $\vec{x}$ and $\vec{N}$.

Proof. (i) By contraposition assume that $M[N/y]$ has Λ-hnf. We will prove that this implies that M has Λ-hnf too. The proof is given by induction on the length p of standard Λ-reduction sequence from $M[N/y]$ to its Λ-hnf. The cases $p = 0, 1$ are trivial. Let $p > 1$ and $M \equiv \lambda\vec{x}.(\lambda z.P)Q\vec{M}$, otherwise M is already in Λ-hnf, and let $R \equiv \lambda\vec{x}.P[Q/z]\vec{M}$. Then

$$M[N/y] \equiv \lambda\vec{x}.(\lambda z.P')Q'\vec{M}' \to_\Lambda \lambda\vec{x}.P'[Q'/z]\vec{M}' \equiv R[N/y],$$

where $P' \equiv P[N/y]$, $Q' \equiv Q[N/y]$ and $\vec{M}' \equiv \vec{M}[N/y]$. Thus $R[N/y]$ has Λ-hnf in less than p steps, so by induction R has Λ-hnf, and by the Church Rosser theorem M has Λ-hnf too.

(ii) We assume that $(\lambda\vec{x}.M)\vec{N}$ has Λ-hnf and we prove that this implies that M has Λ-hnf too. The proof is given by induction on the length p of a standard Λ-reduction sequence from $(\lambda\vec{x}.M)\vec{N}$ to its Λ-hnf. The cases $p = 0, 1$ are trivial. Let $p > 1$ and $M \equiv \lambda\vec{y}.(\lambda z.P)Q\vec{M}$. If $\|\vec{y}\| + \|\vec{x}\| \geq \|\vec{N}\|$ then the proof follows from part (i) of this property and from the confluence property of the Λ-reduction. Otherwise $\exists\vec{N}_1$ such that $\|\vec{N}_1\| = \|\vec{y}\| + \|\vec{x}\|$, $\vec{N} \equiv \vec{N}_1\vec{N}_2$ and $\|\vec{N}_2\| > 1$. In this case, it must be

$$(\lambda\vec{x}.(\lambda\vec{y}.(\lambda z.P)Q\vec{M}))\vec{N}_1\vec{N}_2 \to_\Lambda^* (\lambda z.P')Q'\vec{M}'\vec{N}_2 \to_\Lambda P'[Q'/z]\vec{M}'\vec{N}_2,$$

where $P' \equiv P[\vec{N}_1/\vec{x}\vec{y}]$, $Q' \equiv Q[\vec{N}_1/\vec{x}\vec{y}]$ and $\forall i,\ M_i' \equiv M_i[\vec{N}_1/\vec{x}\vec{y}]$. $P'[Q'/z]\vec{M}'\vec{N}_2$ has Λ-hnf in fewer steps than $(\lambda\vec{x}.M)\vec{N}$, so by induction $\lambda\vec{y}.P[Q/z]\vec{M}$ has Λ-hnf and by confluence $(\lambda\vec{x}.M)\vec{N}$ has Λ-hnf too. □

Note that analogous properties have been proved for the Λ-unsolvable terms (Property 1.2.18).

Now we are able to prove the theorem.

▼ **Proof of Λ-Solvability Theorem** (Theorem 2.1.1 pag. 26).

(⇐) Without loss of generality, we can assume that M is closed. Let $M \equiv \lambda x_1...x_n.x_iM_1...M_m$ $(1 \leq i \leq n)$. Let $P_i \equiv \lambda x_1...x_{m+1}.x_{m+1}$. Then for every sequence $P_1...P_i...P_n$, where P_j is any term, for $i \neq j$,

$$MP_1...P_i...P_n =_\Lambda I.$$

(⇒) If M does not have Λ-hnf, then by Property 2.1.9, for all head contexts $C[.]$, $C[M]$ does not have Λ-hnf; in particular, $C[M]$ cannot be reduced to I. ■

2.1.2 Proof of Böhm's Theorem

The proof will be given in a constructive way, by showing a *separability algorithm.* The algorithm is defined as a formal system, proving statements of the shape:

$$M, N \Rightarrow_\Lambda C[.],$$

where M, N are Λ-normal forms such that $M \neq_{\Lambda\eta} N$ and $C[.]$ is a context. (A very general presentation of formal systems can be found at the beginning of Chap. 5).

The rules of the system are defined by induction on the fact that M, N are Λ-normal forms that are η-different.

Definition 2.1.10. *Let c be a sequence of $n \geq 0$ natural numbers (ϵ denotes the empty sequence) and M, N be Λ-normal forms. $M \not\sim_c N$ if and only if one of following cases arises:*

(i) *if $c \equiv \epsilon$ then either $|p - m| \neq |q - n|$ or $x \not\equiv y$;*
(ii) *if $c \equiv i, c'$ then $M =_\eta \lambda x_1 \ldots x_p . x M_1 \ldots M_m$ and $N =_\eta \lambda x_1 \ldots x_p . y N_1 \ldots N_m$ where $M_i \not\sim_{c'} N_i$ $(1 \leq i \leq m)$.*

Property 2.1.11. Let M and N be Λ-nf's such that $M \neq_{\Lambda\eta} N$. Then there is a sequence c of natural numbers such that $M \not\sim_c N$.

Proof. Easy. □

Some terms will be used extensively in the rest of this section, in particular

$$\begin{aligned} B^n &\equiv \lambda x_1 ... x_{n+1} . x_{n+1} x_1 ... x_n \\ O^n &\equiv \lambda x_1 ... x_{n+1} . x_{n+1} \\ U^i_n &\equiv \lambda x_1 ... x_i . x_n \qquad (i \leq n,\ n \in \mathbb{N}). \end{aligned}$$

A useful structural measure of a term M in Λ-nf is the maximum Λ-degree of its subterms.

Definition 2.1.12. *Let $M \in \Lambda$-NF; $\mathsf{args}(M) \in \mathbb{N}$ is defined inductively as:*

- $\mathsf{args}(x M_1 ... M_m) = \max\{m, \mathsf{args}(M_1), ..., \mathsf{args}(M_n)\}$;
- $\mathsf{args}(\lambda x . M) = \mathsf{args}(M)$.

Example 2.1.13. Let $M \equiv \lambda x . x(\lambda xy . x) x (xu)$; so

$$\mathsf{args}(M) = \max\{3, \mathsf{args}(\lambda xy.x), \mathsf{args}(x), \mathsf{args}(xu)\} = 3.$$

It is easy to check that if N is a subterm of M then $\mathsf{args}(N) \leq \mathsf{args}(M)$.

Definition 2.1.14. *Let M be a term having Λ-normal form. The Λ-normal form of M will be denoted by $\mathsf{nf}_\Lambda(M)$.*

The separability algorithm is presented in Fig. 2.1 (pag. 32). For the sake of simplicity, we assume that all bound and free variables have different names.

The following lemma proves a property on which both the termination and the correctness proofs of the algorithm are based. In fact, rule ($\Lambda 7$) of the algorithm is based on it.

Lemma 2.1.15.
Let $M, N \in \Lambda$-NF, $r \geq \max\{\mathsf{args}(M), \mathsf{args}(N)\}$ and $C^r_x[.] \equiv (\lambda x.[.]) B^r$.

(i) $\exists \bar{M} \in \Lambda$-*NF such that* $C_x^r[M] \to_\Lambda^* \bar{M}$ *and* $r \geq \mathsf{args}(\bar{M})$.
(ii) *If* $M \not\approx_c N$ *then* $\mathsf{nf}_\Lambda(C_x^r[M]) \not\approx_c \mathsf{nf}_\Lambda(C_x^r[N])$.

Proof. (i) By induction on M.
If $M \equiv \lambda z.P$ or $M \equiv zM_1...M_m$ (where $z \not\equiv x$ and $m \leq r$) then the proof follows by induction. Let $M \equiv xM_1...M_m$ ($m \leq r$); so by induction $\forall i \leq m$ there is $\bar{M}_i \in \Lambda$-NF such that $C_x^r[M_i] \to_\Lambda^* \bar{M}_i$ and $r \geq \mathsf{args}(\bar{M}_i)$. Clearly $(\lambda x.M)B^r \to_\Lambda^* \lambda x_{m+1}...x_{r+1}.x_{r+1}\bar{M}_1...\bar{M}_m x_{m+1}...x_r$; hence

$$r \geq \max\{r, \mathsf{args}(\bar{M}_1), ..., \mathsf{args}(\bar{M}_m), \underbrace{0, ...,0}_{r-m}\} = r.$$

Note that $\mathsf{nf}_\Lambda(C_x^r[M])$ is well defined.
(ii) Let $M \equiv \lambda z_1...z_p.zM_1...M_m$ and $N \equiv \lambda y_1...y_q.yN_1...N_n$; we reason by induction on c. Let $c \equiv \epsilon$. Let $z \equiv y$.
If x is different from y, z then the proof is trivial.
In case $|p-m| \neq |q-n|$, let $\bar{M}_i \equiv \mathsf{nf}_\Lambda(C_x^r[M_i])$ and $\bar{N}_i \equiv \mathsf{nf}_\Lambda(C_x^r[N_i])$ for each i; thus

$$\mathsf{nf}_\Lambda(C_x^r[M]) \equiv \lambda z_1...z_p x_{m+1}...x_{r+1}.x_{r+1}\bar{M}_1...\bar{M}_m x_{m+1}...x_r,$$
$$\mathsf{nf}_\Lambda(C_x^r[N]) \equiv \lambda z_1...z_q x_{n+1}...x_{r+1}.x_{r+1}\bar{N}_1...\bar{N}_n x_{n+1}...x_r\,.$$

Since $|p-m| \neq |q-n|$,

$$|(p+(r+1)-m)-r| = |p-m+1| \neq |q-n+1| = |(q+(r+1)-n)-r|.$$

If $z \not\equiv y$ then the proof is simpler.
If $c \equiv i, c'$ (where $i \geq 1$) then the proof follows by induction. □

Example 2.1.16. Let $M \equiv \lambda xyu.x(u(x(yy))(vv))$ and $N \equiv \lambda xyu.x(u(yy)(vv))$. Thus $\mathsf{args}(M) = \mathsf{args}(N) = 2$, so let us pose $r = 2$.
The derivation proving the statement $M, N \Rrightarrow_\Lambda C[.]$ follows:

$$\cfrac{\cfrac{\cfrac{\cfrac{\cfrac{x_3 \not\equiv y \qquad C_5[.] \equiv (\lambda x_3 y.[.])(\lambda x_1 x_2.\tilde{x})(\lambda x_1 x_2 x_3.\tilde{y})}{x_3(yy)x_2\,,\, yyx_2x_3 \Rrightarrow_\Lambda C_5[.]}\,(\Lambda 5)}{\lambda x_2x_3.x_3(yy)x_2\,,\, yy \Rrightarrow_\Lambda C_2[.] \equiv C_5[[.]x_2x_3]}\,(\Lambda 2)}{u(\lambda x_2x_3.x_3(yy)x_2)(vv)\,,\, u(yy)(vv) \Rrightarrow_\Lambda C_6[.] \equiv C_2[(\lambda u.[.])(\lambda z_1z_2.z_1)]}\,(\Lambda 6)}{x(u(x(yy))(vv))\,,\, x(u(yy)(vv)) \Rrightarrow_\Lambda C_7[.] \equiv C_6[(\lambda x.[.])(\lambda x_1x_2x_3.x_3x_1x_2)I(\lambda z_1z_2.z_1)]}\,(\Lambda 7)}{\lambda xyu.x(u(x(yy))(vv))\,,\, \lambda xyu.x(u(yy)(vv)) \Rrightarrow_\Lambda C_7[[.]xyu]}\,(\Lambda 1)$$

where:

$$\begin{array}{lll}
C_5[.] \equiv & (\lambda x_3y.\ [.]\ ) & (\lambda x_1x_2.\tilde{x})(\lambda x_1x_2x_3.\tilde{y}) \\
C_2[.] \equiv & (\lambda x_3y.\ [.]\ x_2x_3) & (\lambda x_1x_2.\tilde{x})(\lambda x_1x_2x_3.\tilde{y}) \\
C_6[.] \equiv & (\lambda x_3y.((\lambda u.\ [.]\ )U_2^1x_2x_3)) & (\lambda x_1x_2.\tilde{x})(\lambda x_1x_2x_3.\tilde{y}) \\
C_7[.] \equiv & (\lambda x_3y.((\lambda u.(\lambda x.\ [.]\ )B^2IU_2^1)U_2^1x_2x_3)) & (\lambda x_1x_2.\tilde{x})(\lambda x_1x_2x_3.\tilde{y}) \\
C[.] \equiv & (\lambda x_3y.((\lambda u.(\lambda x.\ [.]\ xyu)B^2IU_2^1)U_2^1x_2x_3)) & (\lambda x_1x_2.\tilde{x})(\lambda x_1x_2x_3.\tilde{y})\,.
\end{array}$$

So

$$
\begin{aligned}
C[M] \equiv\ & (\lambda x_3 y.((\lambda u.(\lambda x.Mxyu)B^2IU_2^1)U_2^1x_2x_3))(\lambda x_1x_2.\tilde{x})(\lambda x_1x_2x_3.\tilde{y}) \to_\Lambda^* \\
& (\lambda x_3 y.((\lambda u.MB^2yuIU_2^1)U_2^1x_2x_3))(\lambda x_1x_2.\tilde{x})(\lambda x_1x_2x_3.\tilde{y}) \to_\Lambda^* \\
& (\lambda x_3 y.(MB^2yU_2^1IU_2^1x_2x_3))(\lambda x_1x_2.\tilde{x})(\lambda x_1x_2x_3.\tilde{y}) \to_\Lambda^* \\
& (\lambda xyu.x(u(x(yy))(vv)))B^2(\lambda x_1x_2x_3.\tilde{y})U_2^1IU_2^1x_2(\lambda x_1x_2.\tilde{x}) \to_\Lambda^* \\
& B^2(U_2^1(B^2((\lambda x_1x_2x_3.\tilde{y})(\lambda x_1x_2x_3.\tilde{y})))(vv))IU_2^1x_2(\lambda x_1x_2.\tilde{x}) \to_\Lambda^* \\
& U_2^1(U_2^1(B^2((\lambda x_1x_2x_3.\tilde{y})(\lambda x_1x_2x_3.\tilde{y})))(vv))Ix_2(\lambda x_1x_2.\tilde{x}) \to_\Lambda^* \\
& U_2^1(B^2((\lambda x_1x_2x_3.\tilde{y})(\lambda x_1x_2x_3.\tilde{y})))(vv)x_2(\lambda x_1x_2.\tilde{x}) \to_\Lambda^* \\
& B^2((\lambda x_1x_2x_3.\tilde{y})(\lambda x_1x_2x_3.\tilde{y}))x_2(\lambda x_1x_2.\tilde{x}) \to_\Lambda^* \\
& (\lambda x_1x_2.\tilde{x})((\lambda x_1x_2x_3.\tilde{y})(\lambda x_1x_2x_3.\tilde{y}))x_2 \to_\Lambda^* \tilde{x}\,,
\end{aligned}
$$

while on the other hand

$$
\begin{aligned}
C[N] \equiv\ & (\lambda x_3 y.((\lambda u.(\lambda x.Nxyu)B^2IU_2^1)U_2^1x_2x_3))(\lambda x_1x_2.\tilde{x})(\lambda x_1x_2x_3.\tilde{y}) \to_\Lambda^* \\
& (\lambda x_3 y.((\lambda u.NB^2yuIU_2^1)U_2^1x_2x_3))(\lambda x_1x_2.\tilde{x})(\lambda x_1x_2x_3.\tilde{y}) \to_\Lambda^* \\
& (\lambda x_3 y.(NB^2yU_2^1IU_2^1x_2x_3))(\lambda x_1x_2.\tilde{x})(\lambda x_1x_2x_3.\tilde{y}) \to_\Lambda^* \\
& (\lambda xyu.x(u(yy)(vv)))B^2(\lambda x_1x_2x_3.\tilde{y})U_2^1IU_2^1x_2(\lambda x_1x_2.\tilde{x}) \to_\Lambda^* \\
& B^2(U_2^1((\lambda x_1x_2x_3.\tilde{y})(\lambda x_1x_2x_3.\tilde{y}))(vv))IU_2^1x_2(\lambda x_1x_2.\tilde{x}) \to_\Lambda^* \\
& U_2^1(U_2^1((\lambda x_1x_2x_3.\tilde{y})(\lambda x_1x_2x_3.\tilde{y}))(vv))Ix_2(\lambda x_1x_2.\tilde{x}) \to_\Lambda^* \\
& U_2^1((\lambda x_1x_2x_3.\tilde{y})(\lambda x_1x_2x_3.\tilde{y}))(vv)x_2(\lambda x_1x_2.\tilde{x}) \to_\Lambda^* \\
& (\lambda x_1x_2x_3.\tilde{y})(\lambda x_1x_2x_3.\tilde{y})x_2(\lambda x_1x_2.\tilde{x}) \to_\Lambda^* \tilde{y}\,.
\end{aligned}
$$

Now we will prove that the algorithm is correct and complete.

Lemma 2.1.17 (Termination).
If $M, N \in \Lambda$-NF and $M \not\approx_c N$ then $M, N \Rrightarrow_\Lambda C[.]$.

Proof. By induction on c.
Let $c = \epsilon$. Let us consider first the case when M and N have no initial abstractions. If they have different head variables, then axiom ($\Lambda 5$) must be applied, otherwise either axiom ($\Lambda 3$) or axiom ($\Lambda 4$), and then the algorithm stops. If they have initial abstractions, then either rule ($\Lambda 1$) or ($\Lambda 2$) must be applied, and the previous situation is reached.
If $c \neq \epsilon$, either rule ($\Lambda 6$) or ($\Lambda 7$) must be used, and then the result follows, in the first case by induction, in the second one by induction and Lemma 2.1.15. □

Lemma 2.1.18 (Correctness).
Let $M, N \in \Lambda$-NF be such that $M \not\approx_c N$. If $M, N \Rrightarrow_\Lambda C[.]$ then $C[M] =_\Lambda \tilde{x}$ and $C[N] =_\Lambda \tilde{y}$.

Proof. By induction on the derivation of $M, N \Rrightarrow_\Lambda C[.]$, i.e. by cases on the last applied rule.

Let $M, N \in \Lambda$-normal form, $M \not\sim_c N$, $r \geq \max\{\mathsf{args}(M), \mathsf{args}(N)\}$ and $\tilde{x}, \tilde{y}$ be fresh variables such that $\tilde{x} \not\equiv \tilde{y}$.

The rules of the system proving statements $M, N \Rrightarrow_\Lambda C[.]$, are the following:

$$\frac{p \leq q \qquad xM_1...M_m x_{p+1}...x_q, yN_1...N_n \Rrightarrow_\Lambda C[.]}{\lambda x_1...x_p.xM_1...M_m, \lambda x_1...x_q.yN_1...N_n \Rrightarrow_\Lambda C[[.]x_1...x_q]} \ (\Lambda 1)$$

$$\frac{q < p \qquad xM_1...M_m, yN_1...N_n x_{q+1}...x_p \Rrightarrow_\Lambda C[.]}{\lambda x_1...x_p.xM_1...M_m, \lambda x_1...x_q.yN_1...N_n \Rrightarrow_\Lambda C[[.]x_1...x_p]} \ (\Lambda 2)$$

$$\frac{n < m}{xM_1...M_m, xN_1...N_n \Rrightarrow_\Lambda (\lambda x.[.])O^m \underbrace{I.....I}_{m-n-2} K\tilde{x}\tilde{y}} \ (\Lambda 3)$$

$$\frac{m < n}{xM_1...M_m, xN_1...N_n \Rrightarrow_\Lambda (\lambda x.[.])O^n \underbrace{I.....I}_{n-m-2} K\tilde{y}\tilde{x}} \ (\Lambda 4)$$

$$\frac{x \not\equiv y}{xM_1...M_m, yN_1...N_n \Rrightarrow_\Lambda (\lambda xy.[.])(\lambda x_1...x_m.\tilde{x})(\lambda x_1...x_n.\tilde{y})} \ (\Lambda 5)$$

$$\frac{x \notin \mathrm{FV}(M_k) \cup \mathrm{FV}(N_k) \qquad M_k \neq_{\Lambda\eta} N_k \qquad M_k, N_k \Rrightarrow_\Lambda C[.]}{xM_1...M_m, xN_1...N_m \Rrightarrow_\Lambda C[(\lambda x.[.])U^k_m]} \ (\Lambda 6)$$

$$\frac{x \in \mathrm{FV}(M_k) \cup \mathrm{FV}(N_k) \qquad M_k \neq_{\Lambda\eta} N_k \qquad C^r_x[.] \equiv (\lambda x.[.])B^r \qquad \mathsf{nf}_\Lambda(C^r_x[M_k]), \mathsf{nf}_\Lambda(C^r_x[N_k]) \Rrightarrow_\Lambda C[.]}{xM_1...M_m, xN_1...N_m \Rrightarrow_\Lambda C[C^r_x[.] \underbrace{I.....I}_{r-m} U^k_r]} \ (\Lambda 7)$$

Fig. 2.1. Call-by-name separability algorithm.

($\Lambda 1$) By induction $C[xM_1...M_m x_{p+1}...x_q] \rightarrow^*_\Lambda \tilde{x}$ and $C[yN_1...N_n] \rightarrow^*_\Lambda \tilde{y}$; so

$$C[(\lambda x_1...x_p.xM_1...M_m)x_1...x_q] \rightarrow^*_\Lambda C[xM_1...M_m x_{p+1}...x_q] \rightarrow^*_\Lambda \tilde{x}$$
$$C[(\lambda x_1...x_q.yN_1...N_n)x_1...x_q] \rightarrow^*_\Lambda C[yN_1...N_n] \rightarrow^*_\Lambda \tilde{y}.$$

($\Lambda 2$) Similar to ($\Lambda 1$).
($\Lambda 3$) Clearly

$$(\lambda x.xM_1...M_m)O^m \underbrace{I.........I}_{m-n-2} K\tilde{x}\tilde{y} \rightarrow^*_\Lambda$$

$$O^m \underbrace{M_1[O^m/x]...M_m[O^m/x]}_{m} \underbrace{I.........I}_{m-n-2} K\tilde{x}\tilde{y} \rightarrow^*_\Lambda K\tilde{x}\tilde{y} \rightarrow^*_\Lambda \tilde{x}$$

while on the other hand,

$$(\lambda x.xN_1...N_n)O^m \underbrace{I.........I}_{m-n-2} K\tilde{x}\tilde{y} \rightarrow^*_\Lambda$$

$$O^m \underbrace{N_1[O^m/x]...N_n[O^m/x]\ I.........I\ K\ \tilde{x}\, \tilde{y}}_{m} \rightarrow^*_\Lambda \tilde{y}.$$

($\Lambda 4$) Similar to ($\Lambda 3$).
($\Lambda 5$) Easy.
($\Lambda 6$) By induction.
($\Lambda 7$) By induction $C[\mathsf{nf}_\Lambda(C^r_x[M_k])] \rightarrow^*_\Lambda \tilde{x}$ and $C[\mathsf{nf}_\Lambda(C^r_x[N_k]) \rightarrow^*_\Lambda \tilde{y}$, where $C^r_x[.] \equiv (\lambda x.[.])B^r$; thus $C[M_k[B^r/x]] \rightarrow^*_\Lambda \tilde{x}$ and $C[N_k[B^r/x]] \rightarrow^*_\Lambda \tilde{y}$ too.
Hence

$$C[((\lambda x.xM_1...M_m)B^r) \underbrace{I.........I}_{r-m} U^k_r] \rightarrow^*_\Lambda$$

$$C[B^r M_1[B^r/x]...M_m[B^r/x] \underbrace{I.........I}_{r-m} U^k_r] \rightarrow^*_\Lambda$$

$$C[U^k_r M_1[B^r/x]...M_m[B^r/x] \underbrace{I.........I}_{r-m}] \rightarrow^*_\Lambda C[M_k[B^r/x]].$$

The proof of $C[((\lambda x.xN_1...N_m)B^r) \underbrace{I.........I}_{r-m} U^k_r] \rightarrow^*_\Lambda \tilde{y}$ is similar. □

▼ **Proof of Böhm's Theorem** (Theorem 2.1.5 pag. 26).
The proof follows directly from Lemmas 2.1.17 and 2.1.18. ■

Note that $M, N \Rrightarrow_\Lambda C[.]$ does not imply that $C[.]$ is a head context. The original algorithm designed by Böhm produces head contexts. However, the proof of correctness for our version is simpler than that of Böhm. It can be an useful exercise for the reader to modify the algorithm of Fig. 2.1 in such a way that it produces as output a head context.

3. The Call-by-Value λ-Calculus

The more usual programming languages are such that parameters must be evaluated in order to be supplied to a function, and moreover the body of a function is evaluated only when parameters are supplied. The first policy is the so called *call-by-value* parameter passing, and the second policy is called *lazy*-evaluation. In order to mimic this kind of computation with the parametric $\lambda\Delta$-calculus, it is necessary that Δ be a proper subset of Λ, and moreover it contain all the abstraction terms.

So we choose $\Delta = \Gamma$, where $\Gamma = \mathrm{Var} \cup \{\lambda x.M \mid M \in \Lambda\}$ was proved to be a set of input values in Property 1.2.4. The $\lambda\Gamma$-calculus coincides with the $\lambda\beta_v$-calculus, first introduced by Plotkin in [78].

3.1 The Syntax of the $\lambda\Gamma$-Calculus

A term of the $\lambda\Gamma$-calculus is always of the shape: $\lambda x_1 ... x_n.\zeta M_1 ... M_m$, where the head ζ is either a variable or a Γ-redex or a head block (see pag. 8).
A term is in *Γ-normal form* (Γ-nf) if it is of the shape $\lambda x_1 ... x_n.\zeta M_1 ... M_m$, where M_i is in Γ-normal form ($1 \leq i \leq m$) and ζ is either a variable or a head block $(\lambda x.P)Q$, where both P and Q are in Γ-normal form. Γ-NF denotes the set of all Γ-normal forms.

Example 3.1.1. Both xID and $(\lambda x.xI)(yz)w$ are terms in Γ-normal form. DD is a term without Γ-normal form.

Note that, differently from the $\lambda\Lambda$-calculus, here if we want to manipulate some subterms, we need first to transform them into input values. So the notions of Γ-valuable and potentially Γ-valuable terms are important for studying such a calculus.

Definition 3.1.2. (i) *A term M is* Γ-valuable *if and only if there is $N \in \Gamma$ such that $M \to^*_\Gamma N$.*
(ii) *A term M is* potentially Γ-valuable *if and only if there is a substitution* s*, replacing variables by closed terms belonging to Γ, such that* s(M) *is Γ-valuable.*

It is readily verified that a closed term is potentially Γ-valuable if and only if it is valuable. Note that a term can be in Γ-normal form and not potentially Γ-valuable; consider, for example, the term $M \equiv (\lambda z.D)(yI)D$, which is in Γ-normal form. For each term $Q \in \Lambda^0$, the term $M[Q/y] \equiv (\lambda z.D)(QI)D$ is not Γ-valuable; indeed, there are two possible cases:

1. QI is Γ-valuable. Then $M[Q/y] \to_\Gamma^* DD$ and DD is not Γ-valuable, being closed and such that $DD \to_\Gamma DD \notin \Gamma$.
2. QI is not Γ-valuable. Then $(\lambda z.D)Q'D$ is not Γ-valuable, for every Q' such that $QI \to_\Gamma^* Q'$, since $(\lambda z.D)Q'$ is not a Γ-redex.

So to be potentially Γ-valuable is a stronger and more interesting property than to have Γ-normal form.

The class of potentially Γ-valuable terms cannot be characterized through the $\to_\Gamma$ reduction; a new kind of reduction must be defined.

Definition 3.1.3. *Let $\Psi \subseteq \Lambda$.*

(i) *The* lazy Ψ-reduction *($\to_{\Psi\ell}$) is the closure under application of the following rule:*

$$(\lambda x.M)N \to M[N/x] \textit{ if and only if } N \in \Psi;$$

$(\lambda x.M)N$, when it does not occur under the scope of a λ-abstraction, and when $N \in \Psi$, is called a $\Psi\ell$-redex *(or lazy Ψ-redex) and $M[N/x]$ is called its* $\Psi\ell$-contractum *(or lazy Ψ-contractum).*

(ii) *$\to_{\Psi\ell}^*$ and $=_{\Psi\ell}$ are respectively the reflexive and transitive closure of $\to_{\Psi\ell}$ and the symmetric, reflexive and transitive closure of $\to_{\Psi\ell}$.*

(iii) *A term* is in $\Psi\ell$-normal form *($\Psi\ell$-nf) if it has not $\Psi\ell$-redexes and it* has a $\Psi\ell$-normal form, *or it is* $\Psi\ell$-normalizing *if it reduces to a $\Psi\ell$-normal form; the set of $\Psi\ell$-nf is denoted by $\Psi\ell$-NF.*

(iv) *A term is* $\Psi\ell$-strongly normalizing *if it is $\Psi\ell$-normalizing and moreover there is not an infinite $\Psi\ell$-reduction sequence starting from it.*

Let us notice that, in the previous definition, Ψ is not asked to be a set of input values. Moreover, the definition of $\Psi\ell$-reduction, in point (i), does not agree with Definition 1.2.1. In fact, the reduction is defined by closing the reduction rule only under application, while in the standard case the closure is under abstraction too. This allows us to formalize the notion of lazy reduction, where no reduction can be made under the scope of a λ-abstraction.

Potentially Γ-valuable terms will be characterized by the lazy reduction induced by the following subset of Λ.

Definition 3.1.4. *$\Xi \subseteq \Lambda$ is defined as follows:*

$$\Xi = \Gamma \cup \{xM_1 \ldots M_m \mid \forall i \leq m \quad M_i \in \Xi\}.$$

Example 3.1.5. $\lambda x.DD \in \Xi$, $xy(\lambda x.II) \in \Xi$, $I(xy) \notin \Xi$. Note that the last term is in Γ-normal form, while the first two are not.

We will show that terms having $\Xi\ell$-normal forms are all and only the potentially Γ-valuable terms.

Property 3.1.6. Let $M \in \Lambda$.
A term M has $\Xi\ell$-normal form if and only if $M \to^*_{\Xi\ell} P$ for some $P \in \Xi$.

Proof. It is easy to see that $M \in \Xi$ if and only if M is a $\Xi\ell$-normal form. □

Note that Ξ is not a set of input values. In fact, it is easy to see that the contextual reduction $\to_\Xi$ would not be confluent. Let $P \equiv (\lambda x.(\lambda yz.z)(xD))D$. Clearly $P \to_\Xi P_1 \equiv (\lambda yz.z)(DD)$ and $P \to_\Xi P_2 \equiv (\lambda xz.z)D$, but there does not exist a $P_3 \in \Lambda$ such that $P_1 \to^*_\Xi P_3$ and $P_2 \to^*_\Xi P_3$.

Thanks to its "lazy" definition, the $\to_{\Xi\ell}$ reduction enjoys all the good properties we expect.

Theorem 3.1.7. *The $\to_{\Xi\ell}$ reduction enjoys both the confluence and the standardization properties.*

Proof. The proof is in Sect. 3.1.1. □

Moreover, $\to_{\Xi\ell}$ and $\to_\Gamma$ reductions commute as proved by Property 3.1.8.

Property 3.1.8. Let $M \to_{\Xi\ell} P$ and $M \to_\Gamma Q$. Then there is N such that both $Q \to^*_{\Xi\ell} N$ and $P \to^*_\Gamma N$.

Proof. $M \to_{\Xi\ell} P$ implies M is of the shape $\zeta\vec{M}$, where ζ is either a variable, or a Γ-redex, or a head block. Let $M \equiv (\lambda x.R)(z\vec{S})\vec{M}$, since the variable's case is simpler. The proof is given by cases.

1. Let $R \to_\Gamma R'$. It is easy to see that the following diagram commutes:

$$\begin{array}{ccc} (\lambda x.R)(z\vec{S})\vec{M} & \xrightarrow[\Gamma]{} & (\lambda x.R')(z\vec{S})\vec{M} \\ \downarrow{\scriptstyle \Xi\ell} & & \downarrow{\scriptstyle \Xi\ell} \\ R[z\vec{S}/x]\vec{M} & \xrightarrow[\Gamma]{} & R'[z\vec{S}/x]\vec{M} \end{array}$$

2. Let $\vec{S} \equiv S_1...S_j...S_m$ and let $S_j \to_\Gamma S'_j$ $(1 \leq j \leq m)$. $\vec{S}'$ will denote the sequence $S_1...S'_j...S_m$. It is easy to see that the following diagram commutes:

$$\begin{array}{ccc} (\lambda x.R)(z\vec{S})\vec{M} & \xrightarrow[\Gamma]{} & (\lambda x.R)(z\vec{S}')\vec{M} \\ \downarrow{\scriptstyle \Xi\ell} & & \downarrow{\scriptstyle \Xi\ell} \\ R[z\vec{S}/x]\vec{M} & \xrightarrow[\Gamma]{*} & R[z\vec{S}'/x]\vec{M} \end{array}$$

   when a number ≥ 0 of Γ-reductions is needed in order to deal with the copies of $z\vec{S}$ generated by the $\Xi\ell$-reduction.

3. Let $\vec{M} \equiv M_1 ... M_j ... M_m$ and let $M_j \rightarrow_\Gamma M'_j$ $(1 \leq j \leq m)$ and let $\vec{M}'$ denote the sequence $M_1 ... M'_j ... M_m$. It is easy to see that the following diagram commutes:

$$\begin{array}{ccc}
(\lambda x.R)(z\vec{S})\vec{M} & \xrightarrow[\Gamma]{} & (\lambda x.R)(z\vec{S})\vec{M}' \\
\Big\downarrow \Xi\ell & & \Big\downarrow \Xi\ell \\
R[z\vec{S}/x]\vec{M} & \xrightarrow[\Gamma]{} & R[z\vec{S}'/x]\vec{M}'
\end{array}$$

4. The cases when the $\Xi\ell$ and Γ-reductions are made in disjoint subterms of either $\vec{S}$ or $\vec{M}$ are immediate.
5. The cases when the $\Xi\ell$ and Γ-reductions are made in the same subterm of either $\vec{S}$ or $\vec{M}$ can be treated in a similar way as the previous ones.
6. Let $M \equiv (\lambda x.R)S\vec{M}$, where $S \in \Gamma$. Then either $P \equiv Q$, or one of the previous cases applies. □

The $\rightarrow_{\Xi\ell}$-reduction allows a complete characterization of the potentially Γ-valuable terms.

Theorem 3.1.9 (Potential Γ-valuability). [74]
M is potentially Γ-valuable if and only if there is $N \in \Xi$ such that $M \rightarrow^*_{\Xi\ell} N$.

Proof. The proof is in Sect. 3.1.2. □

As an example, let us consider the term $M \equiv (\lambda z.D)(yI)D$, which we proved before to be not potentially Γ-valuable. In fact, $(\lambda z.D)(yI)D \rightarrow_{\Xi\ell} DD$, and clearly DD does not have $\Xi\ell$-normal form, since $DD \rightarrow_{\Xi\ell} DD$.

Now let us study the problem of characterizing the Γ-solvable terms. The next lemma shows us the relationship between the potentially Γ-valuable terms and the Γ-solvable ones.

Lemma 3.1.10. *The class of Γ-solvable terms is properly included in the class of potentially Γ-valuable terms.*

Proof. Let us first prove the inclusion. Let M be Γ-solvable, so there is a head context $(\lambda\vec{x}.[.])\vec{N}$ such that $(\lambda\vec{x}.M)\vec{N} \rightarrow^*_\Gamma I$ (since I is in normal form). Assume $\|\vec{x}\| \leq \|\vec{N}\|$ (otherwise consider the context $(\lambda\vec{x}.[.])\vec{N}\underbrace{I.....I}_{p}$, where $p = \|\vec{x}\| - \|\vec{N}\|$) and $\vec{N} \equiv \vec{N}_1\vec{N}_2$ such that $\|\vec{x}\| = \|\vec{N}_1\|$. So $M[\vec{N}_1/\vec{x}]\vec{N}_2 \rightarrow^*_\Gamma I$. Let $\mathtt{s}$ be a substitution such that $\mathtt{s}(x) \in \Gamma^0$, for each $x \in \mathrm{Var}$. Therefore $\mathtt{s}(M[\vec{N}_1/\vec{x}])\vec{N}_2 \rightarrow^*_\Gamma \mathtt{s}(I) \equiv I$, by Remark 3.1.30 pag. 43, hence $\mathtt{s}(M[\vec{N}/\vec{x}]) \equiv \mathtt{s}(M)[\mathtt{s}(\vec{N})/\vec{x}]$ is Γ-valuable.
The inclusion is proper, since $\lambda x.DD$ is valuable, and so potentially valuable, but clearly Γ-unsolvable. □

In order to characterize the Γ-solvable terms, we need to define a relation between terms, based on the $\rightarrow_{\Xi\ell}$-reduction.

Definition 3.1.11. (i) *The relation* $\searrow\subseteq \Lambda\times\Lambda$ *is defined inductively in the following way:*

- $\lambda x.P \searrow \lambda x.Q$ *if and only if* $P \searrow Q$,
- $xM_1\ldots M_m \searrow xN_1\ldots N_m$ *if and only if* $M_i \rightarrow^*_{\Xi\ell} N_i \in \Xi$ $(1\leq i\leq m)$,
- $(\lambda x.P)QM_1\ldots M_m \searrow R$ *if and only if* $Q \rightarrow^*_{\Xi\ell} \bar{Q} \in \Xi$ *and* $P[\bar{Q}/x]M_1\ldots M_m \searrow R$.

(ii) M is in Γ-head normal form *(Γ-hnf) if and only if* $M \equiv \lambda\vec{x}.xM_1...M_m$, *and for all* $1\leq i\leq m$, $M_i\in\Xi$; *Γ-HNF denotes the set of all Γ-head normal forms.*

(iii) M has Γ-head-normal form *if and only if* $M\searrow \lambda\vec{x}.xM_1...M_m$ *and* $M_i\in\Xi$, *for all* $1\leq i\leq m$. $\|\vec{x}\|$ *is the* Γ-order *and* m *is the* Γ-degree *of* M.

Note that Γ-HNF is a proper subclass of Λ-HNF. In fact, $\lambda x.x(DD)\in\Lambda$-HNF, but $\lambda x.x(DD)\notin\Gamma$-HNF since $DD\notin\Xi$.

The notion of Γ-order (or simply order, when the set of input values is clear from the context) can be extended to terms not having Γ-hnf in the following way:

Definition 3.1.12. (i) *M is of Γ-order 0 if and only if there is no P such that* $M\rightarrow^*_{\Xi\ell}\lambda x.P$;

(ii) *M is of Γ-order $n\geq 1$ if and only if n is the maximum integer such that* $M\rightarrow^*_{\Xi\ell}\lambda x_1.M_1$, $M_i\rightarrow^*_{\Xi\ell}\lambda x_{i+1}.M_{i+1}$ $(1\leq i\leq n)$ *and M_n is Γ-unsolvable of order 0. If such an n does not exists M is of Γ-order ∞.*

Example 3.1.13. DD and $(\lambda zx.xD)(yI)D$ are Γ-unsolvable of order 0, and xy is Γ-solvable of order 0. $(\lambda xy.xx)(\lambda xy.xx)$ is Γ-unsolvable of order ∞.

Theorem 3.1.14 (Γ-Solvability). [74]
A term is Γ-solvable if and only if it has Γ-head-normal form.

Proof. The proof is in Sect. 3.1.2. □

It is also possible to give an operational characterization of the Γ-solvable terms through the notion of $\Xi\ell$-reduction.

Property 3.1.15. M is Γ-solvable if and only if there are terms $M_1,..,M_n$, for some $n\in\mathbb{N}$, such that $M\rightarrow^*_{\Xi\ell}\lambda x_1.M_1$, $M_i\rightarrow^*_{\Xi\ell}\lambda x_{i+1}.M_{i+1}$ $(1\leq i\leq n)$ and $M_n\equiv xP_1...P_m$ where $P_i\in\Xi$ for some $m\in\mathbb{N}$.

Proof. ($\Rightarrow$) By induction on Definition 3.1.11.(i). If M is in Γ-hnf then the proof is trivial. Otherwise, the only not obvious case is when $M\equiv(\lambda x.P)Q\vec{M}$. In this case M is Γ-solvable if and only if $Q\rightarrow^*_{\Xi\ell}Q'$ and $P[Q'/x]\vec{M}\searrow R$, and R is in Γ-hnf. By induction there are terms

$M'_1, .., M'_{n'}$, for some $n \in \mathbb{N}$, such that $P[Q'/x]\vec{M} \to^*_{\Xi\ell} \lambda x_1.M'_1$, $M'_i \to^*_{\Xi\ell} \lambda x_{i+1}.M'_{i+1}$ $(1 \le i \le n')$ and $M'_n \equiv xP'_1...P'_{m'}$, for some m'.
Let $M'_0 \equiv P[Q'/x]\vec{M}$. Since $M \to^*_{\Xi\ell} M_0$, the proof is given.
($\Leftarrow$) By induction on n. If $n = 1$, then $M \to^*_{\Xi\ell} xM_1...M_m$, and so M is Γ-solvable. In all other cases the proof follows easily by induction. □

Differently from the call-by-name case, in the $\lambda\Gamma$-calculus the notion of Γ-nf is not semantically meaningful; in fact, we have seen that a term in Γ-nf can be not potentially valuable, and so is Γ-unsolvable. Moreover, consider the two terms $(\lambda z.D)(yI)D$ and $(\lambda z.D)(yK)D$: they are Γ-normal forms, and they are $\neq_\Gamma$, but they are both Γ-unsolvable of Γ-order 0. We will see that all the Γ-unsolvable terms of Γ-order 0 can be consistently equated.

Nevertheless, Λ-normal forms maintain a semantic importance also in this calculus, as the next theorem shows. Note that a Λ-normal form is a particular case of a $\Xi\ell$-normal form.

Theorem 3.1.16 (Γ-Separability). [72]
Let $M, N \in \Lambda$-NF. If $M \neq_{\Lambda\eta} N$ then M and N are Γ-separable.

Proof. The proof is in Sect. 3.1.3. □

The Γ-separability theorem has an interesting semantical consequence.

Corollary 3.1.17. *Let M, N be two Λ-normal forms, and let $M \neq_{\Lambda\eta} N$. For every Γ-theory $\mathcal{T}$, if $M =_{\mathcal{T}} N$ then $\mathcal{T}$ is input inconsistent.*

Proof. The proof is identical to the proof of Property 1.3.6, just putting $\Delta = \Gamma$. □

We proved that, for every $\lambda\Lambda$-theory $\mathcal{T}$, $I =_{\mathcal{T}} E$ if and only if $\mathcal{T}$ is fully extensional. In the case of the $\lambda\Gamma$-calculus, we can prove only a weaker property.

Property 3.1.18. Let $M \in \Lambda$ and $x \notin \mathrm{FV}(M)$.
$I =_{\mathcal{T}} E$ if and only if $M =_{\mathcal{T}} \lambda x.Mx$, for every $M \in \Gamma$.

Proof. ($\Rightarrow$) $I =_{\mathcal{T}} E$ implies $IM =_{\mathcal{T}} EM$, for all M. Thus, if $M \in \Gamma$ then $M =_{\mathcal{T}} \lambda x.Mx$ $(x \notin \mathrm{FV}(M))$.
($\Leftarrow$) If $M =_{\mathcal{T}} \lambda x.Mx$, for each $M \in \Gamma$, then $x =_{\mathcal{T}} \lambda y.xy$, so $\lambda x.x =_{\mathcal{T}} \lambda xy.xy$. □

The notion of fixed point can be easily extended in the call-by-value setting, in the sense that N is a call-by-value fixed point of M if and only if $MN =_\Gamma N$.

Theorem 3.1.19. *M is Γ-valuable implies that M has a call-by-value fixed point, i.e. there is N such that $MN =_\Gamma N$.*

Proof. Let $M \to_\Gamma^* M' \in \Gamma$ and let Y be defined as in the proof of Theorem 2.1.8. Then $YM \to_\Gamma^* YM' \to_\Gamma^* (\lambda y.M'(yy))(\lambda y.M'(yy)) =_\Gamma M'(YM') =_\Gamma M(YM)$. □

Let us call *call-by-value fixed point operator* terms Z such that if M is Γ-valuable then ZM is a call-by-value fixed point of M.

We will see, using denotational tools, that in the call-by-value setting, the notion of fixed point is in some sense meaningless, since every fixed point operator Z is such that ZM is not potentially Γ-valuable, for every M.

A more useful notion related to this one is the notion of *call-by-value recursion operator.* In fact, it will be used in Sect. 16.2, for expressing the recursive functions in a call-by-value setting. A call-by-value recursion operator is a term Z such that $ZM =_\Gamma M(\lambda z.ZMz)$, for all Γ-valuable terms M. The following theorem holds.

Theorem 3.1.20. *A call-by-value recursion operator exists.*

Proof. The term $\lambda x.(\lambda y.x(\lambda z.yyz))(\lambda y.x(\lambda z.yyz))$ has the desired behaviour. □

3.1.1 $\Xi\ell$-Confluence and $\Xi\ell$-Standardization

The confluence property for the reduction $\to_{\Xi\ell}$ follows directly from the fact that it enjoys the diamond property, as proved in the next lemma.

Lemma 3.1.21 ($\Xi\ell$-Diamond property).
Let $M, N_0, N_1 \in \Lambda$ and $N_0 \not\equiv N_1$. If $M \to_{\Xi\ell} N_0$ and $M \to_{\Xi\ell} N_1$ then there is $Q \in \Lambda$ such that $N_0 \to_{\Xi\ell} Q$ and $N_1 \to_{\Xi\ell} Q$.

Proof. We will prove only the most difficult case, i.e. $M \equiv (\lambda x.P)M_1...M_m$, by induction on M.

- If $N_0 \equiv (\lambda x.P)M_1...M'_k...M_m$ such that $M_k \to_{\Xi\ell} M'_k$, for some k, and $N_1 \equiv (\lambda x.P)M_1...M'_h...M_m$ such that $M_h \to_{\Xi\ell} M'_h$, for some $k \neq h$, then $Q \equiv (\lambda x.P)M_1...M'_h...M'_k...M_m$.
- Let $N_0 \equiv (\lambda x.P)M_1...M'_k...M_m$ and $N_1 \equiv (\lambda x.P)M_1...M''_k...M_m$ such that $M'_k \not\equiv M''_k$, $M_k \to_{\Xi\ell} M'_k$ and $M_k \to_{\Xi\ell} M''_k$, for some k. By induction on M_k there is Q' such that $M'_k \to_{\Xi\ell} Q'$ and $M''_k \to_{\Xi\ell} Q'$, thus $Q \equiv (\lambda x.P)M_1...Q'...M_m$.
- Let $M_1 \in \Xi$, so both $(\lambda x.P)$ and M_1 are $\Xi\ell$-normal forms. Let $N_0 \equiv P[M_1/x]M_2...M_m$ and $N_1 \equiv (\lambda x.P)M_1...M'_k...M_m$ such that $M_k \to_{\Xi\ell} M'_k$, for some k. Clearly, $Q \equiv P[M_1/x]M_2...M'_k...M_m$. □

Theorem 3.1.22 ($\Xi\ell$-Confluence).
Let $M, N_0, N_1 \in \Lambda$. If $M \to_{\Xi\ell}^* N_0$ and $M \to_{\Xi\ell}^* N_1$ then there is $P \in \Lambda$ such that $N_0 \to_{\Xi\ell}^* P$ and $N_1 \to_{\Xi\ell}^* P$.

Proof. By Lemma 3.1.21, following the same reasoning as in Theorem 1.2.5. □

Let $M \to^*_{\Xi\ell} N$; by the $\Xi\ell$-confluence theorem, M has $\Xi\ell$-normal form if and only if N has $\Xi\ell$-normal form.

Corollary 3.1.23. *The $\Xi\ell$-normal form of a term, if it exists, is unique.*

In order to state a standardization theorem for $\to_{\Xi\ell}$, we need to redefine some notions already stated for the $\lambda\Delta$-calculus. The fact that Ξ is not a set of input values forces this redefinition.

Definition 3.1.24. (i) *A symbol λ in a term M is $\Xi\ell$-*active *if and only if it is the first symbol of a $\Xi\ell$-redex of M.*

(ii) *The $\Xi\ell$-*degree *of a $\Xi\ell$-redex R in M is the numbers of λ's that both are active in M and occur on the left of R.*

(iii) *The* principal *$\Xi\ell$-redex of M, if it exists, is the redex of M with minimum degree.*

(iv) *A sequence $M \equiv P_0 \to_{\Xi\ell} P_1 \to_{\Xi\ell} \ldots \to_{\Xi\ell} P_n \to_{\Xi\ell} N$ is* standard *if and only if the $\Xi\ell$-degree of the redex contracted in P_i is less than or equal to the degree of the redex contracted in P_{i+1}, for every $i < n$. We denote by $M \to^{\circ}_{\Xi\ell} N$ a standard reduction sequence from M to N.*

It can be easily checked that the definition of $\Xi\ell$-degree of a redex, given in the definition before, can be obtained by specializing the general notion of sequentialization given in Definition 1.2.7: its simplification is due to the laziness of the reduction.

If $M \to_{\Xi\ell} N$ by reducing a $\Xi\ell$-redex of degree $k \in \mathbb{N}$, then we use the notation $M \stackrel{k}{\to}_{\Xi\ell} N$.

Lemma 3.1.25. *Let $P_0 \stackrel{k}{\to}_{\Xi\ell} P_1 \stackrel{h}{\to}_{\Xi\ell} P_2$ and $k > h$.*
There is $n \in \mathbb{N}$ and $P'_1 \in \Lambda$ such that $P_0 \stackrel{h}{\to}_{\Xi\ell} P'_1 \stackrel{n}{\to}_{\Xi\ell} P_2$ and $n \geq h$.

Proof. By induction on P_0.
We will prove only the most difficult case, when $P_0 \equiv (\lambda x.P)QM_1...M_m$ ($m \in \mathbb{N}$). Note that $k > h$ implies $k \geq 1$, so the principal redex cannot be reduced in $P_0 \stackrel{k}{\to}_{\Xi\ell} P_1$; thus either $P_1 \equiv (\lambda x.P)Q'M_1...M_m$ where $Q \stackrel{k}{\to}_{\Xi\ell} Q'$ or $P_1 \equiv (\lambda x.P)QM_1...M'_j...M_m$ where $M_j \to_{\Xi\ell} M'_j$ ($1 \leq j \leq m$).

- In the first case, $k > h$ implies there is $Q'' \in \Lambda$ such that $Q' \to_{\Xi\ell} Q''$, $P_2 \equiv (\lambda x.P)Q''M_1...M_m$. The proof follows by induction on Q.
  Note that $Q \stackrel{k}{\to}_{\Xi\ell} Q'$ implies $Q \notin \Xi$; moreover, $Q \stackrel{k}{\to}_{\Xi\ell} Q'$ and $k > h$ imply $Q' \notin \Xi$, since the reduction is not principal.
- In the last case:
  1. either $P_2 \equiv (\lambda x.P)QM_1...M''_j...M_m$ where $M \stackrel{k'}{\to}_{\Xi\ell} M' \stackrel{h'}{\to}_{\Xi\ell} M''$ and $k' \geq h'$;
  2. or $P_2 \equiv (\lambda x.P)QM_1...M'_j...M'_r...M_m$ where $r > j$.

  In case 1 the proof follows by induction on M_j, in case 2 we take the reduction sequence $P_0 \to_{\Xi\ell} (\lambda x.P)QM_1...M'_j...M_m \to_{\Xi\ell} P_2$.

□

Corollary 3.1.26. *If* $P_0 \stackrel{k}{\rightarrow}_{\Xi\ell} P_1 \stackrel{0}{\rightarrow}_{\Xi\ell} P_2$ *and* $k \geq 1$, *then there are* $P_1' \in \Lambda$ *and* $h \in \mathbb{N}$ *such that* $P_0 \stackrel{0}{\rightarrow}_{\Xi\ell} P_1' \stackrel{h}{\rightarrow}_{\Xi\ell} P_2$.

Proof. By the Lemma 3.1.25, just putting $h = 0$. □

Now we can state the standardization theorem.

Theorem 3.1.27 ($\Xi\ell$-Standardization).
If $M \rightarrow^*_{\Xi\ell} N$ *then there is a standard reduction sequence from* M *to* N.

Proof. By induction on M. Let $M \equiv xM_1...M_m$, thus N must be of the shape $xP_1...P_m$, where $M_i \rightarrow^*_{\Xi\ell} P_i$. By induction there is a standard reduction sequence $M_i \rightarrow^{\circ}_{\Xi\ell} P_i$, and so the desired standard sequence is: $M \rightarrow^{\circ}_{\Xi\ell} xP_1M_2...M_m \rightarrow^{\circ}_{\Xi\ell} xP_1P_2...M_m \rightarrow^{\circ}_{\Xi\ell} xP_1...P_m$ $(1 \leq i \leq m)$.
If $M \equiv \lambda x.M'$ then it must be that $M \equiv N$, and the empty reduction sequence is trivially standard.
Let $M \equiv (\lambda x.P)M_1...M_m$ $(m \geq 1)$. The proof follows by induction on the length of the reduction $M \rightarrow^*_{\Xi\ell} N$, by using the previous corollary. □

The principal reduction is normalizing.

Corollary 3.1.28. $M \rightarrow^*_{\Xi\ell} N \in \Xi$ *if and only if* $M \rightarrow^{\circ}_{\Xi\ell} N \in \Xi$.

Proof. Trivial. □

3.1.2 Proof of Potential Γ-Valuability and Γ-Solvability Theorems

In order to prove the theorems, we need to introduce a measure for carrying out some inductive proofs.

Definition 3.1.29. *The* weight $\langle \_ \rangle : \Lambda \longrightarrow \mathbb{N}$ *is the partial function defined as follows:*

- $\langle \lambda x.M' \rangle = 0$.
- $\langle xM_1...M_m \rangle = 1 + \langle M_1 \rangle + ... + \langle M_m \rangle$.
- $\langle (\lambda x.M_0)M_1...M_m \rangle = 1 + \langle M_1 \rangle + \langle M_0[M_1/x]M_2...M_m \rangle$.

In Sect. 3.2, we will show that the weight of a term M is defined if and only if M has $\Xi\ell$-normal form.

The following remark will be extensively used in what follows.

Remark 3.1.30. Let $M, N, P, Q \in \Delta$, where Δ is a set of input values.
If $M \rightarrow_{\Delta} N$ and $P \rightarrow_{\Delta} Q$ then $M[P/z] \rightarrow_{\Delta} N[Q/z]$.

Proof. Easy, by induction on M. □

Lemma 3.1.31. *Let $Q \in \Xi$, $P \in \Lambda$ and $C[.]$ be a context.*

(i) *If $Q \in \Xi$ then $\langle Q \rangle$ is defined.*
(ii) *If $M \rightarrow_{\Xi\ell} N$ and $\langle N \rangle$ is defined then $\langle M \rangle$ is defined.*
(iii) *If M has $\Xi\ell$-normal form then $\langle M \rangle$ is defined.*

Proof. (i) By induction on Q.
(ii) By induction on $\langle N \rangle$.
If $\langle N \rangle = 0$ then $N \equiv \lambda x.N'$, so $M \equiv (\lambda z.P)Q$ and $Q \in \Xi$; hence, $\langle M \rangle = 1 + \langle Q \rangle + \langle N \rangle$ where $\langle Q \rangle$ is defined by the previous point of this lemma.
Let $\langle N \rangle \geq 1$; there are many cases.
1. Let $M \equiv xM_1...M_m \rightarrow_{\Xi\ell} xN_1...N_m \equiv N$ $(m \geq 1)$, where there is a unique $k \leq m$ such that $M_k \rightarrow_{\Xi\ell} N_k$ while $M_h \equiv N_h$ if $h \neq k$. The proof follows easily by induction.
2. Let $M \equiv (\lambda z.P)QM_1...M_m \rightarrow_{\Xi\ell} RM_1...M_m \equiv N$ $(m \geq 1)$, where $Q \in \Xi$ and $(\lambda z.P)Q \rightarrow_{\Xi\ell} R$; hence, $\langle M \rangle = 1 + \langle Q \rangle + \langle N \rangle$ where $\langle Q \rangle$ is defined by the point (i) of this lemma.
3. The case $M \equiv \lambda x.P$ is not possible, since $\rightarrow_{\Xi\ell}$ is lazy.

(iii) By induction on the length of the sequence to $\Xi\ell$-normal form, by using the previous points of this lemma. □

The weight of a term allows us to induce on the length of reduction sequences with respect to different notions of reduction.

Remark 3.1.32. If $M \in \Lambda^0$ is Γ-valuable then $M \equiv (\lambda z.P)QM_1...M_m$, for some $m \in \mathbb{N}$; moreover, M_i is closed and also Γ-valuable $(1 \leq i \leq m)$.

Property 3.1.33. Let $M, N \in \Lambda^0$.

(i) If M is Γ-valuable then $\langle M \rangle$ is defined.
(ii) $M \rightarrow_{\Lambda}^{*} N$ and $\langle M \rangle$ is defined imply $\langle N \rangle$ is defined and $\langle M \rangle \geq \langle N \rangle$.
(iii) Let either $M \rightarrow_{\Gamma}^{*} N$ or $M \rightarrow_{\Xi\ell}^{*} N$.
If $\langle M \rangle$ is defined then $\langle N \rangle$ is defined and $\langle M \rangle \geq \langle N \rangle$.

Proof. (i) Γ is a standard set of input values and $M \rightarrow_{\Gamma}^{*} N \in \Gamma$ imply that there is $N' \in \Gamma$ such that $M \rightarrow_{\Gamma}^{*p} N' \in \Gamma$; moreover, since M is closed there is M' such that $M \rightarrow_{\Gamma}^{p} M_1 \rightarrow_{\Gamma}^{p} M_2 ... \rightarrow_{\Gamma}^{p} M_r \rightarrow_{\Gamma}^{p} \lambda z.M' \rightarrow_{\Gamma}^{p*} N' \in \Gamma$, where M_i is not an abstraction, for all i $(1 \leq i \leq r)$.
Then $M \rightarrow_{\Gamma\ell} \lambda z.M'$ by definition of principal reduction. Clearly $M \rightarrow_{\Gamma\ell} \lambda z.M' \in \Gamma$ implies $M \rightarrow_{\Xi\ell} \lambda z.M' \in \Xi$, so the proof follows by Lemma 3.1.31.(iii).
(ii) Let $\langle M \rangle = k$ and let p be the number of steps of the standard reduction sequence $M \rightarrow_{\Lambda}^{*} N$. The proof is given by induction on the pair (k, p), ordered according to the lexicographical order.
The cases where either $\langle M \rangle = 0$ or $p = 0$ are trivial. $M \equiv xM_1 ... M_m$ is not possible, since $M \in \Lambda^0$ by hypothesis. Let $M \equiv (\lambda x.M_0)M_1 ... M_m$, $h' = \langle M_1 \rangle$ and $h'' = \langle M_0[M_1/x]M_2 ... M_m \rangle$, thus $k = 1 + h' + h''$.

Let the reduction path be: $M \to_\Lambda R_1 \to_\Lambda ..... \to_\Lambda R_p \equiv N$ $(p > 0)$. There are three cases:

1. If $R_1 \equiv M_0[M_1/x]M_2...M_m$ then $\langle R_1 \rangle = h'' < k$, so the proof follows by induction.
2. Let $R_1 \equiv (\lambda x.N_0)M_1N_2...N_m$ where $\exists! j \in \mathbb{N}$ such that $M_j \to_\Lambda N_j$, while $\forall i \neq j$, $M_i \equiv N_i$ $(0 \leq i \leq m$ and $i \neq 1)$.
   Hence $M_0[M_1/x]M_2...M_m \to_\Lambda N_0[M_1/x]N_2...N_m$ and $h'' < k$ imply $\langle N_0[M_1/x]N_2...N_m \rangle \leq h''$, by induction.
   Thus $\langle R_1 \rangle = 1 + \langle M_1 \rangle + \langle N_0[M_1/x]N_2...N_m \rangle \leq k$ and the proof follows by induction.
3. Let $R_1 \equiv (\lambda x.M_0)N_1M_2...M_m$, where $M_1 \to_\Lambda N_1$. Thus by induction on $M_0[M_1/x]M_2...M_m \to_\Lambda^* M_0[N_1/x]M_2...M_m$ and $h'' < k$, $\langle M_0[N_1/x]M_2...M_m \rangle \leq h''$. Again, by induction $\langle M_1 \rangle \geq \langle N_1 \rangle$. Thus the conclusion follows by definition of weight and by induction.

(iii) By the previous point of this Property, since $M \to_\Psi N$ implies $M \to_\Lambda N$, for each $\Psi \subseteq \Lambda$. □

The Lemma 3.1.34 proves that if a term is potentially Γ-valuable, then it has $\Xi\ell$-normal form.

Lemma 3.1.34. *Let $M \in \Lambda$, $\mathrm{FV}(M) \subseteq \{x_1...x_n\}$ and let $\mathbf{s}$ be a substitution such that $\mathbf{s}(x_i) = P_i \in \Gamma^0$. If $\mathbf{s}(M) \to_\Gamma^* \bar{M} \in \Gamma$ then there is $N \in \Xi$ such that both $M \to_{\Xi\ell}^* N$ and $\mathbf{s}(N) \to_\Gamma^* \bar{M}$.*

Proof. The proof is carried out by induction on $k = \langle \mathbf{s}(M) \rangle$, where we assume $\mathbf{s}(x_i) = P_i$ $(1 \leq i \leq n)$ and $\mathbf{s}(M) \equiv M[P_1/x_1, ..., P_n/x_n]$.

- $k = 0$. Thus $\mathbf{s}(M)$ is an abstraction; there are two cases:
  1. $M \equiv x_j$ and $P_j \equiv \lambda z.P \in \Lambda^0$, so $N \equiv x_j$.
  2. $M \equiv \lambda z.P$, so $N \equiv \lambda z.P$.

  In both cases the proof is immediate.
- $k > 0$. $\mathbf{s}(M) \in \Lambda^0$, so $\mathbf{s}(M) \equiv (\lambda u.R_0)R_1...R_r$ $(r \geq 1)$. Two cases are possible, according to the shape of M:
  1. $M \equiv x_jM_1...M_m$ $(j \leq n,\ 1 \leq m)$. Assume $P_j \equiv (\lambda z.P')$ (indeed $P_j \in \Gamma^0$); then $\mathbf{s}(M) \equiv P_j\mathbf{s}(M_1)...\mathbf{s}(M_m) \to_\Gamma^* \bar{M} \in \Gamma$.
     Since $\mathbf{s}(M)$ is Γ-valuable, there are $\bar{M}_i$ such that $\mathbf{s}(M_i) \to_\Gamma^* \bar{M}_i \in \Gamma$ and $\langle \mathbf{s}(M_i) \rangle < \langle \mathbf{s}(M) \rangle$; hence, by induction there are $N_i \in \Xi$ such that $M_i \to_{\Xi\ell} N_i$ and $\mathbf{s}(N_i) \to_\Gamma^* \bar{M}_i$ $(1 \leq i \leq m)$.
     Let $N \equiv x_jN_1...N_m \in \Xi$, thus $x_jM_1...M_m \to_{\Xi\ell}^* x_jN_1...N_m$ and $P_j\mathbf{s}(N_1)...\mathbf{s}(N_m) \to_\Gamma^* P_j\bar{M}_1...\bar{M}_m \to_\Gamma^* \bar{M}$ $(1 \leq i \leq m)$.
  2. $M \equiv (\lambda z.P)QM_1...M_m$ $(m \geq 0)$. Since $\mathbf{s}(M)$ is Γ-valuable, there is $\bar{Q}$ such that $\mathbf{s}(Q) \to_\Gamma^* \bar{Q} \in \Gamma$ and $\mathbf{s}(P[\bar{Q}/z])\mathbf{s}(M_1)...\mathbf{s}(M_m) \to_\Gamma^* \bar{M}$. Moreover, $\langle \mathbf{s}(Q) \rangle < \langle \mathbf{s}(M) \rangle$, so by induction there is $R \in \Xi$ such that $Q \to_{\Xi\ell}^* R$ and $\mathbf{s}(R) \to_\Gamma^* \bar{Q}$.
     But from $\mathbf{s}(P[R/z])\mathbf{s}(M_1)...\mathbf{s}(M_m) \to_\Gamma^* \mathbf{s}(P[\bar{Q}/z])\mathbf{s}(M_1)...\mathbf{s}(M_m)$ together with $\mathbf{s}(P[\bar{Q}/z])\mathbf{s}(M_1)...\mathbf{s}(M_m) \to_\Gamma^* \bar{M}$ it follows that:
     $\mathbf{s}(P[R/z])\mathbf{s}(M_1)...\mathbf{s}(M_m) \to_\Gamma^* \bar{M}$ and, by Property 3.1.33.(iii)

$$\langle \mathbf{s}(P[\mathbf{s}(R)/z])\mathbf{s}(M_1)...\mathbf{s}(M_m)\rangle \leq \langle \mathbf{s}(P[\bar{Q}/z])\mathbf{s}(M_1)...\mathbf{s}(M_m)\rangle < \langle \mathbf{s}(M)\rangle.$$

Then, by induction, there is $T \in \Xi$ such that $P[R/z]M_1...M_m \to^*_{\Xi\ell} T$ and $\mathbf{s}(T) \to^*_{\Gamma} \bar{M}$. Let $N \equiv T$; clearly $M \equiv (\lambda z.P)QM_1...M_m \to^*_{\Xi\ell} (\lambda z.P)RM_1...M_m \to_{\Xi\ell} P[R/z]M_1...M_m \to^*_{\Xi\ell} N$, so the proof is given.

□

The Lemma 3.1.35 proves that if a term has $\Xi\ell$-normal form then it is potentially Γ-valuable.

Lemma 3.1.35. *Let $M \in \Lambda$, $\mathrm{FV}(M) \subseteq \{x_1,...,x_n\}$.*
$M \to^*_{\Xi\ell} N \in \Xi$ implies that $\exists h \in \mathbb{N}$ such that $\forall r \geq h$, $\exists \bar{M}^r \in \Gamma$

$$M[O^r/x_1,...,O^r/x_n] \to^*_{\Gamma} \bar{M}^r\,,$$
$$N[O^r/x_1,...,O^r/x_n] \to^*_{\Gamma} \bar{M}^r\,,$$

where $O^r \equiv \lambda x_1...x_{r+1}.x_{r+1}$.

Proof. Let $\mathbf{p}_r$ be the substitution such that $\mathbf{p}_r(y) = O^r$, for all $y \in \mathrm{Var}$ and $r \geq 0$; so $\mathbf{p}_r(M) = M[O^r/x_1,...,O^r/x_n]$.
The proof will be given by induction on $\langle M\rangle$.

- Let $\langle M\rangle = 0$, so M is an abstraction and the proof is trivial.
- Let $\langle M\rangle \geq 1$. If $M \equiv xM_1...M_m$ ($m \in \mathbb{N}$) then by induction, $\forall i \leq m$ there are $h_i \in \mathbb{N}$ such that $\forall r \geq \max\{m, h_1,...,h_m\}$, $\mathbf{p}_r(M_i) \to^*_{\Gamma} \bar{M}^r_i \in \Gamma$ and the proof is immediate.
  Otherwise, let $M \equiv (\lambda z.P)QM_1...M_m$ ($m \in \mathbb{N}$); M has $\Xi\ell$-normal form implies that there is $R \in \Xi$ such that $Q \to^*_{\Xi\ell} R$. Hence $\langle Q\rangle < \langle M\rangle$ and this implies by induction, that there is $h_0 \in \mathbb{N}$ such that $\forall r \geq h_0$ $\exists \bar{Q}^r \in \Gamma$, $\mathbf{p}_r(Q) \to^*_{\Gamma} \bar{Q}^r$ and $\mathbf{p}_r(R) \to^*_{\Gamma} \bar{Q}^r$. Clearly $P[R/z]M_1...M_m \to^*_{\Xi\ell} N$ too.
  By Property 3.1.33.(iii) $\langle P[R/z]M_1...M_m\rangle \leq \langle P[Q/z]M_1...M_m\rangle < \langle M\rangle$ then, by induction, there is $h_1 \in \mathbb{N}$ such that $\forall r \geq h_1$ $\exists \bar{P}^r \in \Gamma$ satisfying $\mathbf{p}_r(P[R/z]M_1...M_m) \to^*_{\Gamma} \bar{P}^r$ and $\mathbf{p}_r(N) \to^*_{\Gamma} \bar{P}^r$.
  $\forall r \geq \max\{h_0, h_1\}$, $\exists \bar{Q}^r \in \Lambda^0$ $\mathbf{p}_r(R) \to^*_{\Gamma} \bar{Q}^r$ implies, by the confluence theorem,

  $$\mathbf{p}_r(P[R/z])\mathbf{p}_r(M_1)...\mathbf{p}_r(M_m) \to^*_{\Gamma} \mathbf{p}_r(P[\bar{Q}^r/z])\mathbf{p}_r(M_1)...\mathbf{p}_r(M_m) \to^*_{\Gamma} \bar{P}^r.$$

  Since $\mathbf{p}_r(M) \to^*_{\Gamma} \mathbf{p}_r(P[\bar{Q}^r/z])\mathbf{p}_r(M_1)...\mathbf{p}_r(M_m)$, the proof is done.

□

▼ **Proof of Potential Γ-Valuability Theorem** (Theorem 3.1.9 pag. 38).
The proof of the (only if) part follows directly from Lemma 3.1.34, while the proof of the (if) part follows directly from Lemma 3.1.35. ■

The Lemma 3.1.36 implies as an immediate corollary that, if $M \in \Lambda$ has Γ-head normal form, then M is Γ-solvable.

Lemma 3.1.36. *If M has Γ-head normal form and* $\mathrm{FV}(M) = \{x_1, ..., x_n\}$ *then $\exists s \in \mathbb{N}$, $\forall r \geq s$, $\exists k \in \mathbb{N}$ such that* $(\lambda x_1...x_n.M)\underbrace{O^r...O^r}_{r} \to^*_\Gamma O^k$.

Proof. Let $\mathbf{p}_r$ (where $r \in \mathbb{N}$) be the substitution such that $\mathbf{p}_r(y) = O^r$, for each $y \in \mathrm{Var}$. The proof is done by induction on the minimum number q of steps necessary to prove that $M \searrow N$, for some N in Γ-head normal form. If M is an abstraction, the proof follows directly by induction. Let $M \equiv xM_1...M_m$, where M_i have $\Xi\ell$-normal forms $(1 \leq i \leq m)$. By Lemma 3.1.35, $\exists s_i \in \mathbb{N}$ such that $\forall r \geq s_i$, $M_i[O^r/x_1, ..., O^r/x_n] \to^*_\Gamma \bar{M}_i \in \Gamma$ $(1 \leq i \leq m)$. Let $r \geq \max\{m, n, s_1, ..., s_m\}$; thus for some $k \in \mathbb{N}$,

$$(\lambda x_1...x_n.M)\underbrace{O^r...O^r}_{r} \to^*_\Gamma O^r\bar{M}_1...\bar{M}_m\underbrace{O^r...O^r}_{r-n} \to^*_\Gamma O^{r-m}\underbrace{O^r...O^r}_{r-n} \to^*_\Gamma O^k.$$

If $r - m \geq r - n$ then $n \geq m$ and $k = (r - m) - (r - n) = n - m$; otherwise $r - m < r - n$ and $O^{r-m}\underbrace{O^r.....O^r}_{r-n} \to^*_\Gamma \underbrace{O^r.....O^r}_{r-n-(r-m)}$, thus $r - n - (r - m) = m - n$ and $k = r - (m - n - 1) = r + 1 + n - m$.
Let $M \equiv (\lambda x.P)QM_1...M_m$ $(m \geq 1)$. By definition $Q \to^*_{\Xi\ell} R \in \Xi$ and $M \searrow P[R/x]M_1...M_m$, which has Γ-head normal form. $Q \to^*_{\Xi\ell} R \in \Xi$ implies that $\exists h_0 \in \mathbb{N}$ such that $\forall r \geq h_0$, $\exists\bar{Q}^r \in \Gamma$, $\mathbf{p}_r(Q) \to^*_\Gamma \bar{Q}^r$ and $\mathbf{p}_r(R) \to^*_\Gamma \bar{Q}^r$, by Lemma 3.1.35. By induction $\exists h \in \mathbb{N}$ such that $\forall r \geq h$, $(\lambda x_1...x_n.P[R/x]M_1...M_m)\underbrace{O^r...O^r}_{r} \to^*_\Gamma O^k$, for some $k \in \mathbb{N}$.

Let $r \geq \max\{h_0, h\}$, so for some $v \in \mathbb{N}$,

$$\begin{aligned}(\lambda x_1...x_n.M)\underbrace{O^r...O^r}_{r} &\to^*_\Gamma (\mathbf{p}_r(\lambda x.P)\mathbf{p}_r(Q)\mathbf{p}_r(M_1)...\mathbf{p}_r(M_m))\underbrace{O^r...O^r}_{r-n}\\ &\to^*_\Gamma (\mathbf{p}_r(\lambda x.P)\bar{Q}^r\mathbf{p}_r(M_1)...\mathbf{p}_r(M_m))\underbrace{O^r...O^r}_{r-n}\\ &\to_\Gamma (\mathbf{p}_r(P[\bar{Q}^r/x])\mathbf{p}_r(M_1)...\mathbf{p}_r(M_m))\underbrace{O^r...O^r}_{r-n} \to^*_\Gamma O^v.\end{aligned}$$

But $\mathbf{p}_r\big(P[\mathbf{p}_r(R)/x]\big)\mathbf{p}_r(M_1)...\mathbf{p}_r(M_m) \to^*_\Gamma \mathbf{p}_r\big(P[\bar{Q}^r/x]\big)\mathbf{p}_r(M_1)...\mathbf{p}_r(M_m)$, thus by the confluence theorem, it follows that $(\lambda x_1...x_n.M)\underbrace{O^r...O^r}_{r} \to^*_\Gamma O^v$ too. □

The following lemma implies as an immediate corollary that, if $M \in \Lambda$ is Γ-solvable, then M has Γ-head normal form.

Lemma 3.1.37. *Let $M \in \Lambda$,* $\mathrm{FV}(M) \subseteq \{x_1, ..., x_n\}$ *and $P_1, ..., P_k \in \Gamma^0$. If $(\lambda x_1...x_n.M)P_1...P_k \to^*_\Gamma I$ then there is N in Γ-head-normal form such that $M \searrow N$ and $(\lambda x_1...x_n.N)P_1...P_k \to^*_\Gamma I$.*

Proof. Let $\mathbf{s}$ be a substitution such that $\mathbf{s}(x_i) = P_i \in \Gamma^0$.
Let $M \equiv \lambda x_{n+1}...x_r.\xi M_1...M_m$ $(m, r \in \mathbb{N}, n \leq r)$ where either $\xi \equiv x_j$ $(j \leq r)$ or $\xi \equiv (\lambda x.P)Q$, for some $P, Q \in \Lambda$, and let $C[.] \equiv (\lambda x_1...x_n.[.])P_1...P_k$, so

$$C[M] \equiv (\lambda x_1 ... x_r.\xi M_1 ... M_m) P_1 ... P_k.$$

Note that $r \leq k+1$, otherwise $C[M] \to^*_{\Gamma} \lambda x_{k+1} ... x_r.S' \neq_{\Gamma} I$. The proof is given for induction on $\langle C[M] \rangle$e by taking into account all possible shapes of the term M.

- $M \equiv \lambda x_{n+1} ... x_r.x_j M_1 ... M_m$. If $m = 0$ then the proof is trivial by putting $N \equiv M$, so let $m \geq 1$. There are 2 cases.
  1. If $r \leq k$ then

$$\begin{aligned} C[M] &\equiv (\lambda x_1 ... x_r.x_j M_1 ... M_m) P_1 ... P_k \\ &\to^*_{\Gamma} P_j \mathbf{s}(M_1) ... \mathbf{s}(M_m) P_r + 1 ... P_k \to^*_{\Gamma} I. \end{aligned}$$

  By Remark 3.1.32, let $\mathbf{s}(M_i) \to^*_{\Gamma} \bar{M}_i \in \Gamma$ $(1 \leq i \leq m)$; so by Lemma 3.1.34 $M_i \to^*_{\Xi\ell} N_i \in \Xi$ and $\mathbf{s}(N_i) \equiv N_i[P_1/x_1, ..., P_n/x_n] \to^*_{\Gamma} \bar{M}_i$. Let $N \equiv \lambda x_{n+1} ... x_r.x_j N_1 ... N_m$; so $\lambda x_{n+1} ... x_r.x_j M_1 ... M_m \searrow N$ and

$$\begin{aligned} (\lambda x_1 ... x_r.x_j N_1 ... N_m) P_1 ... P_k &\to^*_{\Gamma} P_j \mathbf{s}(N_1) ... \mathbf{s}(N_m) P_{r+1} ... P_k \\ &=_{\Gamma} (\lambda x_1 ... x_r.x_j M_1 ... M_m) P_1 ... P_k \\ &\to^*_{\Gamma} P_j \bar{M}_1 ... \bar{M}_m P_{r+1} ... P_k \\ &=_{\Gamma} (\lambda x_1 ... x_n.M) P_1 ... P_k \to^*_{\Gamma} I. \end{aligned}$$

  2. If $r = k+1$ then the proof is similar to that of the previous case, since

$$\begin{aligned} C[M] &\equiv (\lambda x_1 ... x_r.x_j M_1 ... M_m) P_1 ... P_k \\ &\to^*_{\Gamma} \lambda x_r.P_j \mathbf{s}(M_1) ... \mathbf{s}(M_m) \to^*_{\Gamma} \lambda x_r.x_r. \end{aligned}$$

- $M \equiv \lambda x_{n+1} ... x_r.(\lambda z.P) Q M_1 ... M_m$ $(m \geq 0,\ r \geq n)$; there are 2 cases.
  1. If $r \leq k$ then

$$\begin{aligned} C[M] &\equiv (\lambda x_1 ... x_r.(\lambda z.P) Q M_1 ... M_m) P_1 ... P_k \\ &\to^*_{\Gamma} \mathbf{s}(\lambda z.P) \mathbf{s}(Q) \mathbf{s}(M_1) ... \mathbf{s}(M_m) P_{r+1} ... P_k \\ &\to^*_{\Gamma} \mathbf{s}(\lambda z.P)\, \bar{Q}\, \mathbf{s}(M_1) ... \mathbf{s}(M_m) P_{r+1} ... P_k \to^*_{\Gamma} I, \end{aligned}$$

  where $\mathbf{s}(Q) \to^*_{\Gamma} \bar{Q} \in \Gamma$. Hence, by Lemma 3.1.34 $Q \to^*_{\Xi\ell} R \in \Xi$ and $\mathbf{s}(R) \to^*_{\Gamma} \bar{Q} \in \Gamma$. Moreover

$$\begin{aligned} &\mathbf{s}(P[R/z]) \mathbf{s}(M_1) ... \mathbf{s}(M_m) P_{r+1} ... P_k \to^*_{\Gamma} \\ &\mathbf{s}(P[\bar{Q}/z]) \mathbf{s}(M_1) ... \mathbf{s}(M_m) P_{r+1} ... P_k \to^*_{\Gamma} I. \end{aligned}$$

  Let $U \equiv \lambda x_{n+1} ... x_r.P[R/z] M_1 ... M_m$, thus $C[U] \to^*_{\Gamma} I$. Remember that $P_1, \ldots, P_k \in \Gamma^0$; then

$$\begin{aligned} &\langle C[M] \rangle = \\ &= r + \langle P_1 \rangle + ... + \langle P_r \rangle + \langle \mathbf{s}(\lambda z.P) \mathbf{s}(Q) \mathbf{s}(M_1) ... \mathbf{s}(M_m) P_{r+1} ... P_k \rangle = \\ &= r + \underbrace{0 + ... + 0}_{r} + 1 + \langle \mathbf{s}(Q) \rangle + \langle \mathbf{s}(P[\mathbf{s}(Q)/z]) \mathbf{s}(M_1) ... \mathbf{s}(M_m) P_{r+1} ... P_k \rangle, \end{aligned}$$

and by Property 3.1.33.(ii)

$$\langle \mathbf{s}(P[\mathbf{s}(Q)/z])\mathbf{s}(M_1)...\mathbf{s}(M_m)P_{r+1}...P_k\rangle \geq \langle \mathbf{s}(P[\mathbf{s}(R)/z])\mathbf{s}(M_1)...\mathbf{s}(M_m)P_{r+1}...P_k\rangle.$$

Hence $\langle C[U]\rangle < \langle C[M]\rangle$, and by induction, we can state $U \searrow T$ and $C[U] \to_\Gamma^* I$, for some T in Γ-head normal form.
Let N be T; so $C[N] \to_\Gamma^* I$ and

$$M \equiv \lambda x_{n+1}...x_r.(\lambda z.P)QM_1...M_m \searrow \lambda x_{n+1}...x_r.P[R/z]M_1...M_m \searrow T.$$

2. The case $r = k+1$ is similar to the previous one. □

▼ **Proof of Γ-Solvability Theorem** (Theorem 3.1.14 pag. 39).
Let $\mathrm{FV}(M) = \{x_1,\ldots,x_n\}$.

($\Rightarrow$) By Lemma 3.1.36 for some r and h:

$$(\lambda x_1...x_n.M)\underbrace{O^r.....O^r}_{r} \to_\Gamma^* O^h \ (h \geq 0).$$

Let $R_1,\ldots,R_h \in \Gamma$, thus $(\lambda x_1...x_n.M)\underbrace{O^r.....O^r}_{r} R_1...R_h \to_\Gamma^* I$.

($\Leftarrow$) By Lemma 3.1.37. ■

3.1.3 Proof of Γ-Separability Theorem

The proof will be given in a constructive way, by showing a *separability algorithm*. The algorithm is defined as a formal system, proving statements of the shape

$$M, N \Rrightarrow_\Gamma C[.],$$

where M, N are Λ-normal forms such that $M \neq_{\Lambda\eta} N$.
Differently from the call-by-name case, the context $C[.]$ generated from the algorithm is not yet the separating one. More precisely, it is a separating context if $M, N \in \Lambda^0$, but for open terms some additional work must be done.

Let $B^n \equiv \lambda x_1...x_{n+1}.x_{n+1}x_1...x_n$, $O^n \equiv \lambda x_1...x_{n+1}.x_{n+1}$ and $U_i^n \equiv \lambda x_1...x_n.x_i$ ($i \leq n$, $n \in \mathbb{N}$).
Furthermore, if $S \subseteq \mathrm{Var}$ then let $X_{x,S}^n \equiv \begin{cases} \lambda z_1\ldots z_n.\, x_i z_1\ldots z_n & \text{if } x \notin S, \\ x & \text{otherwise.} \end{cases}$

The notions of args, nf_Λ and $\simeq_c$ are defined respectively in Definitions 2.1.12, 2.1.14 and 2.1.10.

Remark 3.1.38. It is easy to check that every Λ-normal form M is a potentially Γ-valuable term; so every subterm of a Γ-nf, being in turn a Γ-nf, is potentially Γ-valuable.

A first problem in the development of a separability theorem for the Γ-calculus is the transformation of potentially valuable terms (subterms) in valuable ones. As proved in the next Lemma, the solution is to substitute to free variables some values with a suitable number of initial abstractions.

Lemma 3.1.39. *Let $M \in \Lambda$-NF, $\mathrm{FV}(M) = \{x_1, ..., x_n\}$ and $r \geq \mathsf{args}(M)$. If $\forall j \leq n$, $Q_j^r \equiv \lambda x_1...x_r.Q_j$ and $Q_j \in \Gamma$ then*

$$M[Q_1^r/x_1, ..., Q_n^r/x_n] \to_\Gamma^* \bar{M} \in \Gamma.$$

Proof. By induction on M. □

Lemma 3.1.40 proves an important result on which the inductive rule ($\Gamma 7$) of the algorithm is based. The relation $\not\sim_c$ was defined in Definition 2.1.10.

Lemma 3.1.40. *Let M, N be Λ-NF, $r \geq \max\{\mathsf{args}(M), \mathsf{args}(N)\}$, $y, z \in \mathrm{Var}$ and $C_k^r[.] \equiv (\lambda x_1...x_k.[.])X^r_{x_1,\{y,z\}}...X^r_{x_k,\{y,z\}}$, for some $k \in \mathbb{N}$.*

(i) *$\exists \bar{M} \in \Lambda$-NF such that $C_k^r[M] \to_\Lambda^* \bar{M}$ and $r \geq \mathsf{args}(\bar{M})$.*
(ii) *If $M \not\sim_c N$ then $\mathsf{nf}_\Lambda(C_k^r[M]) \not\sim_c \mathsf{nf}_\Lambda(C_k^r[N])$.*

Proof. (i) By induction on M.
If $M \equiv \lambda u.P$ or $M \equiv uM_1...M_m$ (where $u \notin \{x_1, ..., x_k\}$ and $m \leq r$) then the proof follows by induction. Let $M \equiv x_j M_1...M_m$ ($1 \leq j \leq k$, $m \leq r$); so by induction $\forall i \leq m$ there is $\bar{M}_i \in \Lambda$-NF such that $C_k^r[M_i] \to_\Lambda^* \bar{M}_i$ and $r \geq \mathsf{args}(\bar{M}_i)$. If $x_j \in \{y, z\}$ the proof is immediate, since $X^r_{x_j,\{y,z\}} \equiv x_j$. Let $x_j \notin \{y, z\}$; clearly

$$C_k^r[M] \equiv (\lambda x_1...x_k.x_j M_1...M_m)X^r_{x_1,\{y,z\}}...X^r_{x_k,\{y,z\}} \to_\Lambda^*$$
$$(\lambda u_1...u_r.x_j u_1...u_r)\bar{M}_1...\bar{M}_m \to_\Lambda^* \lambda u_{m+1}...u_r.x_j\bar{M}_1...\bar{M}_m u_{m+1}...u_r;$$

so

$$r \geq \max\{r, \mathsf{args}(\bar{M}_1), ..., \mathsf{args}(\bar{M}_m), \underbrace{0, ..., 0}_{r-m}\} = r.$$

Note that $\mathsf{nf}_\Lambda(C_k^r[M])$ is well defined.

(ii) Let $M \equiv \lambda u_1...u_p.uM_1...M_m$ and $N \equiv \lambda v_1...v_q.vN_1...N_n$; we reason by induction on c. Let $c \equiv \epsilon$, $|p - m| \neq |q - n|$ and $v \equiv u$.
$\forall i$ $\bar{M}_i \equiv \mathsf{nf}_\Lambda(C_k^r[M_i])$ and $\bar{N}_i \equiv \mathsf{nf}_\Lambda(C_k^r[N_i])$. Let $u \equiv v \equiv x_j$, for some $j \leq k$ and $u, v \notin \{y, z\}$; otherwise the proof is simpler. Thus

$$\mathsf{nf}_\Lambda(C_k^r[M]) \equiv \lambda u_1...u_p w_{m+1}...w_r.x_j\bar{M}_1...\bar{M}_m w_{m+1}...w_r ,$$
$$\mathsf{nf}_\Lambda(C_k^r[N]) \equiv \lambda v_1...v_q w_{n+1}...w_r.x_j\bar{N}_1...\bar{N}_n w_{n+1}...w_r.$$

$|(p+(r-m))-(m+(r-m))| \neq |(q+(r-n))-(n+(r-n))|$, since $|p-m| \neq |q-n|$.

If $u \not\equiv v$ the proof is similar.

If $c \equiv i, c'$ (where $i \geq 1$) then the proof follows by induction. □

The call-by-value separability algorithm is presented in fig. 3.1 (pag. 52). For sake of simplicity, in the algorithm description we assume that different bound variables have different names. The algorithm follows essentially the same pattern as the call-by-name separability algorithm, but the context in the conclusion replaces variables by terms having enough initial abstractions to assure that subterms become Γ-valuable, by using the result of Lemma 3.1.39. Note that, in rules (Γ3)-(Γ6), every occurrence of the term B^r in the context could be safely replaced by I, thereby following an approach similar to the call-by-name case. In fact such terms are erased when the context, filled by one of the two input terms, is Γ-reduced. Using B^r allows for an easier correctness proof.

Example 3.1.41. Let us consider the same terms as in Example 2.1.16, i.e. let $M \equiv \lambda xyu.x(u(x(yy))(vv))$ and $N \equiv \lambda xyu.x(u(yy)(vv))$.

Clearly $\mathsf{args}(M) = \mathsf{args}(N) = 2$, so let $r = 2$. The derivation proving the statement $M, N \Rrightarrow_\Gamma C[.]$ is the following:

$$\dfrac{\dfrac{\dfrac{\dfrac{\dfrac{x_3 \not\equiv y \qquad C_5[.] \equiv (\lambda x_3 y.[.])(\lambda x_1x_2x_3x_4.\tilde{x})(\lambda x_1x_2x_3x_4x_5.\tilde{y})B^2B^2}{x_3(yy)(\lambda v_1v_2.x_2v_1v_2)\ ,\ yy(\lambda v_1v_2.x_2v_1v_2)x_3 \Rrightarrow_\Gamma C_5[.]}(\Gamma 5)}{\lambda x_2x_3.x_3(yy)x_2\ ,\ yy \Rrightarrow_\Gamma C_2[.] \equiv C_5[[.](\lambda v_1v_2.x_2v_1v_2)x_3]}(\Gamma 2)}{u(\lambda x_2x_3.x_3(yy)x_2)(vv)\ ,\ u(yy)(vv) \Rrightarrow_\Gamma C_6[.] \equiv C_2[(\lambda u.[.])(\lambda z_1z_2.z_1)]}(\Gamma 6)}{x(u(x(yy))(vv))\ ,\ x(u(yy)(vv)) \Rrightarrow_\Gamma C_7[.] \equiv C_6[(\lambda x.[.])(\lambda x_1x_2x_3.x_3x_1x_2)B^2(\lambda z_1z_2.z_1)]}(\Gamma 7)}{\lambda xyu.x(u(x(yy))(vv))\ ,\ \lambda xyu.x(u(yy)(vv)) \Rrightarrow_\Gamma C_7[[.]xy(\lambda u_1u_2.uu_1u_2)]}(\Gamma 1)$$

where, if $\tilde{X}_4 \equiv (\lambda x_1x_2x_3x_4.\tilde{x})$ and $\tilde{Y}_5 \equiv (\lambda x_1x_2x_3x_4x_5.\tilde{y})$:

$$\begin{array}{lll}
C_5[.] \equiv & (\lambda x_3y.\ [.]\ ) & \tilde{X}_4\tilde{Y}_5B^2B^2 \\
C_2[.] \equiv & (\lambda x_3y.\ [.]\ (\lambda v_1v_2.x_2v_1v_2)x_3) & \tilde{X}_4\tilde{Y}_5B^2B^2 \\
C_6[.] \equiv & (\lambda x_3y.((\lambda u.\ [.]\ )U^1_2(\lambda v_1v_2.x_2v_1v_2)x_3)) & \tilde{X}_4\tilde{Y}_5B^2B^2 \\
C_7[.] \equiv & (\lambda x_3y.((\lambda u.(\lambda x.\ [.]\ )B^2B^2U^1_2)U^1_2(\lambda v_1v_2.x_2v_1v_2)x_3)) & \tilde{X}_4\tilde{Y}_5B^2B^2 \\
C[.] \equiv & (\lambda x_3y.((\lambda u.(\lambda x.\ [.]\ xy(\lambda u_1u_2.uu_1u_2))B^2B^2U^1_2)U^1_2(\lambda v_1v_2.x_2v_1v_2)x_3)) & \tilde{X}_4\tilde{Y}_5B^2B^2
\end{array}$$

We can check that $(C[M])[O^2/v] \rightarrow^*_\Gamma \tilde{x}$; in fact,

Let $M, N \in \Lambda$-NF, $M \not\simeq_c N$, $r \geq \max\{\mathsf{args}(M), \mathsf{args}(N)\}$ and $\tilde{x}, \tilde{y}$ be fresh variables such that $\tilde{x} \not\equiv \tilde{y}$.

The rules of the system proving statements $M, N \Rrightarrow_\Gamma C[.]$, are the following:

$$\frac{p \leq q \qquad C^r_k[.] \equiv (\lambda x_1 ... x_k.[.]) X^r_{x_1,\{x,y\}} ... X^r_{x_k,\{x,y\}} \quad (k \in \{p,q\}) \qquad \left.\begin{array}{r} x\mathsf{nf}_\Lambda(C^r_p[M_1])...\mathsf{nf}_\Lambda(C^r_p[M_m]) X^r_{x_{p+1},\{x,y\}} ... X^r_{x_q,\{x,y\}}, \\ y\mathsf{nf}_\Lambda(C^r_q[N_1])...\mathsf{nf}_\Lambda(C^r_q[N_n]) \end{array}\right\} \Rrightarrow_\Gamma C[.]}{\lambda x_1 \ldots x_p.xM_1 \ldots M_m, \lambda x_1 \ldots x_q.yN_1 \ldots N_n \Rrightarrow_\Gamma C[[.] X^r_{x_1,\{x,y\}} ... X^r_{x_q,\{x,y\}}]} \ (\Gamma 1)$$

$$\frac{p > q \qquad C^r_k[.] \equiv (\lambda x_1 ... x_k.[.]) X^r_{x_1,\{x,y\}} ... X^r_{x_k,\{x,y\}} \quad (k \in \{p,q\}) \qquad \left.\begin{array}{r} x\mathsf{nf}_\Lambda(C^r_p[M_1])...\mathsf{nf}_\Lambda(C^r_p[M_m]), \\ y\mathsf{nf}_\Lambda(C^r_q[N_1])...\mathsf{nf}_\Lambda(C^r_q[N_n]) X^r_{x_{q+1},\{x,y\}} ... X^r_{x_p,\{x,y\}} \end{array}\right\} \Rrightarrow_\Gamma C[.]}{\lambda x_1 \ldots x_p.xM_1 \ldots M_m, \lambda x_1 \ldots x_q.yN_1 \ldots N_n \Rrightarrow_\Gamma C[[.] X^r_{x_1,\{x,y\}} ... X^r_{x_p,\{x,y\}}]} \ (\Gamma 2)$$

$$\frac{n < m}{xM_1 \ldots M_m, xN_1 \ldots N_n \Rrightarrow_\Gamma (\lambda x.[.]) O^{r+n} \underbrace{B^r .....B^r}_{r+n-m} (\lambda x_1 \ldots x_{m-n}.\tilde{x}) \underbrace{\tilde{y} .....\tilde{y}}_{m-n}} \ (\Gamma 3)$$

$$\frac{m < n}{xM_1 \ldots M_m, xN_1 \ldots N_n \Rrightarrow_\Gamma (\lambda x.[.]) O^{r+m} \underbrace{B^r .....B^r}_{r+m-n} (\lambda x_1 \ldots x_{n-m}.\tilde{y}) \underbrace{\tilde{x} .....\tilde{x}}_{n-m}} \ (\Gamma 4)$$

$$\frac{x \not\equiv y}{xM_1 \ldots M_m, yN_1 \ldots N_n \Rrightarrow_\Gamma (\lambda xy.[.])(\lambda x_1 \ldots x_{r+m}.\tilde{x})(\lambda x_1 \ldots x_{r+n}.\tilde{y}) \underbrace{B^r .....B^r}_{r}} \ (\Gamma 5)$$

$$\frac{M_k \neq_{\Lambda\eta} N_k \qquad x \notin \mathrm{FV}(M_k) \cup \mathrm{FV}(N_k) \qquad M_k, N_k \Rrightarrow_\Gamma C[.]}{xM_1 \ldots M_m, xN_1 \ldots N_m \Rrightarrow_\Gamma C[(\lambda x.[.]) U^r_k \underbrace{B^r .....B^r}_{r-m}]} \ (\Gamma 6)$$

$$\frac{\begin{array}{c} M_k \neq_{\Lambda\eta} N_k \qquad x \in \mathrm{FV}(M_k) \cup \mathrm{FV}(N_k) \\ C^r_x[.] \equiv (\lambda x.[.]) B^r \qquad \mathsf{nf}_\Lambda(C^r_x[M_k]), \mathsf{nf}_\Lambda(C^r_x[N_k]) \Rrightarrow_\Gamma C[.] \end{array}}{xM_1 \ldots M_m, xN_1 \ldots N_m \Rrightarrow_\Gamma C[C^r_x[.] \underbrace{B^r .....B^r}_{r-m} U^r_k]} \ (\Gamma 7)$$

Fig. 3.1. Call-by-value separability algorithm

$$(\lambda x_3 y.((\lambda u.(\lambda x.M[O^2/v]xy(\lambda u_1u_2.uu_1u_2))B^2B^2U_2^1)U_2^1(\lambda v_1v_2.x_2v_1v_2)x_3))\tilde{X}_4\tilde{Y}_5B^2B^2 \to_\Gamma^*$$
$$(\lambda x_3 y.((\lambda u.M[O^2/v]B^2y(\lambda u_1u_2.uu_1u_2)B^2U_2^1)U_2^1(\lambda v_1v_2.x_2v_1v_2)x_3))\tilde{X}_4\tilde{Y}_5B^2B^2 \to_\Gamma^*$$
$$(\lambda x_3 y.(M[O^2/v]B^2y(\lambda u_1u_2.U_2^1u_1u_2)B^2U_2^1(\lambda v_1v_2.x_2v_1v_2)x_3))\tilde{X}_4\tilde{Y}_5B^2B^2 \to_\Gamma^*$$
$$(\lambda xyu.x(u(x(yy))(vv)))[O^2/v]B^2\tilde{Y}_5(\lambda u_1u_2.U_2^1u_1u_2)B^2U_2^1(\lambda v_1v_2.x_2v_1v_2)\tilde{X}_4B^2B^2 \to_\Gamma^*$$
$$(\lambda xyu.x(u(x(yy))(O^2O^2)))B^2\tilde{Y}_5(\lambda u_1u_2.u_1)B^2U_2^1(\lambda v_1v_2.x_2v_1v_2)\tilde{X}_4B^2B^2 \to_\Gamma^*$$
$$B^2((\lambda u_1u_2.u_1)(B^2(\tilde{Y}_5\tilde{Y}_5))(O^2O^2))B^2U_2^1(\lambda v_1v_2.x_2v_1v_2)\tilde{X}_4B^2B^2 \to_\Gamma^*$$
$$B^2((\lambda u_1u_2.u_1)(B^2(\lambda x_2x_3x_4x_5.\tilde{y}))O^1)B^2U_2^1(\lambda v_1v_2.x_2v_1v_2)\tilde{X}_4B^2B^2 \to_\Gamma^*$$
$$B^2((\lambda u_1u_2.u_1)(\lambda x_2x_3.x_3(\lambda x_2x_3x_4x_5.\tilde{y})x_2)O^1)B^2U_2^1(\lambda v_1v_2.x_2v_1v_2)\tilde{X}_4B^2B^2 \to_\Gamma^*$$
$$B^2(\lambda x_2x_3.x_3(\lambda x_2x_3x_4x_5.\tilde{y})x_2)B^2U_2^1(\lambda v_1v_2.x_2v_1v_2)\tilde{X}_4B^2B^2 \to_\Gamma^*$$
$$U_2^1(\lambda x_2x_3.x_3(\lambda x_2x_3x_4x_5.\tilde{y})x_2)B^2(\lambda v_1v_2.x_2v_1v_2)\tilde{X}_4B^2B^2 \to_\Gamma^*$$
$$(\lambda x_2x_3.x_3(\lambda x_2x_3x_4x_5.\tilde{y})x_2)(\lambda v_1v_2.x_2v_1v_2)\tilde{X}_4B^2B^2 \to_\Gamma^*$$
$$\tilde{X}_4(\lambda x_2x_3x_4x_5.\tilde{y})(\lambda v_1v_2.x_2v_1v_2)B^2B^2 \to_\Gamma^* \tilde{x}\,.$$

While $(C[N])[O^2/v] \to_\Gamma^* \tilde{y}$; in fact,

$$(\lambda x_3 y.((\lambda u.(\lambda x.N[O^2/v]xy(\lambda u_1u_2.uu_1u_2))B^2B^2U_2^1)U_2^1(\lambda v_1v_2.x_2v_1v_2)x_3))\tilde{X}_4\tilde{Y}_5B^2B^2 \to_\Gamma^*$$
$$(\lambda x_3 y.((\lambda u.N[O^2/v]B^2y(\lambda u_1u_2.uu_1u_2)B^2U_2^1)U_2^1(\lambda v_1v_2.x_2v_1v_2)x_3))\tilde{X}_4\tilde{Y}_5B^2B^2 \to_\Gamma^*$$
$$(\lambda x_3 y.N[O^2/v]B^2y(\lambda u_1u_2.U_2^1u_1u_2)B^2U_2^1(\lambda v_1v_2.x_2v_1v_2)x_3)\tilde{X}_4\tilde{Y}_5B^2B^2 \to_\Gamma^*$$
$$(\lambda xyu.x(u(yy)(O^2O^2)))B^2\tilde{Y}_5(\lambda u_1u_2.U_2^1u_1u_2)B^2U_2^1(\lambda v_1v_2.x_2v_1v_2)\tilde{X}_4B^2B^2 \to_\Gamma^*$$
$$B^2((\lambda u_1u_2.U_2^1u_1u_2)(\tilde{Y}_5\tilde{Y}_5)(O^2O^2))B^2U_2^1(\lambda v_1v_2.x_2v_1v_2)\tilde{X}_4B^2B^2 \to_\Gamma^*$$
$$B^2((\lambda u_1u_2.u_1)(\lambda x_2x_3x_4x_5.\tilde{y})O^1)B^2U_2^1(\lambda v_1v_2.x_2v_1v_2)\tilde{X}_4B^2B^2 \to_\Gamma^*$$
$$B^2(\lambda x_2x_3x_4x_5.\tilde{y})B^2U_2^1(\lambda v_1v_2.x_2v_1v_2)\tilde{X}_4B^2B^2 \to_\Gamma^*$$
$$U_2^1(\lambda x_2x_3x_4x_5.\tilde{y})B^2(\lambda v_1v_2.x_2v_1v_2)\tilde{X}_4B^2B^2 \to_\Gamma^*$$
$$(\lambda x_2x_3x_4x_5.\tilde{y})(\lambda v_1v_2.x_2v_1v_2)\tilde{X}_4B^2B^2 \to_\Gamma^* \tilde{y}.$$

Lemma 3.1.42 (Termination).
If $M, N \in \Lambda$-NF and $M \not\approx_c N$ then $M, N \Rrightarrow_\Gamma C[.]$.

Proof. By induction on c. Similar to the termination proof (Lemma 2.1.17) of the Böhm Theorem, by using Lemma 3.1.40. □

The next Lemma is necessary for proving the correctness.

Lemma 3.1.43. *Let $zP_1...P_m, T \in \Lambda$-NF, $\mathsf{args}(zP_1...P_m) \leq r$, $\mathsf{args}(T) \leq r$. Let $D[.] \equiv (\lambda u_1...u_k.[.])R_1...R_h$ be a context ($k \leq h$) such that both $R_j \equiv \lambda x_1...x_r.\bar{R}_j$ ($1 \leq j \leq h$) and $\mathsf{nf}_\Lambda(D[T]) \equiv zP_1...P_m$.*
Let $\mathbf{b}$ denote a substitution such that, if $y \in \mathrm{FV}(D[T])$ then there is $Q^y \in \Lambda$ and $\mathbf{b}(y) = \lambda x_1...x_r.Q^y$. In particular, $\mathbf{b}(z) = \lambda x_1...x_r.Q^z$.
If $x_1, ..., x_r \notin \mathrm{FV}(Q^z)$ then $\mathbf{b}(D[T]) \to_\Gamma^* \lambda x_{r-m}...x_r.Q^z$.

Proof. The proof is given by induction on h.

$h = 0$. Clearly $D[.] \equiv [.]$, so T must be $zP_1...P_m$.
$\mathbf{b}(P_j) \to_\Gamma^* \bar{P}_j \in \Gamma$ ($1 \leq j \leq m$), by Lemma 3.1.39; so, $\mathbf{b}(zP_1...P_m) \to_\Gamma^* (\lambda x_1...x_r.Q^z)\mathbf{b}(P_1)...\mathbf{b}(P_m) \to_\Gamma^* (\lambda x_1...x_r.Q^z)\bar{P}_1...\bar{P}_m \to_\Gamma^* \lambda x_{r-m}...x_r.Q^z$.

$h \geq 1$. If $k = 0$ then $TR_1...R_h \equiv zP_1...P_m$, so the proof is similar to the previous case.
Let $k \geq 1$, let $D'[.]$ be $[.]R_{k+1}...R_h$ and $\mathbf{b}'$ denote a substitution such that $\mathbf{b}'(u_j) = R_j$ $(1 \leq j \leq k)$, while $\mathbf{b}'(y) = \mathbf{b}(y)$ for all other variables. Hence, $\mathbf{b}'(D'[T]) \to_\Gamma^* \lambda x_{r-m}...x_r.Q^z$ by case $k = 0$; the proof follows, since $\mathbf{b}(D[zP_1...P_m]) \to_\Gamma^* \mathbf{b}'(D'[T])$. □

In the next lemma, $\tilde{x}, \tilde{y}$ are the pair of fresh variables considered by the algorithm.

Lemma 3.1.44 (Correctness).
Let M, N be different Λ-normals form such that $r \geq \max\{\mathsf{args}(M), \mathsf{args}(N)\}$. If $M, N \Rrightarrow_\Gamma C[.]$ and $\mathrm{FV}(C[M]) \cup \mathrm{FV}(C[N]) = \{u_1, ..., u_n\} - \{\tilde{x}, \tilde{y}\}$ then $(\lambda u_1...u_n.C[M])\underbrace{O^r...O^r}_n \to_\Gamma^* \tilde{x}$ and $(\lambda u_1...u_n.C[N])\underbrace{O^r...O^r}_n \to_\Gamma^* \tilde{y}$.

Proof. Let $\mathbf{e}$ denote a substitution such that $\mathbf{e}(\tilde{x}) = \tilde{x}$, $\mathbf{e}(\tilde{y}) = \tilde{y}$, while $\forall z \in \mathrm{Var} - \{\tilde{x}, \tilde{y}\}$, $\mathbf{e}(z) = O^r$.
We will prove that $\mathbf{e}(C[M]) \to_\Gamma^* \tilde{x}$ and $\mathbf{e}(C[N]) \to_\Gamma^* \tilde{y}$;
moreover, let $T \in \Lambda$-NF, let $\mathsf{args}(T) \leq r$, let $D[.] \equiv (\lambda u_1...u_k.[.])R_1...R_h$ be a context $(k \leq h)$ such that either $R_j = B^r$ or $R_j = U_t^r$ or $R_j = X_{u,S}^r$ (where $1 \leq j \leq h$, $t \leq r$, $u \in \mathrm{Var}$ and $S \subseteq \mathrm{Var}$) and let $\tilde{x}, \tilde{y} \notin \mathrm{FV}(D[T])$, so:

- if $\mathsf{nf}_\Lambda(D[T]) \equiv M$ then $\mathbf{e}(C[D[T]]) \to_\Gamma^* \tilde{x}$,
- if $\mathsf{nf}_\Lambda(D[T]) \equiv N$ then $\mathbf{e}(C[D[T]]) \to_\Gamma^* \tilde{y}$.

The proof is given by induction on the derivation proving $M, N \Rrightarrow_\Gamma C[.]$.

($\Gamma 1$) Let $C_p^r[.] \equiv (\lambda x_1...x_p.[.])X_{x_1,\{x,y\}}^r...X_{x_p,\{x,y\}}^r$ and

$$\left.\begin{array}{r} x\mathsf{nf}_\Lambda(C_p^r[M_1])...\mathsf{nf}_\Lambda(C_p^r[M_m])X_{x_{p+1},\{x,y\}}^r...X_{x_q,\{x,y\}}^r, \\ y\mathsf{nf}_\Lambda(C_q^r[N_1])...\mathsf{nf}_\Lambda(C_q^r[N_n]) \end{array}\right\} \Rrightarrow_\Gamma C[.]\,;$$

so the two inductive hypothesis follow.

- $\mathbf{e}(C[x\mathsf{nf}_\Lambda(C_p^r[M_1])...\mathsf{nf}_\Lambda(C_p^r[M_m])X_{x_{p+1},\{x,y\}}^r...X_{x_q,\{x,y\}}^r]) \to_\Gamma^* \tilde{x}$.
- Moreover, if $T \in \Lambda$-NF, $\mathsf{args}(T) \leq r$, $D[.] \equiv (\lambda u_1...u_k.[.])R_1...R_h$ is a context $(k \leq h)$, where $R_j \in \{B^r, U_t^r, X_{u,S}^r \mid t \in \mathbb{N} \wedge u \in \mathrm{Var} \wedge S \subseteq \mathrm{Var}\}$ $(1 \leq j \leq h)$,

$$\mathsf{nf}_\Lambda(D[T]) \equiv x\mathsf{nf}_\Lambda(C_p^r[M_1])...\mathsf{nf}_\Lambda(C_p^r[M_m])X_{x_{p+1},\{x,y\}}^r...X_{x_q,\{x,y\}}^r$$

and $\tilde{x}, \tilde{y} \notin \mathrm{FV}(D[T])$ then $\mathbf{e}(C[D[T]]) \to_\Gamma^* \tilde{x}$.

Let $T \equiv xM_1...M_m$ and $D[.] \equiv C_p^r[.]$; so, $\mathsf{args}(T) \leq r$, $\tilde{x}, \tilde{y} \notin \mathrm{FV}(D[T])$ and $\mathsf{nf}_\Lambda(C_p^r[T]) \equiv x\mathsf{nf}_\Lambda(C_p^r[M_1])...\mathsf{nf}_\Lambda(C_p^r[M_m])X_{x_{p+1},\{x,y\}}^r...X_{x_q,\{x,y\}}^r$ imply, by induction, $\mathbf{e}\Big(C[D[xM_1...M_m]]\Big) \to_\Gamma^* \tilde{x}$, so

$$\mathbf{e}\Big(C[(\lambda x_1...x_p.xM_1...M_m)X^r_{x_1,\{x,y\}}...X^r_{x_q,\{x,y\}}]\Big) \to^*_\Gamma \tilde{x}.$$

We must yet prove that, if $T^* \in \Lambda$-NF, $D^*[.] \equiv (\lambda u^*_1...u^*_{k^*}.[.])R^*_1...R^*_{h^*}$ is a context ($k^* \leq h^*$) such that either $R^*_j = B^r$ or $R^*_j = U^r_t$ or $R^*_j = X^r_{u,S}$ (where $1 \leq j \leq h^*$, $t \in \mathbb{N}$, $u \in$ Var and $S \subseteq$ Var), $\mathsf{args}(T^*) \leq r$, $\tilde{x}, \tilde{y} \notin \mathrm{FV}(D^*[T^*])$ and $\mathsf{nf}_\Lambda(D^*[T^*]) \equiv \lambda x_1...x_p.xM_1...M_m$ then $\mathbf{e}(C^*[D^*[T^*]]) \to^*_\Gamma \tilde{x}$, where $C^*[.] \equiv C[[.]X^r_{x_1,\{x,y\}}...X^r_{x_q,\{x,y\}}]$.
Let $T \equiv T^*$ and $D[.] \equiv D^*[.]X^r_{x_1,\{x,y\}}...X^r_{x_q,\{x,y\}}$; thus, both

$$\begin{aligned}&\mathsf{nf}_\Lambda(D^*[T^*]X^r_{x_1,\{x,y\}}...X^r_{x_q,\{x,y\}})\\&\qquad\equiv x\mathsf{nf}_\Lambda(C^r_p[M_1])...\mathsf{nf}_\Lambda(C^r_p[M_m])X^r_{x_{p+1},\{x,y\}}...X^r_{x_q,\{x,y\}},\end{aligned}$$

$\mathsf{args}(T) \leq r$ and $\tilde{x}, \tilde{y} \notin \mathrm{FV}(D[T^*])$. By induction,

$$\mathbf{e}\Big(C[D[T^*]]\Big) \equiv \mathbf{e}\Big(C^*[D^*[T^*]]\Big) \to^*_\Gamma \tilde{x}.$$

The proof for the term on the right is similar.

(Γ2) Similar to (Γ1).

(Γ3) Let $n < m$ and let $xM_1...M_m, xN_1...N_n \Rrightarrow_\Gamma C[.]$ where

$$C[.] \equiv (\lambda x.[.])O^{r+n}\underbrace{B^r.....B^r}_{r+n-m}(\lambda x_1...x_{m-n}.\tilde{x})\underbrace{\tilde{y}.....\tilde{y}}_{m-n}.$$

We will prove that, if $T \in \Lambda$-NF, $\mathsf{args}(T) \leq r$, $D[.] \equiv (\lambda u_1...u_k.[.])R_1...R_h$ is a context ($h, k \in \mathbb{N}$), where

$$R_j \in \{B^r, U^r_t, X^r_{u,S} \mid t \in \mathbb{N} \wedge u \in \mathrm{Var} \wedge S \subseteq \mathrm{Var}\} \qquad (1 \leq j \leq h),$$

$\tilde{x}, \tilde{y} \notin \mathrm{FV}(D[T])$ and $\mathsf{nf}_\Lambda(D[T]) \equiv xM_1...M_m$ then $\mathbf{e}(C[D[T]]) \to^*_\Gamma \tilde{x}$.
Let $\mathbf{b}(x) = O^{r+n}$ and $\mathbf{b}(y) = \mathbf{e}(y)$ for all other variables; therefore, by Lemma 3.1.43, $\mathbf{b}(D[T])) \to^*_\Gamma O^{r+n-m}$. Hence

$$\begin{aligned}&\mathbf{e}\Big((\lambda x.D[T])O^{r+n}\underbrace{B^r.....B^r}_{r+n-m}(\lambda x_1...x_{m-n}.\tilde{x})\underbrace{\tilde{y}.....\tilde{y}}_{m-n}\Big) \to^*_\Gamma\\&O^{r+n-m}\underbrace{B^r.....B^r}_{r+n-m}(\lambda x_1...x_{m-n}.\tilde{x})\underbrace{\tilde{y}.....\tilde{y}}_{m-n} \to^*_\Gamma \tilde{x}.\end{aligned}$$

On the other hand, we will prove that, if $T \in \Lambda$-NF, $\mathsf{args}(T) \leq r$, $D[.] \equiv (\lambda u_1...u_k.[.])R_1...R_h$ is a context ($h, k \in \mathbb{N}$), where

$$R_j \in \{B^r, U^r_t, X^r_{u,S} \mid t \in \mathbb{N} \wedge u \in \mathrm{Var} \wedge S \subseteq \mathrm{Var}\} \qquad (1 \leq j \leq h),$$

$\tilde{x}, \tilde{y} \notin \mathrm{FV}(D[T])$ and $\mathsf{nf}_\Lambda(D[T]) \equiv xN_1...N_n$ then $\mathbf{e}(C[D[T]]) \to^*_\Gamma \tilde{y}$.
Let $\mathbf{b}(x) = O^{r+n}$ and $\mathbf{b}(y) = \mathbf{e}(y)$ for all other variables; therefore, by Lemma 3.1.43, $\mathbf{b}(D[T])) \to^*_\Gamma O^r$. Hence

$$\mathbf{e}\Big((\lambda x.D[T])O^{r+n}\underbrace{B^r.....B^r}_{r+n-m}(\lambda x_1...x_{m-n}.\tilde{x})\underbrace{\tilde{y}.....\tilde{y}}_{m-n}\Big) \to^*_\Gamma$$
$$O^r \underbrace{B^r.....B^r\ (\lambda x_1...x_{m-n}.\tilde{x})\ \tilde{y}.....\tilde{y}}_{r+1} \to^*_\Gamma \tilde{y}.$$

(Γ4) Symmetric to (Γ3).
(Γ5) Let $xM_1...M_m, yN_1...N_n \Rrightarrow_\Gamma C[.]$ where

$$C[.] \equiv (\lambda xy.[.])(\lambda x_1...x_{r+m}.\tilde{x})(\lambda x_1...x_{r+n}.\tilde{y})\underbrace{B^r.....B^r}_{r}.$$

We will prove that, if $T \in \Lambda$-NF, $\mathsf{args}(T) \le r$, $D[.] \equiv (\lambda u_1...u_k.[.])R_1...R_h$ is a context ($h, k \in \mathbb{N}$), where

$$R_j \in \{B^r, U^r_t, X^r_{u,S} \mid t \in \mathbb{N} \wedge u \in \mathrm{Var} \wedge S \subseteq \mathrm{Var}\} \qquad (1 \le j \le h),$$

$\tilde{x}, \tilde{y} \notin \mathrm{FV}(D[T])$ and $\mathsf{nf}_\Lambda(D[T]) \equiv xM_1...M_m$ then $\mathbf{e}(C[D[T]]) \to^*_\Gamma \tilde{x}$. Let $\mathbf{b}(x) = \lambda x_1...x_{r+m}.\tilde{x}$, $\mathbf{b}(y) = \lambda x_1...x_{r+n}.\tilde{y}$ and $\mathbf{b}(z) = \mathbf{e}(z)$ for each other variable z; thus, by Lemma 3.1.43, $\mathbf{b}(D[T]) \to^*_\Gamma \lambda x_1...x_r.\tilde{x}$. Hence

$$\mathbf{e}\Big((\lambda xy.xM_1...M_m)(\lambda x_1...x_{r+m}.\tilde{x})(\lambda x_1...x_{r+n}.\tilde{y})\underbrace{B^r.....B^r}_{r}\Big) \to^*_\Gamma \tilde{x}.$$

On the other hand, we will prove that, if $T \in \Lambda$-NF, $\mathsf{args}(T) \le r$, and $D[.] \equiv (\lambda u_1...u_k.[.])R_1...R_h$ is a context ($h, k \in \mathbb{N}$) where

$$R_j \in \{B^r, U^r_t, X^r_{u,S} \mid t \in \mathbb{N} \wedge u \in \mathrm{Var} \wedge S \subseteq \mathrm{Var}\} \qquad (1 \le j \le h),$$

$\tilde{x}, \tilde{y} \notin \mathrm{FV}(D[T])$ and $\mathsf{nf}_\Lambda(D[T]) \equiv yN_1...N_n$ then $\mathbf{e}(C[D[T]]) \to^*_\Gamma \tilde{y}$. Let $\mathbf{b}(x) = \lambda x_1...x_{r+m}.\tilde{x}$, $\mathbf{b}(y) = \lambda x_1...x_{r+n}.\tilde{y}$ and $\mathbf{b}(z) = \mathbf{e}(z)$ for each other variable z; thus, by Lemma 3.1.43, $\mathbf{b}(D[T])) \to^*_\Gamma \lambda x_1...x_r.\tilde{y}$. Hence

$$\mathbf{e}\Big((\lambda xy.yN_1...N_n)(\lambda x_1...x_{r+m}.\tilde{x})(\lambda x_1...x_{r+n}.\tilde{y})\underbrace{B^r.....B^r}_{r}\Big) \to^*_\Gamma \tilde{y}.$$

(Γ6) Let $M_k \neq_{\Lambda\eta} N_k$, $x \notin \mathrm{FV}(M_k) \cup \mathrm{FV}(N_k)$ and $M_k, N_k \Rrightarrow_\Gamma C[.]$. The two inductive hypothesis follow.
- $\mathbf{e}(C[M_k]) \to^*_\Gamma \tilde{x}$.
- If $T \in \Lambda$-NF, $\mathsf{args}(T) \le r$, $D[.] \equiv (\lambda u_1...u_k.[.])R_1...R_h$ is a context ($k \le h$), where

$$R_j \in \{B^r, U^r_t, X^r_{u,S} \mid t \in \mathbb{N} \wedge u \in \mathrm{Var} \wedge S \subseteq \mathrm{Var}\} \qquad (1 \le j \le h),$$

$\mathsf{nf}_\Lambda(D[T]) \equiv M_k$ and $\tilde{x}, \tilde{y} \notin \mathrm{FV}(D[T])$ then $\mathbf{e}(C[D[T]]) \to^*_\Gamma \tilde{x}$.

Let $T \equiv xM_1...M_m$ and $D[.] \equiv (\lambda x.[.])U^r_k \underbrace{B^r.....B^r}_{r-m}$; thus, $\mathsf{args}(T) \le r$, $\tilde{x}, \tilde{y} \notin \mathrm{FV}(D[T])$ and $\mathsf{nf}_\Lambda(D[T]) \equiv M_k$ imply, by induction,

$$\mathbf{e}\Big(C[D[T]]\Big) \equiv \mathbf{e}\Big(C[(\lambda x.xM_1...M_m)U^r_k \underbrace{B^r.....B^r}_{r-m}]\Big) \to^*_\Gamma \tilde{x}.$$

We must yet prove that, if $T^* \in \Lambda$-NF, $D^*[.] \equiv (\lambda u_1^*...u_{k^*}^*.[.])R_1^*...R_{h^*}^*$ is a context ($k^* \leq h^*$) such that either $R_j^* = B^r$ or $R_j^* = U_t^r$ or $R_j^* = X_{u,S}^r$ (where $1 \leq j \leq h^*$, $t \in \mathbb{N}$, $u \in$ Var and $S \subseteq$ Var), $\mathsf{args}(T^*) \leq r$, $\tilde{x}, \tilde{y} \notin \mathrm{FV}(D^*[T^*])$ and $\mathsf{nf}_\Lambda(D^*[T^*]) \equiv xM_1...M_m$, then

$\mathbf{e}(C^*[D^*[T^*]]) \to_\Gamma^* \tilde{x}$, where $C^*[.] \equiv C[(\lambda x.[.])U_k^r \underbrace{B^r.....B^r}_{r-m}]$.

Let $D[.] \equiv (\lambda x u_1^*...u_{k^*}^*.[.])U_k^r R_1^*[U_k^r/x]...R_{h^*}^*[U_k^r/x] \underbrace{B^r.....B^r}_{r-m}$ and $T \equiv T^*$; therefore $R_j^*[U_k^r/x] \in \{B^r, U_t^r, X_{u,S}^r \mid t \in \mathbb{N} \wedge u \in \mathrm{Var} \wedge S \subseteq \mathrm{Var}\}$ ($1 \leq j \leq h^*$), $\mathsf{nf}_\Lambda(D[T]) \equiv M_k$ (since $x \notin \mathrm{FV}(M_k)$), $\mathsf{args}(T) \leq r$ and $\tilde{x}, \tilde{y} \notin \mathrm{FV}(D[T])$. So, by induction,

$$\mathbf{e}\Big(C[(\lambda x.D^*[T^*])U_k^r \underbrace{B^r.....B^r}_{r-m}]\Big) =_\Gamma \mathbf{e}\Big(C[D[T]]\Big) \to_\Gamma^* \tilde{x}.$$

The proof for the term on the right is similar.

($\Gamma 7$) Let $M_k \neq_{\Lambda\eta} N_k$ and $x \in \mathrm{FV}(M_k) \cup \mathrm{FV}(N_k)$; furthermore, let $C_x^r[.] \equiv (\lambda x.[.])B^r$ and $\mathsf{nf}_\Lambda(C_x^r[M_k]), \mathsf{nf}_\Lambda(C_x^r[N_k]) \Rrightarrow_\Gamma C[.]$.
The two inductive hypothesis follow.
- $\mathbf{e}(C[\mathsf{nf}_\Lambda(C_x^r[M_k])]) \to_\Gamma^* \tilde{x}$.
- moreover, if $T \in \Lambda$-NF, $\mathsf{args}(T) \leq r$, and $D[.] \equiv (\lambda u_1...u_k.[.])R_1...R_h$ is a context ($k \leq h$), where

$$R_j \in \{B^r, U_t^r, X_{u,S}^r \mid t \in \mathbb{N} \wedge u \in \mathrm{Var} \wedge S \subseteq \mathrm{Var}\} \qquad (1 \leq j \leq h),$$

$\mathsf{nf}_\Lambda(D[T]) \equiv \mathsf{nf}_\Lambda(C_x^r[M_k])$ and $\tilde{x}, \tilde{y} \notin \mathrm{FV}(D[T])$, then $\mathbf{e}(C[D[T]]) \to_\Gamma^* \tilde{x}$.

Let $T \equiv xM_1...M_m$ and $D[.] \equiv (\lambda x.[.])B^r \underbrace{B^r.....B^r}_{r-m} U_k^r$; so, $\mathsf{args}(T) \leq r$, $\tilde{x}, \tilde{y} \notin \mathrm{FV}(D[T])$ and $\mathsf{nf}_\Lambda(D[T]) \equiv \mathsf{nf}_\Lambda(C_x^r[M_k])$ imply, by induction,

$$\mathbf{e}\Big(C[D[T]]\Big) \equiv \mathbf{e}\Big(C[(\lambda x.xM_1...M_m)B^r \underbrace{B^r.....B^r}_{r-m} U_k^r]\Big) \to_\Gamma^* \tilde{x}.$$

We must yet prove that, if $T^* \in \Lambda$-NF, $D^*[.] \equiv (\lambda u_1^*...u_{k^*}^*.[.])R_1^*...R_{h^*}^*$ is a context ($k^* \leq h^*$) such that either $R_j^* = B^r$ or $R_j^* = U_t^r$ or $R_j^* = X_{u,S}^r$ (where $1 \leq j \leq h^*$, $t \in \mathbb{N}$, $u \in$ Var and $S \subseteq$ Var), $\mathsf{args}(T^*) \leq r$, $\tilde{x}, \tilde{y} \notin \mathrm{FV}(D^*[T^*])$ and $\mathsf{nf}_\Lambda(D^*[T^*]) \equiv xM_1...M_m$, then

$\mathbf{e}(C^*[D^*[T^*]]) \to_\Gamma^* \tilde{x}$, where $C^*[.] \equiv C[(\lambda x.[.])B^r \underbrace{B^r.....B^r}_{r-m} U_k^r]$.

Let $D[.] \equiv (\lambda x u_1^*...u_{k^*}^*.[.])B^r R_1^*[B^r/x]...R_{h^*}^*[B^r/x] \underbrace{B^r.....B^r}_{r-m} U_k^r$ and let $T \equiv T^*$. Thus $R_j^*[B^r/x] \in \{B^r, U_t^r, X_{u,S}^r \mid t \in \mathbb{N} \wedge u \in \mathrm{Var} \wedge S \subseteq \mathrm{Var}\}$ where $1 \leq j \leq h^*$, $\mathsf{nf}_\Lambda(D[T]) \equiv \mathsf{nf}_\Lambda(C_x^r[M_k])$, $\mathsf{args}(T) \leq r$ and $\tilde{x}, \tilde{y} \notin \mathrm{FV}(D[T])$. By induction,

$$\mathbf{e}\Big(C[(\lambda x.D^*[T^*])B^r \underbrace{B^r.....B^r}_{r-m} U_k^r]\Big) =_\Gamma \mathbf{e}\Big(C[D[T]]\Big) \rightarrow_\Gamma^* \tilde{x}.$$

The proof for the term on the right is similar. □

▼ **Proof of Γ-Separability Theorem** (Theorem 3.1.16 pag. 40).
The proof follows directly from Lemmas 3.1.42 and 3.1.44. ■

3.2 Potentially Γ-Valuable Terms and Λ-Reduction

In this section the relation between the call-by-value $\lambda\Gamma$-calculus and the call-by-name $\lambda\Lambda$-calculus is explored. In particular, we show that the potentially Γ-valuable terms, which were characterized through the notion of $\Xi\ell$-reduction, introduced on purpose, coincide with the strongly normalizing terms with respect to the $\Lambda\ell$-reduction. The notion of $\Lambda\ell$-reduction is a particular case of Definition 3.1.3. In order to prove the result, the notion of weight of a term, introduced in Definition 3.1.29, will be used.

Lemma 3.2.1. *If $\langle M\rangle$ is defined then M has $\Xi\ell$-normal form.*

Proof. By induction on $\langle M\rangle$.

$\langle M\rangle = 0$. Trivial, since M is an abstraction.
$\langle M\rangle \geq 1$. If $M \equiv xM_1...M_m$ $(m \in \mathbb{N})$ the proof follows by induction. Let $M \equiv (\lambda x.P)QM_1...M_m$ $(m \in \mathbb{N})$. By induction $P[Q/x]M_1...M_m$ and Q have $\Xi\ell$-normal forms, so let $Q \rightarrow_{\Xi\ell}^* R \in \Xi$; by Property 3.1.33.(ii)

$$\langle P[R/x]M_1...M_m\rangle \leq \langle P[Q/x]M_1...M_m\rangle < \langle M\rangle,$$

so the proof follows by induction. □

Corollary 3.2.2. *M has $\Xi\ell$-normal form if and only if $\langle M\rangle$ is defined.*

Proof. By the previous lemma and Lemma 3.1.31.(iii). □

The next lemma proves that the notion of weight also works well for the $\Lambda\ell$-reduction.

Lemma 3.2.3. *Let $M \in \Lambda$ and $\langle M\rangle$ be defined.*
If $M \rightarrow_{\Lambda\ell} N$ then $\langle N\rangle$ is defined and $\langle N\rangle < \langle M\rangle$.

Proof. The proof is given by induction on $k = \langle M\rangle$.
The case where $k = 0$ is not possible, since M is an abstraction and it cannot be $\Lambda\ell$-reduced; so let $k \geq 1$.
If $M \equiv xM_1...M_m$ then the proof follows by induction.
Let $M \equiv (\lambda x.M_0)M_1...M_m$, $h' = \langle M_1\rangle$ and $h'' = \langle M_0[M_1/x]M_2...M_m\rangle$, thus $k = 1 + h' + h''$. There are only three cases, by the laziness of $\rightarrow_{\Lambda\ell}$.

(i) If $N \equiv M_0[M_1/x]M_2...M_m$ then the proof follows from the definition of weight.

(ii) Let $N \equiv (\lambda x.M_0)M_1N_2...N_m$ where there is a unique $j \geq 2$ such that $M_j \rightarrow_{\Lambda\ell} N_j$, while $\forall i \neq j \;\; M_i \equiv N_i$ $(0 \leq i \leq m$ and $i \neq 1)$.
$M_0[M_1/x]M_2...M_m \rightarrow_{\Lambda\ell} M_0[M_1/x]N_2...N_m$ and $h'' < k$ imply, by induction, $\langle N_0[M_1/x]N_2...N_m\rangle < h''$.
Thus $\langle N\rangle = 1 + \langle M_1\rangle + \langle N_0[M_1/x]N_2...N_m\rangle < k$.

(iii) Let $N \equiv (\lambda x.M_0)N_1M_2...M_m$, where $M_1 \rightarrow_{\Lambda\ell} N_1$. By Property 3.1.33.(ii) we can state $\langle M_0[N_1/x]M_2...M_m\rangle \leq h''$, since $M_0[M_1/x]M_2...M_m \rightarrow^*_{\Lambda} M_0[N_1/x]M_2...M_m$. Again, by induction $\langle M_1\rangle < \langle N_1\rangle$; so the proof follows by the definition of weight.

□

Theorem 3.2.4.
M has $\Xi\ell$-normal form if and only if M is $\Lambda\ell$-strongly normalizing.

Proof. ($\Leftarrow$) Trivial, since $M \rightarrow_{\Xi\ell} N$ implies $M \rightarrow_{\Lambda\ell} N$.
($\Rightarrow$) By Corollary 3.2.2, $\langle M\rangle$ is defined. Let N be such that $M \rightarrow_{\Lambda\ell} N$; thus by Lemma 3.2.3, both $\langle N\rangle$ is defined and $\langle N\rangle < \langle M\rangle$. This implies that there is not an infinite sequence of $\Lambda\ell$-reductions starting from M, and so M is $\Lambda\ell$-strongly normalizing, by Definition 3.1.3.(iv). □

Corollary 3.2.5 shows the desired result.

Corollary 3.2.5. *Let $M \in \Lambda^0$.*
M is Γ-valuable if and only if M is $\Lambda\ell$-strongly normalizing.

Proof. From Theorems 3.1.9 and 3.2.4. □

4. Further Reading

Λ-separability. The separability property of Λ-normal forms was extended to finite sets of different $\Lambda\eta$-normal forms in [18], and the separability of infinite sets of Λ-normal forms was studied in [84]. An algebraic analysis of the technique used by Böhm for proving his theorem was developed in [76]. More refined notions of separability were studied in [20] and [22].

Call-by-value λ-calculus. Extensions of the $\lambda\Gamma$-calculus for studying imperative and control features in the call-by-value setting were introduced respectively in [45] and [46]. Moggi [70], starting from the $\lambda\Gamma$-calculus, developed a further paradigmatic language for reasoning about the call-by-value computation, called the partial λ-calculus.

Call-by-value versus call-by-name. Some interesting observations on the relationship between call-by-value and call-by-name computation can be found in [33], in a typed setting, where it was shown that call-by-value is the De Morgan dual of call-by-name. This idea was further developed in [96].

Part II

Operational Semantics

5. Parametric Operational Semantics

In this part we will study the evaluation of terms and the induced operational semantics. Our notion of operational semantics is inspired by the structured operational semantics (SOS) developed by Plotkin [80] and by Kahn [55].

In Sect. 1.2, we introduced in an informal way the notion of evaluation, by saying that a possible way of evaluating a term is to apply the reduction rule to it until a normal from is reached. Clearly, such evaluation can never stop, for example, in the case when $D \in \Delta$ and the Δ-reduction is applied to the term DD, which do not have Δ-normal form.

But the normal forms are not the only terms we can reasonably consider as output results. For example, we defined the notion of head normal form, both in the $\lambda\Lambda$ and in the $\lambda\Gamma$ setting. It is natural to ask if such terms can be considered as output values, and so if it is possible to check, through an evaluation, whether or not a term possesses head normal form.

Hence, in order to study the evaluation of terms, we need to introduce behind the notion of input values, that of output values. The definition of a set of output results is parametric with respect to the set of input values.

Definition 5.0.1. *Let Δ be a set of input values.*
A set of output values *with respect to Δ is any set $\Theta \subseteq \Lambda$ such that:*

(i) *Θ contains all the Δ-normal forms,*
(ii) *if $M =_\Delta N$ and $N \in \Theta$ then there is $P \in \Theta$ such that $M \to_\Delta^{*p} P$* (principality condition).

The first condition of the previous definition takes into account the fact that the set of normal forms is in some sense the most "natural" set of output values, corresponding to the complete evaluation of terms. Remember that Corollary 1.2.13 assures us that, to reach the normal form of a term, if it exists, it is sufficient to perform at every step the principal redex. So the second condition simply says that we are interested in those evaluations that are an initial step of the complete one. As we show in the following, each evaluation of interest is of this kind.

Lemma 5.0.2. *Let $\Theta \subseteq \Lambda$ be such that Δ-NF $\subseteq \Theta$.*
If Θ is closed under $\to_\Delta$ and the set $\{M \in \Lambda \mid M \notin \Theta\}$ is closed under $\to_\Delta^i$, then Θ is a set of output values with respect to Δ.

Proof. We must prove that Θ satisfies the principality condition.
If $M =_\Delta N \in \Theta$ then there is a term $M' \in \Lambda$ such that $M \to^*_\Delta M'$ and $N \to^*_\Delta M'$ by the confluence theorem; so $M' \in \Theta$ by the fact that output values are closed under $\to_\Delta$. By the standardization theorem, there is a standard reduction sequence $M \to^{*p}_\Delta M'' \to^{*i}_\Delta M'$; hence $M' \in \Theta$ implies $M'' \in \Theta$, by the fact that $\{M \in \Lambda \mid M \notin \Theta\}$ is closed under $\to^i_\Delta$. □

The next property shows some examples of sets of output values.

Property 5.0.3. 1. Λ, Λ-NF, Λ-HNF and Λ-LHNF are sets of output values with respect to Λ.
2. Λ and Γ-NF are sets of output values with respect to Γ.
3. The set of Γ-lazy blocked normal forms (Γ-lbnf's), namely $\Gamma\text{-LBNF} = \{\lambda x.M \mid M \in \Lambda\} \cup \{xM_1...M_m \mid M_i \in \Lambda\ ,\ m \in \mathbb{N}\} \cup \{(\lambda x.P)QM_1...M_m \mid P, M_i \in \Lambda\ ,\ Q \notin \Gamma\ ,\ Q \in \Gamma\text{-LBNF}\ ,\ m \in \mathbb{N}\}$, is a set of output values with respect to Γ.
4. Γ is not a set of output values with respect to either Λ nor Γ.
5. Ξ is a set of output values with respect to Λ, but not with respect to Γ.
6. $\Gamma\ell$-NF is not a set of output values with respect to either Λ nor Γ.

Proof. 1. The case for Λ is trivial. In case $\Theta \in \{\Lambda\text{-NF}, \Lambda\text{-HNF}, \Lambda\text{-LHNF}\}$, the proof follows by Lemma 5.0.2. In fact,
- $\Lambda\text{-NF} \subseteq \Lambda\text{-HNF} \subseteq \Lambda\text{-LHNF}$,
- if $P \in \Theta$ and $P \to_\Lambda Q$ then $Q \in \Theta$,
- if $P \to^i_\Lambda Q$ and $P \notin \Theta$ then $Q \notin \Theta$.

2. The case for Λ is trivial.
Γ-NF is a set of output values with respect to Γ by Corollary 1.2.13.
3. The proof follows by Lemma 5.0.2. In fact,
- $\Gamma\text{-NF} \subseteq \Gamma\text{-LBNF}$,
- if $P \in \Gamma\text{-LBNF}$ and $P \to_\Gamma Q$ then $Q \in \Gamma\text{-LBNF}$,
- if $P \to^i_\Gamma Q$ and $P \notin \Gamma\text{-LBNF}$ then $Q \notin \Gamma\text{-LBNF}$.

4. In fact, $xI \in \Lambda$-NF and $xI \in \Gamma$-NF, but $xI \notin \Gamma$.
5. It is easy to see that $\Xi \equiv \Lambda\ell$-NF (see Definition 3.1.3 and Property 3.1.6), thus $\Lambda\text{-NF} \subseteq \Xi$; so Ξ is a set of output values with respect to Λ, by Lemma 5.0.2. But it is not a set of output values with respect to Γ; in fact, $I(xI) \in \Gamma$-NF, but $I(xI) \notin \Xi$.
6. Let $\Delta \in \{\Lambda, \Gamma\}$; thus $(\lambda x.DD)(xI)(II) =_{\Delta\ell} (\lambda x.DD)(xI)I \in \Gamma\ell$-NF; nevertheless, there is no $P \in \Delta\ell$-NF such that $(\lambda x.DD)(xI)(II) \to^{*p}_\Delta P$, against the principality condition. □

In what follows, Θ always denotes a generic set of output values.

Definition 5.0.4. *Let Θ be a set of output values with respect to Δ.*

(i) *An* evaluation relation $\mathbf{O}$ *on the $\lambda\Delta$-calculus with respect to Θ is any subset of $\Lambda \times \Theta$, such that $(M, N) \in \mathbf{O}$ implies $M \to^*_\Delta N$.*

(ii) *$\mathcal{E}(\Delta, \Theta)$ denotes the class of all evaluation relations $\mathbf{O}$ on the $\lambda\Delta$-calculus with respect to Θ.*

Evaluation relations are denoted by bold capital letters.

Example 5.0.5. It is easy to see that the following evaluation relations are well defined.

1. Let $\mathbf{N_{nd}} \in \mathcal{E}(\Lambda, \Lambda\text{-NF})$ be $\{(M, N) \in \Lambda \times \Lambda\text{-NF} \mid M \rightarrow^*_\Lambda N\}$.
2. Let $\mathbf{N} \in \mathcal{E}(\Lambda, \Lambda\text{-NF})$ be $\{(M, N) \in \Lambda \times \Lambda\text{-NF} \mid M \rightarrow^{*p}_\Lambda N\}$.
3. Let $\mathbf{H_{nd}} \in \mathcal{E}(\Lambda, \Lambda\text{-HNF})$ be $\{(M, N) \in \Lambda \times \Lambda\text{-HNF} \mid M \rightarrow^*_\Lambda N\}$.
4. Let $\mathbf{H} \in \mathcal{E}(\Lambda, \Lambda\text{-HNF})$ be $\{(M, N) \in \Lambda \times \Lambda\text{-HNF} \mid M \rightarrow^p_\Lambda M_1 \rightarrow^p_\Lambda \ldots$
$\ldots \rightarrow^p_\Lambda M_r \rightarrow^p_\Lambda N$, and $M_i \notin \Lambda\text{-HNF}(1 \leq i \leq r)\}$.
5. Let $\mathbf{L_{nd}} \in \mathcal{E}(\Lambda, \Lambda\text{-LHNF})$ be $\{(M, N) \in \Lambda \times \Lambda\text{-LHNF} \mid M \rightarrow^*_\Lambda N\}$.
6. Let $\mathbf{L} \in \mathcal{E}(\Lambda, \Lambda\text{-LHNF})$ be $\{(M, N) \in \Lambda \times \Lambda\text{-LHNF} \mid M \rightarrow^p_\Lambda M_1 \rightarrow^p_\Lambda \ldots$
$\ldots \rightarrow^p_\Lambda M_r \rightarrow^p_\Lambda N$, and $M_i \notin \Lambda\text{-LHNF}(1 \leq i \leq r)\}$.
7. Let $\mathbf{G_{nd}} \in \mathcal{E}(\Gamma, \Gamma\text{-NF})$ be $\{(M, N) \in \Lambda \times \Gamma\text{-NF} \mid M \rightarrow^*_\Gamma N\}$.
8. Let $\mathbf{G} \in \mathcal{E}(\Gamma, \Gamma\text{-NF})$ be $\{(M, N) \in \Lambda \times \Gamma\text{-NF} \mid M \rightarrow^{*p}_\Gamma N\}$.
9. Let $\mathbf{V_{nd}} \in \mathcal{E}(\Gamma, \Gamma\text{-LBNF})$ be $\{(M, N) \in \Lambda \times \Gamma\text{-LBNF} \mid M \rightarrow^*_\Gamma N\}$.
10. Let $\mathbf{V} \in \mathcal{E}(\Gamma, \Gamma\text{-LBNF})$ be $\{(M, N) \in \Lambda \times \Gamma\text{-LBNF} \mid M \rightarrow^p_\Gamma M_1 \rightarrow^p_\Gamma \ldots$
$\ldots \rightarrow^p_\Gamma M_r \rightarrow^p_\Lambda N$, and $M_i \notin \Gamma\text{-LBNF}(1 \leq i \leq r)\}$.

An evaluation relation can be presented by using a formal system.

A *logical rule*, or briefly rule, has the following shape:

$$\frac{\mathfrak{P}_1 \;\ldots\ldots\; \mathfrak{P}_m}{\mathfrak{C}}\ \mathsf{name}$$

where the *premises* $\mathfrak{P}_i$ $(1 \leq i \leq m)$ and the *conclusion* $\mathfrak{C}$ are logical judgments (written using metavariables); while name is the name of the rule.

The intended meaning of a rule is that, for every instance s of the metavariables in the rule, $s(\mathfrak{C})$ is implied by the logical AND of $s(\mathfrak{P}_i)$ $(1 \leq i \leq m)$.

For sake of simplicity, we will use the syntax of terms for denoting the metaterms in the logical rules.

A *derivation* is a finite tree of logical rules, such that each leaf is an axiom, each intermediate node has as premises the consequences of its son nodes and its consequence is one of the premises of its father node. The conclusion of the root node is the proved judgment. The *size* of a derivation is the number of nodes in it.

A formal system defining an evaluation relation $\mathbf{O} \in \mathcal{E}(\Delta, \Theta)$ is a set of logical rules for establishing judgments of the shape $M \Downarrow_{\mathbf{O}} N$, whose meaning is $(M, N) \in \mathbf{O}$. We will denote with $M \Downarrow_{\mathbf{O}}$ the fact that the judgment $M \Downarrow_{\mathbf{O}} N$

can be proved in the system for some N, i.e. $(M,N) \in \mathbf{O}$. We will denote with $M \Uparrow_{\mathbf{O}}$ the fact that there is no $N \in \Theta$ such that $M \Downarrow_{\mathbf{O}} N$.

The evaluation relation $\mathbf{O} \in \mathcal{E}(\Delta,\Theta)$ is *deterministic* if, in case $M \Downarrow_{\mathbf{O}}$, there is a unique term N such that $M \Downarrow_{\mathbf{O}} N$ (i.e. the evaluation relation is a partial function). All the evaluation relations we show in this part of the book are deterministic.

Example 5.0.6. The evaluation relations $\mathbf{N_{nd}}$ and $\mathbf{N}$, defined in Example 5.0.5, coincide and are both deterministic by Corollaries 1.2.6 and 1.2.13. The same is true for the evaluation relations $\mathbf{G_{nd}}$ and $\mathbf{G}$.
The evaluation relations $\mathbf{H}$, $\mathbf{L}$ and $\mathbf{V}$ are deterministic, while $\mathbf{H_{nd}}$, $\mathbf{L_{nd}}$ and $\mathbf{V_{nd}}$ are not deterministic. For example, both $(\lambda x.x(II), \lambda x.x(II)) \in \mathbf{H_{nd}}$ and $(\lambda x.x(II), \lambda x.xI) \in \mathbf{H_{nd}}$; the same two pairs of terms are in $\mathbf{L_{nd}}$ and $\mathbf{V_{nd}}$.

A formal system establishing judgments of the shape $M \Downarrow_{\mathbf{O}} N$ can be viewed as a logical representation of a *reduction machine.* In particular the evaluation process of the machine is simulated by a derivation in the logical system. In the terminology of reduction machines, $M \Downarrow_{\mathbf{O}} N$ means that "on input M, the reduction machine $\mathbf{O}$ stops and gives as output N"; $M \Downarrow_{\mathbf{O}}$ means that "on input M, the reduction machine $\mathbf{O}$ stops"; while $M \Uparrow_{\mathbf{O}}$ means "on input M, the reduction machine never stops".

In the rest of the book, we will use the metavariable $\mathbf{O}$ to denote an evaluation relation actually defined by a formal system.

Definition 5.0.7. *An evaluation relation $\mathbf{O} \in \mathcal{E}(\Delta,\Theta)$ induces naturally an* operational semantics, *i.e. a preorder relation on terms denoted by $\preceq_{\mathbf{O}}$. The* operational preorder induced by $\mathbf{O}$ *is defined as:*

$$M \preceq_{\mathbf{O}} N$$
$$\textit{if and only if}$$
$$\forall C[.] \textit{ such that } C[M], C[N] \in \Lambda^0 \quad (C[M] \Downarrow_{\mathbf{O}} \textit{ implies } C[N] \Downarrow_{\mathbf{O}}).$$

$\prec_{\mathbf{O}}$ denotes the strict version of $\preceq_{\mathbf{O}}$, while $\approx_{\mathbf{O}}$ is the equivalence relation on terms induced by $\preceq_{\mathbf{O}}$.
If $M \approx_{\mathbf{O}} N$ then M and N are $\mathbf{O}$-operationally equivalent.

This operational equivalence amounts to the Leibniz Equality Principle for programs, i.e. a criterion for establishing equivalence on the basis of the behaviour of programs regarded as black boxes. It is natural to model a program by a closed term. So a context can be viewed as a *partially specified* program, where every occurrence of the hole denotes a place that must be filled by a subprogram, while a generic term can be viewed as a subprogram. So two terms are equivalent if they can be replaced by each other in the same program without changing its behaviour (with respect to an evaluation relation $\mathbf{O}$).

Since we are considering "pure" calculi, i.e. calculi without constants, the only behaviour we can observe on terms is the termination, and this justifies the previous definition of operational semantics.

> In the presence of constants, a subset of them (the "basic constants") will be the possible results of a computation, and the definition would change in the following way:
>
> $$M \preceq N$$
> if and only if
> $\forall C[.]$ such that $C[M], C[N] \in \Lambda^0$, for all basic constants a, $(C[M] \Downarrow a$ implies $C[N] \Downarrow a)$.

Definition 5.0.8. *Let $\mathbf{O} \in \mathcal{E}(\Delta, \Theta)$ be an evaluation relation.*

(i) *The $\lambda\Delta$-calculus is* correct *with respect to the $\mathbf{O}$-operational semantics if and only if $M =_\Delta N$ implies $M \approx_{\mathbf{O}} N$.*
(ii) *The $\lambda\Delta$-calculus is* complete *with respect to the $\mathbf{O}$-operational semantics if and only if $M \approx_{\mathbf{O}} N$ implies $M =_\Delta N$.*

The $\lambda\Delta$-calculus is correct with respect to the $\mathbf{O}$-operational semantics if and only if $\approx_{\mathbf{O}}$ is a Δ-theory. In fact, it is easy to check that $\approx_{\mathbf{O}}$ is always a congruence relation (see Definition 1.3.1).

Example 5.0.9. The evaluation relations of Example 5.0.5 are correct with respect to their respective set of input values. Some counterexamples to the correctness follow:

(i) Let $\mathbf{J} \in \mathcal{E}(\Lambda, \Lambda\text{-NF})$ be $\{(M, N) \in \Lambda \times \Lambda\text{-NF} \mid M \rightarrow_{\Lambda}^{*i} N\}$.
The $\lambda\Lambda$-calculus is not correct with respect to $\mathbf{J}$. In fact, $KI(DD) =_\Lambda I$ but $KI(DD) \not\approx_{\mathbf{J}} I$, since $KI(DD) \Uparrow_{\mathbf{J}}$ while $I \Downarrow_{\mathbf{J}}$.
(ii) Let $\mathbf{W} \in \mathcal{E}(\Gamma, \Gamma\text{-LBNF})$ be $\{(M, N) \in \Gamma \times \Gamma\text{-LBNF} \mid M \rightarrow_{\Gamma}^{*i} N\}$. The $\lambda\Gamma$-calculus is not correct with respect to $\mathbf{W}$. In fact, $KI(\lambda x.DD) =_\Gamma I$ but $KI(\lambda x.DD) \not\approx_{\mathbf{W}} I$, since $KI(\lambda x.DD) \Uparrow_{\mathbf{W}}$ while $I \Downarrow_{\mathbf{W}}$.

The notion of $\mathbf{O}$-relevant context, introduced in the next definition, is a technical tool that is useful for proving operational equivalences.

Definition 5.0.10. *Let $\mathbf{O} \in \mathcal{E}(\Delta, \Theta)$.*

(i) *A context $C[.]$ is* $\mathbf{O}$-relevant *if and only if there are $M, N \in \Lambda^0$ such that $C[M] \Downarrow_{\mathbf{O}}$ and $C[N] \Uparrow_{\mathbf{O}}$.*
(ii) *Let $M, N \in \Lambda$. A context $C[.]$ is said a* discriminating context for M and N *if and only if $C[M] \Downarrow_{\mathbf{O}}$ and $C[N] \Uparrow_{\mathbf{O}}$, or vice versa.*

This notion of relevance is inspired by the corresponding one of relevant context, introduced by Plotkin in order to study the operational behaviour of the paradigmatic programming language PCF (see [79]).

5.1 The Universal Δ-Reduction Machine

The fact that the set of output values satisfies the principality condition allows us to define a universal evaluation relation, parametric both in the set of input and output values, from which many interesting evaluation relations can be derived by suitable instantiations. Such an evaluation relation is based on a formal system, defining the principal evaluation of a term of the $\lambda\Delta$-calculus.

Definition 5.1.1. *A formal system proving statements of the kind*

$$M \to_{\Delta}^{p} N$$

where $M, N \in \Lambda$, is formalized in Fig. 5.1.
The notation $M \to_{\Delta}^{p} N$ is defined in Definition 1.2.7 (N is obtained from M by reducing its principal redex).

$$\frac{M \to_{\Delta}^{p} N}{\lambda x.M \to_{\Delta}^{p} \lambda x.N}\ p1$$

$$\frac{i = \min\{j \leq m | M_i \notin \Delta\text{-nf}\} \qquad M_i \to_{\Delta}^{p} N_i}{xM_1...M_m \to_{\Delta}^{p} xM_1...N_i...M_m}\ p2$$

$$\frac{Q \in \Delta}{(\lambda x.P)QM_1...M_m \to_{\Delta}^{p} P[Q/x]M_1...M_m}\ p3$$

$$\frac{Q \notin \Delta \qquad Q \notin \Delta\text{-nf} \qquad Q \to_{\Delta}^{p} Q'}{(\lambda x.P)QM_1...M_m \to_{\Delta}^{p} (\lambda x.P)Q'M_1...M_m}\ p4$$

$$\frac{Q \notin \Delta \qquad Q \in \Delta\text{-nf} \qquad P \notin \Delta\text{-nf} \qquad P \to_{\Delta}^{p} P'}{(\lambda x.P)QM_1...M_m \to_{\Delta}^{p} (\lambda x.P')QM_1...M_m}\ p5$$

$$\frac{Q \notin \Delta \qquad P, Q \in \Delta\text{-nf} \qquad i = \min\{j \leq m | M_i \notin \Delta\text{-nf}\} \qquad M_i \to_{\Delta}^{p} N_i}{(\lambda x.P)QM_1...M_m \to_{\Delta}^{p} (\lambda x.P)QM_1...N_i...M_m}\ p6$$

Fig. 5.1. Principal reduction machine

The machine described in Fig. 5.1 is "step-by-step", since each of its rules describes just one application of the reduction rule.

Definition 5.1.2 (Universal evaluation relation).
Let Θ be a set of output values with respect to the set of input value Δ.

(i) $\mathcal{U}_\Theta^\Delta \in \mathcal{E}(\Delta, \Theta)$ *is the evaluation relation defined through the following rules:*

$$\frac{M \in \Theta}{M \Downarrow_{\mathcal{U}_\Theta^\Delta} M}\ (axiom) \qquad\qquad \frac{M \to_\Delta^p P \qquad P \Downarrow_{\mathcal{U}_\Theta^\Delta} N}{M \Downarrow_{\mathcal{U}_\Theta^\Delta} N}\ (eval)$$

(ii) $M \preceq_{\mathcal{U}_\Theta^\Delta} N$ *if and only if, for all contexts* $C[.]$ *such that* $C[M], C[N] \in \Lambda^0$, $(C[M] \Downarrow_{\mathcal{U}_\Theta^\Delta}$ *implies* $C[N] \Downarrow_{\mathcal{U}_\Theta^\Delta})$.

(iii) $M \approx_{\mathcal{U}_\Theta^\Delta} N$ *if and only if* $M \preceq_{\mathcal{U}_\Theta^\Delta} N$ *and* $N \preceq_{\mathcal{U}_\Theta^\Delta} M$.

It is easy to check that the previous definition is well posed, i.e. $M \Downarrow_{\mathcal{U}_\Theta^\Delta} N$ implies $M \to_\Delta^* N$. Furthermore, the evaluation relation $\mathcal{U}_\Theta^\Delta$ is deterministic for all Δ, Θ.

Theorem 5.1.3 proves that the evaluation relation $\mathcal{U}_\Theta^\Delta$ is universal, in the sense that it subsumes all deterministic evaluation relations obtained by instantiating Δ and Θ in a correct way.

Theorem 5.1.3. *If* $M \to_\Delta^* N \in \Theta$ *then* $M \Downarrow_{\mathcal{U}_\Theta^\Delta}$.

Proof. Since Θ satisfies the principality condition, $M \to_\Delta^* N \in \Theta$ implies there is $N' \in \Theta$ such that $M \to_\Delta^{*p} N'$. Then the proof follows by induction on the length of the reduction sequence $M \to_\Delta^{*p} N' \in \Theta$. If $M \in \Theta$, then the proof follows by rule (*axiom*) of the formal system defining $\mathcal{U}_\Theta^\Delta$. Otherwise, $M \to_\Delta^{*p} N'$ means $M \to_\Delta^p N'' \to_\Delta^{*p} N'$, so the proof follows by induction. □

For each choice of the sets of the input and output values, the $\lambda\Delta$-calculus is *correct* with respect to the $\mathcal{U}_\Theta^\Delta$ operational semantics, as proved in Theorem 5.1.4.

Theorem 5.1.4 ($\mathcal{U}_\Theta^\Delta$-Correctness).
The $\lambda\Delta$-calculus is correct with respect to the $\mathcal{U}_\Theta^\Delta$-operational semantics.

Proof. $M =_\Delta N$ implies $C[M] =_\Delta C[N]$, for all contexts $C[.]$.
If there is $P \in \Theta$ such that $C[M] \to_\Delta^* P$, then $C[M] \Downarrow_{\mathcal{U}_\Theta^\Delta}$, by Theorem 5.1.3. Clearly $P =_\Delta C[N]$; thus, by principality, there is $P' \in \Theta$ such that $C[N] \to_\Delta^{*p} P'$, so $C[N] \Downarrow_{\mathcal{U}_\Theta^\Delta}$.
In case there is not such a P, both $C[M] \Uparrow_{\mathcal{U}_\Theta^\Delta}$ and $C[N] \Uparrow_{\mathcal{U}_\Theta^\Delta}$. □

So, $\mathcal{U}_\Theta^\Delta$-operational semantics induce a Δ-theory; as far as completeness is concerned, it depends on the choice of the set of output values. But all operational semantics of interest are not complete, as we will see in the following.

Example 5.1.5. Let $\mathcal{U}^{\Lambda}_{\Lambda\text{-HNF}}$ be the universal evaluation relation, where Λ is the set of input values and Λ-HNF is the set of output values.

(i) Let $M_0 \equiv \lambda x.(\lambda uv.xuv)I(DD)$, $M_1 \equiv \lambda x.(\lambda v.xIv)(DD)$ and $M_2 \equiv \lambda x.xI(DD)$; note that $M_0 \to^p_\Lambda M_1 \to^p_\Lambda M_2 \in \Lambda$-HNF.
$\lambda x.(\lambda uv.xuv)I(DD) \Downarrow_{\mathcal{U}^{\Lambda}_{\Lambda\text{-HNF}}} \lambda x.xI(DD)$; in fact, we can build the following derivation:

$$\dfrac{M_0 \to^p_\Lambda M_1 \qquad \dfrac{M_1 \to^p_\Lambda M_2 \qquad \dfrac{M_2 \in \Lambda\text{-HNF}}{M_2 \Downarrow_{\mathcal{U}^{\Lambda}_{\Lambda\text{-HNF}}} M_2}\,(axiom)}{M_1 \Downarrow_{\mathcal{U}^{\Lambda}_{\Lambda\text{-HNF}}} M_2}\,(eval)}{\lambda x.(\lambda uv.xuv)I(DD) \Downarrow_{\mathcal{U}^{\Lambda}_{\Lambda\text{-HNF}}} \lambda x.xI(DD)}\,(eval)$$

(ii) It is possible to check that there is not a derivation proving $\lambda x.DD \Downarrow_{\mathcal{U}^{\Lambda}_{\Lambda\text{-HNF}}}$, i.e. $\lambda x.DD \Uparrow_{\mathcal{U}^{\Lambda}_{\Lambda\text{-HNF}}}$.
Every derivation proving $\lambda x.DD \Downarrow_{\mathcal{U}^{\Lambda}_{\Lambda\text{-HNF}}}$ must be of the following shape:

$$\dfrac{\lambda x.DD \to^p_\Lambda \lambda x.DD \qquad \dfrac{\lambda x.DD \to^p_\Lambda \lambda x.DD \qquad d}{\lambda x.DD \Downarrow_{\mathcal{U}^{\Lambda}_{\Lambda\text{-HNF}}} R}\,(eval)}{\lambda x.DD \Downarrow_{\mathcal{U}^{\Lambda}_{\Lambda\text{-HNF}}} R}\,(eval)$$

for some $R \in \Lambda$ and some derivation d proving $\lambda x.DD \Downarrow_{\mathcal{U}^{\Lambda}_{\Lambda\text{-HNF}}} R$. Since all derivations are applications of a finite number of rules, d cannot exist, and so also the whole derivation.

In the remainder of this part of the book we will present four different operational semantics: three for the call-by-name λ-calculus and one for the call-by-value calculus. They formalize the deterministic evaluation relations given in Example 5.0.5, except for **G**. We will not develop such a semantics, since the notion of Γ-normal form is semantically meaningless, as already noted.

Each one of the operational semantics we are interested in can be derived from the "universal Δ-reduction machine" by instantiating the sets of input and output values in a suitable way. But we choose to present the various operational semantics independently, both for clarity and for technical reasons. In fact, while the universal reduction machine is based on a step-by-step description of the evaluation relation, the reduction machines we will present supply an input-output description of it, and this makes the proofs easier.

6. Call-by-Name Operational Semantics

6.1 H-Operational Semantics

$\mathbf{H} \in \mathcal{E}(\Lambda, \Lambda\text{-HNF})$ is the first evaluation relation that we will study; it is the universal evaluation relation $\mathcal{U}^{\Lambda}_{\Lambda\text{-HNF}}$ (see Example 5.1.5).

In this setting, the converging terms represent computations that can always produce a given output value when applied to suitable arguments. In fact, the set of terms having Λ-HNF coincides with the set of Λ-solvable terms.

Definition 6.1.1 (H-Operational semantics).

(i) $\mathbf{H} \in \mathcal{E}(\Lambda, \Lambda\text{-}HNF)$ *is the evaluation relation induced by the formal system proving judgments of the shape*

$$M \Downarrow_{\mathbf{H}} N$$

where $M \in \Lambda$ *and* $N \in \Lambda\text{-}HNF$. *It consists of the following rules:*

$$\frac{m \geq 0}{xM_1 \ldots M_m \Downarrow_{\mathbf{H}} xM_1 \ldots M_m}\ (var)$$

$$\frac{M \Downarrow_{\mathbf{H}} N}{\lambda x.M \Downarrow_{\mathbf{H}} \lambda x.N}\ (abs)$$

$$\frac{P[Q/x]M_1 \ldots M_m \Downarrow_{\mathbf{H}} N}{(\lambda x.P)QM_1 \ldots M_m \Downarrow_{\mathbf{H}} N}\ (head)$$

(ii) $M \preceq_{\mathbf{H}} N$ *if and only if, for all contexts* $C[.]$ *such that* $C[M], C[N] \in \Lambda^0$, $(C[M] \Downarrow_{\mathbf{H}}$ *implies* $C[N] \Downarrow_{\mathbf{H}})$.

(iii) $M \approx_{\mathbf{H}} N$ *if and only if* $M \preceq_{\mathbf{H}} N$ *and* $N \preceq_{\mathbf{H}} M$.

As we already noticed, **H** is deterministic.

Example 6.1.2. (i) $\lambda x.(\lambda uv.xuv)I(DD) \Downarrow_{\mathbf{H}} \lambda x.xI(DD)$. In fact, we can build the following derivation:

$$\cfrac{\cfrac{\cfrac{\cfrac{}{xI(DD) \Downarrow_{\mathbf{H}} xI(DD)}\,(var)}{(\lambda v.xIv)(DD) \Downarrow_{\mathbf{H}} xI(DD)}\,(head)}{(\lambda uv.xuv)I(DD) \Downarrow_{\mathbf{H}} xI(DD)}\,(head)}{\lambda x.(\lambda uv.xuv)I(DD) \Downarrow_{\mathbf{H}} \lambda x.xI(DD)}\,(abs)$$

where the unique leaf is the axiom (var) and the conclusion of the root node is the judgment $\lambda x.(\lambda uv.xuv)I(DD) \Downarrow_{\mathbf{H}} \lambda x.xI(DD)$. Note that, in the particular case of the system $\Downarrow_{\mathbf{H}}$, every derivation is such that each node has a unique son.

(ii) It is possible to check that there is no derivation proving $\lambda x.DD \Downarrow_{\mathbf{H}}$. In fact, if a such derivation exists then it must be of the following shape:

$$\cfrac{\cfrac{d}{DD \Downarrow_{\mathbf{H}} R}\,(head)}{\lambda x.DD \Downarrow_{\mathbf{H}} \lambda x.R}\,(abs)$$

for some R, and some derivation d. But the rule (head) implies that the derivation d must be in its turn of the shape

$$\cfrac{d}{DD \Downarrow_{\mathbf{H}} R}\,(head)$$

Since all derivations are the application of a finite number of rules, d cannot exist, and so also the whole derivation.

The system $\Downarrow_{\mathbf{H}}$ characterizes completely the class of terms having Λ-head normal forms, as shown in Theorem 6.1.3.

Theorem 6.1.3. (i) *$M \Downarrow_{\mathbf{H}} N$ implies $M \to_{\Lambda}^{*p} N$ and N is in Λ-hnf.*
(ii) *$M \Downarrow_{\mathbf{H}}$ if and only if M has a Λ-hnf.*

Proof. (i) By induction on the definition of $\Downarrow_{\mathbf{H}}$.
(ii) ($\Rightarrow$) The proof is a consequence of (i).
($\Leftarrow$) M has Λ-hnf means that there is $N \in \Lambda$-HNF such that $M =_{\Lambda} N$. But Λ-HNF is a set of output values with respect to Λ, by Property 5.0.3; so there is a reduction sequence $M \to_{\Lambda}^{*p} M' \in \Lambda$-HNF.
The proof is done by induction on the length of the reduction sequence $M \to_{\Lambda}^{*p} M'$. Let $M \equiv \lambda x_1 \ldots x_n.\zeta M_1 \ldots M_m$ $(n, m \in \mathbb{N})$.
If ζ is a variable then M is already in Λ-hnf. In fact $M \Downarrow_{\mathbf{H}} M$, by n applications of rule (abs) and one application of the rule (var).

If $\zeta \equiv (\lambda x.P)Q$ then by induction, $P[Q/x]M_1 \ldots M_m \Downarrow_{\mathbf{H}} N$, for some N; thus $M \Downarrow_{\mathbf{H}} \lambda x_1 \ldots x_n.N$, by n applications of rule (abs) and one application of the rule (head). □

The following property will be quite useful in the Sect. 15.2.

Property 6.1.4. Let $M, N, T, U \in \Lambda$ and $M \Downarrow_{\mathbf{H}} N$.

(i) $M[T/z] \Downarrow_{\mathbf{H}} U$ if and only if $N[T/z] \Downarrow_{\mathbf{H}} U$.
(ii) $MT \Downarrow_{\mathbf{H}} U$ if and only if $NT \Downarrow_{\mathbf{H}} U$.

Proof. (i) Clearly $M[T/z] =_\Lambda N[T/z]$, so by the confluence theorem $M[T/z]$ has Λ-hnf if and only if $N[T/z]$ has Λ-hnf; hence let $M[T/z] \Downarrow_{\mathbf{H}} U_0$ and $N[T/z] \Downarrow_{\mathbf{H}} U_1$, for some $U_0, U_1 \in \Lambda$.
We show that $U_0 \equiv U_1$ by induction on the derivation of $M \Downarrow_{\mathbf{H}} N$. The case (var) is trivial. The case (abs) follows by induction. The more complex case is (head); if $P[Q/x]M_1 \ldots M_m \Downarrow_{\mathbf{O}} N$ then by induction, $(P[Q/x])[T/z]M_1[T/z]...M_m[T/z] \Downarrow_{\mathbf{H}} V$ and $N[T/z] \Downarrow_{\mathbf{H}} V$, thus $(\lambda x.P[T/z])Q[T/z]M_1[T/z]...M_m[T/z] \Downarrow_{\mathbf{H}} V$ too and the proof is done.
(ii) Since $M =_\Lambda N$ implies $MT =_\Lambda NT$, we can assume $MT \Downarrow_{\mathbf{H}} U_0$ if and only if $NT \Downarrow_{\mathbf{H}} U_1$. We show that $U_0 = U_1$ by induction on the derivation of $M \Downarrow_{\mathbf{H}} N$. The case (var) is trivial. The case (abs) follows by using the previous point. The case (head) follows by induction. □

$\approx_{\mathbf{H}}$ is a Λ-theory, as proved by Theorem 6.1.5.

Theorem 6.1.5 (H-Correctness).
The $\lambda\Lambda$-calculus is correct with respect to the **H***-operational semantics.*

Proof. We must prove that $M =_\Lambda N$ implies $M \approx_{\mathbf{H}} N$, by definition of correctness. Let $M =_\Lambda N$ and let $C[.]$ be a context such that $C[M], C[N] \in \Lambda^0$. By definition of $=_\Lambda$, $C[M] =_\Lambda C[N]$. So the proof follows from Theorem 6.1.3.(ii), since by the confluence theorem the property of having Λ-hnf is closed under $=_\Lambda$. □

The $\lambda\Lambda$-calculus, nevertheless, is not complete with respect to the **H**-operational semantics. To show the incompleteness, the notion of **H**-relevant context is used, which is the specialization to **H** of the general notion presented in Definition 5.0.10. Lemma 6.1.6 shows a syntactical characterization of **H**-relevant context.

Lemma 6.1.6 (H-Relevance).
A context $C[.]$ is **H**-relevant *whenever there is a context $C'[.] \equiv [.]C_1[.]...C_m[.]$ ($m \in \mathbb{N}$) such that for all $M \in \Lambda^0$, $C[M] \Downarrow_{\mathbf{H}}$ if and only if $C'[M] \Downarrow_{\mathbf{H}}$.*

Proof. ($\Rightarrow$) Assume that $C[.]$ is **H**-relevant, namely there are $M, N \in \Lambda^0$ such that $C[M] \Downarrow_{\mathbf{H}}$ and $C[N] \Uparrow_{\mathbf{H}}$. By induction on $C[M] \Downarrow_{\mathbf{H}}$ we will prove that there is a context $C'[.]$ satisfying the statement.

If the last applied rule is (var) then either $C[.] \equiv xC_1[.]...C_m[.]$ ($m \in \mathbb{N}$) or $C[.] \equiv [.]C_1[.]...C_m[.]$ ($m \in \mathbb{N}$). In the first case the context is not relevant, while the second case is not possible, since $M \in \Lambda^0$.
If the last applied rule is (abs) then either $C[.] \equiv [.]$ or $C[.] \equiv \lambda z.C''[.]$. The first case is immediate, while the second follows by induction.
If the last applied rule is (head) then either $C[.] \equiv [.]C_1[.]...C_m[.]$ ($m \in \mathbb{N}$) or $C[.] \equiv (\lambda z.C_0[.])C_1[.]...C_m[.]$ ($m \geq 1$). The first case is trivial, while the second follows by induction; in fact, the context $C_0[.][C_1[.]/z]C_2[.]...C_m[.]$ is discriminating M and N and so is **H**-relevant too.

($\Leftarrow$) Let $C'[.]$ be a context satisfying the statement of this Lemma, so $C'[M] \Downarrow_{\mathbf{H}}$ if and only if $C[M] \Downarrow_{\mathbf{H}}$, for each $M \in \Lambda$. Thus $M \equiv DD$ and $N \equiv \lambda x_1...x_m z.z$ are witnesses of the **H**-relevance of $C[.]$. □

By observing the details of the proof, it is easy to see that actually, for all $M \in \Lambda^0$, if $C[M] \Downarrow_{\mathbf{H}}$ then in the derivation of $C[M] \Downarrow_{\mathbf{H}}$ there are contexts $C_1[.], ..., C_m[.]$ ($m \in \mathbb{N}$) and there is a subderivation proving $MC_1[M]...C_m[M] \Downarrow_{\mathbf{H}}$.

Lemma 6.1.7. *Let $C[.]$ be* **H**-*relevant. If $M \in \Lambda^0$ and $C[M] \Downarrow_{\mathbf{H}}$ then $M \Downarrow_{\mathbf{H}}$.*

Proof. By induction on the derivation of $C[M] \Downarrow_{\mathbf{H}}$.
If the last applied rule is (var) then either $C[.] \equiv xC_1[.]...C_m[.]$ ($m \in \mathbb{N}$) or $C[.] \equiv [.]C_1[.]...C_m[.]$ ($m \in \mathbb{N}$). In the first case the context is not relevant, while the second case is not possible, since $M \in \Lambda^0$.
If the last applied rule is (abs) then either $C[.] \equiv [.]$ or $C[.] \equiv \lambda z.C''[.]$. The first case is immediate, while the second follows by induction.
If the last applied rule is (head) then either $C[.] \equiv [.]C_1[.]...C_m[.]$ ($m \in \mathbb{N}$) or $C[.] \equiv (\lambda z.C_0[.])C_1[.]...C_m[.]$ ($m \geq 1$). The first case follows by Property 2.1.9.(ii), while the second follows by induction on the derivation proving $C_0[M][C_1[M]/z]C_2[M]...C_m[M]$. □

We can check that all closed Λ-unsolvable terms are equated in the **H**-operational semantics.

Theorem 6.1.8 (H-Incompleteness).
The λ-calculus is incomplete with respect to the **H**-*operational semantics.*

Proof. Let P and Q be two closed Λ-unsolvable terms such that $M \neq_{\Lambda} N$. A non-**H**-relevant context cannot discriminate P and Q. Let $C[.]$ be a **H**-relevant context: P, Q have not Λ-hnf (by Theorem 2.1.1), thus both $P \Uparrow_{\mathbf{H}}$ and $Q \Uparrow_{\mathbf{H}}$. Thus $C[P] \Uparrow_{\mathbf{H}}$ and $C[Q] \Uparrow_{\mathbf{H}}$ by Lemma 6.1.7. Hence, $P \approx_{\mathbf{H}} Q$. □

The proof of the following property is an example of a useful technique for proving operational equality between terms.

Property 6.1.9. $I \approx_{\mathbf{H}} E$.

Proof. By absurdum assume that the two terms can be discriminated. This means that there is a context $C[.]$ discriminating them. Let $C[.]$ be such that $C[I] \Downarrow_{\mathbf{H}}$ while $C[E] \Uparrow_{\mathbf{H}}$. Clearly $C[.]$ must be **H**-relevant.
Let $C[.]$ be a minimal discriminating context for I and E, in the sense that the derivation of $C[I] \Downarrow_{\mathbf{H}}$ has a minimal size between all the proofs of $C'[I] \Downarrow_{\mathbf{H}}$, for every $C'[.]$ discriminating between I and E in such a way that $C'[I] \Downarrow_{\mathbf{H}}$ while $C'[E] \Uparrow_{\mathbf{H}}$. The proof is done by considering the last applied rule in the derivation proving $C[I] \Downarrow_{\mathbf{H}}$.

The last used rule cannot be (var), since $C[.] \equiv xC_1[.]...C_m[.]$ ($m \in \mathbb{N}$) is not **H**-relevant. If the last used rule is (abs) then either $C[.] \equiv \lambda x.C'[.]$ or $C[.] \equiv [.]$. In the former case, $C'[.]$ would be a discriminating context such that the derivation of $C'[I] \Downarrow_{\mathbf{H}}$ has smaller size than the derivation of $C[I] \Downarrow_{\mathbf{H}}$, against the hypothesis. The latter case is not possible, since clearly $[.]$ is not a discriminating context for I and E.
Let the last used rule be (head); thus either $C[.] \equiv (\lambda x.C_0[.])C_1[.]...C_m[.]$ ($m \geq 1$) or $C[.] \equiv [.]C_1[.]...C_m[.]$ ($m \in \mathbb{N}$). Let consider the former case. By the rule (head), $C[I] \Downarrow_{\mathbf{H}}$ if and only if $C_0[I][C_1[I]/x]C_2[I]...C_m[I] \Downarrow_{\mathbf{H}}$. But in this case $C_0[.][C_1[.]/x]C_2[.]...C_m[.]$ would be a discriminating context for M and N with a derivation having smaller size than $C[.]$, against the hypothesis that $C[.]$ is minimum.
The case $C[.] \equiv [.]C_1[.]...C_m[.]$ leads to a similar contradiction. In fact, in this case $C_1[.]...C_m[.]$ would be a discriminating context "smaller" than $C[.]$.

The case when $C[I] \Uparrow_{\mathbf{H}}$ and $C[E] \Downarrow_{\mathbf{H}}$ is symmetric. □

Theorem 6.1.10. *The theory* **H** *is fully extensional.*

Proof. By Properties 2.1.7 and 6.1.9. □

6.2 N-Operational Semantics

$\mathbf{N} \in \mathcal{E}(\Lambda, \Lambda\text{-NF})$ is the evaluation relation studied in this section; it is the universal evaluation relation $\mathcal{U}^{\Lambda}_{\Lambda\text{-NF}}$.

In some sense, **N** induces the most natural operational semantics for the $\lambda\Lambda$-calculus; in fact, converging terms represent the completely terminating computations.

Definition 6.2.1 (N-Operational semantics).

(i) $\mathbf{N} \in \mathcal{E}(\Lambda, \Lambda\text{-}NF)$ *is the evaluation relation induced by the formal system proving judgments of the shape*

$$M \Downarrow_{\mathbf{N}} N$$

where $M \in \Lambda$ *and* $N \in \Lambda\text{-}NF$. *It consists of the following rules:*

$$\frac{(M_i \Downarrow_{\mathbf{N}} N_i)_{(i \leq m)}}{xM_1 \ldots M_m \Downarrow_{\mathbf{N}} xN_1 \ldots N_m}\ (var)$$

$$\frac{M \Downarrow_{\mathbf{N}} N}{\lambda x.M \Downarrow_{\mathbf{N}} \lambda x.N}\ (abs)$$

$$\frac{P[Q/x]M_1 \ldots M_m \Downarrow_{\mathbf{N}} N}{(\lambda x.P)QM_1 \ldots M_m \Downarrow_{\mathbf{N}} N}\ (head)$$

(ii) $M \preceq_{\mathbf{N}} N$ *if and only if, for all contexts* $C[.]$ *such that* $C[M], C[N] \in \Lambda^0$, $(C[M] \Downarrow_{\mathbf{N}}$ *implies* $C[N] \Downarrow_{\mathbf{N}})$.

(iii) $M \approx_{\mathbf{N}} N$ *if and only if* $M \preceq_{\mathbf{N}} N$ *and* $N \preceq_{\mathbf{N}} M$.

As is true for **H**, the relation **N** is also deterministic.

Example 6.2.2. $\lambda x_1 x_2.x_1(ID)((\lambda uv.u)(II)x_2) \Downarrow_{\mathbf{N}} \lambda x_1 x_2.x_1 DI$, as shown by the following derivation.

$$\dfrac{\dfrac{\dfrac{\dfrac{\dfrac{\dfrac{}{x \Downarrow_{\mathbf{N}} x}\,(var)}{xx \Downarrow_{\mathbf{N}} xx}\,(var)}{\lambda x.xx \Downarrow_{\mathbf{N}} \lambda x.xx}\,(abs)}{ID \Downarrow_{\mathbf{N}} D}\,(head) \qquad \dfrac{\dfrac{\dfrac{\dfrac{\dfrac{}{x \Downarrow_{\mathbf{N}} x}\,(var)}{\lambda x.x \Downarrow_{\mathbf{N}} \lambda x.x}\,(abs)}{II \Downarrow_{\mathbf{N}} I}\,(head)}{\lambda v.II \Downarrow_{\mathbf{N}} \lambda v.I}\,(abs)}{(\lambda uv.u)(II) \Downarrow_{\mathbf{N}} \lambda v.I}\,(head)}{x_1(ID)((\lambda uv.u)(II)) \Downarrow_{\mathbf{N}} x_1 D(\lambda v.I)}\,(var)}{\lambda x_2.x_1(ID)((\lambda uv.u)(II)) \Downarrow_{\mathbf{N}} \lambda x_2.x_1 D(\lambda v.I)}\,(abs)}{\lambda x_1 x_2.x_1(ID)((\lambda uv.u)(II)) \Downarrow_{\mathbf{N}} \lambda x_1 x_2.x_1 D(\lambda v.I)}\,(abs)$$

The system $\Downarrow_{\mathbf{N}}$ characterizes completely, from an operational point of view, the class of Λ-normal forms.

Theorem 6.2.3. (i) $M \Downarrow_{\mathbf{N}} N$ *implies* $M \to_{\Lambda}^{*p} N$ *and* N *is in* Λ*-nf.*
(ii) $M \Downarrow_{\mathbf{N}}$ *if and only if* M *has* Λ*-nf.*

Proof. (i) By induction on the definition of $\Downarrow_{\mathbf{N}}$.
(ii) $(\Rightarrow)$ Directly from (i).
$(\Leftarrow)$ If $M \to_{\Lambda}^{*} N \in \Lambda$-NF then $M \to_{\Lambda}^{*p} N$, by Corollary 1.2.13. The proof follows by induction on the pair (M, p), where p is the length of the reduction sequence $M \to_{\Lambda}^{*p} N$ ordered in a lexicographic way.
Let $M \equiv \lambda x_1 \ldots x_n.\zeta M_1 \ldots M_m$.

If ζ is a variable then $N \equiv \lambda x_1 \ldots x_n.\zeta\mathsf{nf}_\Lambda(M_1)\ldots\mathsf{nf}_\Lambda(M_m)$. By induction $M_i \Downarrow_{\mathbf{N}}$ $(1 \leq i \leq m)$, thus $M \Downarrow_{\mathbf{N}}$ by rule (var) having as premises the derivation proving $M_i \Downarrow_{\mathbf{N}}$ and n instances of (abs).
If $\zeta \equiv (\lambda x.P)Q$ then $\mathsf{nf}_\Lambda(M) \equiv \lambda x_1 \ldots x_n.\mathsf{nf}_\Lambda(P[Q/x]M_1 \ldots M_m)$; so, by induction, $P[Q/x]M_1 \ldots M_m \Downarrow_{\mathbf{N}} R$, for some R; hence $(\lambda x.P)QM_1 \ldots M_m \Downarrow_{\mathbf{N}} N$, by applying rule (head) and $M \Downarrow_{\mathbf{N}} \lambda x_1 \ldots x_n.N$ by n instances of (abs). □

The following property will be quite useful in Sect. 15.2.

Property 6.2.4. Let $M \Downarrow_{\mathbf{N}} N$. $MT \Downarrow_{\mathbf{N}} U$ if and only if $NT \Downarrow_{\mathbf{N}} U$.

Proof. Clearly $MT =_\Lambda NT$, so the by confluence theorem MT has Λ-nf if and only if NT has Λ-nf; hence let $MT \Downarrow_{\mathbf{N}} U_0$ and $NT \Downarrow_{\mathbf{N}} U_1$. By Corollary 1.2.6, it is easy to show that $U_0 \equiv U_1$. □

An immediate consequence of the Theorem 6.2.3.(ii) is that $M \Downarrow_{\mathbf{N}}$ implies $M \Downarrow_{\mathbf{H}}$. Moreover, $\approx_{\mathbf{N}}$ is a Λ-theory as proved in Theorem 6.2.5.

Theorem 6.2.5 (N-Correctness).
The $\lambda\Lambda$-calculus is correct with respect to the **N***-operational semantics.*

Proof. By definition of correctness, we must prove that $M =_\Lambda N$ implies $M \approx_{\mathbf{N}} N$. Let $M =_\Lambda N$ and let $C[.]$ be a context such that $C[M], C[N] \in \Lambda^0$. By definition of $=_\Lambda$, $C[M] =_\Lambda C[N]$. So the proof follows from Theorem 6.2.3, since by the confluence theorem the property of having Λ-nf is closed under $=_\Lambda$. □

We will prove that the $\lambda\Lambda$-calculus is not complete with respect to the **N**-operational semantics by using a syntactical characterization of **N**-relevant context.

Lemma 6.2.6 (N-Relevance).
A context $C[.]$ is **N**-relevant *whenever there are $n \geq 1$ contexts $C^i[.] \equiv [.]C^i_1[.]\ldots C^i_{m_i}[.]$ $(m_i \in \mathbb{N},\ 1 \leq i \leq n)$ such that for all $M \in \Lambda^0$, $C[M] \Downarrow_{\mathbf{N}}$ if and only if $\forall i \leq n$, $C^i[M] \Downarrow_{\mathbf{N}}$.*

Proof. $(\Rightarrow)$ Assume that $C[.]$ is **N**-relevant; namely, there are $M, N \in \Lambda^0$ such that $C[M] \Downarrow_{\mathbf{N}}$ and $C[N] \Uparrow_{\mathbf{N}}$. By induction on $C[M] \Downarrow_{\mathbf{N}}$ we will prove that there is at least one context satisfying the statement.
If the last applied rule is (var) then either $C[.] \equiv xC_1[.]\ldots C_m[.]$ $(m \in \mathbb{N})$ or $C[.] \equiv [.]C_1[.]\ldots C_m[.]$ $(m \in \mathbb{N})$. In the first case the **N**-relevance implies $m \geq 1$, and $C[M] \Downarrow_{\mathbf{N}}$ implies that $C_j[M] \Downarrow_{\mathbf{N}}$, for each $1 \leq j \leq m$. Let $\{D_1[.], \ldots D_h[.]\}$ $(h \leq m)$ be the subset of all relevant contexts in $\{C_1[.]\ldots C_m[.]\}$; it is not empty by the hypothesis that $C[.]$ is **N**-relevant. So the proof follows by induction on contexts $D_i[.]$. The second case is not possible, since $M \in \Lambda^0$.
If the last applied rule is (abs) then either $C[.] \equiv [.]$ or $C[.] \equiv \lambda z.C''[.]$. The first case is immediate, while the second follows by induction.

If the last applied rule is (head) then either $C[.] \equiv [.]C_1[.]...C_m[.]$ ($m \in \mathbb{N}$) or $C[.] \equiv (\lambda z.C_0[.])C_1[.]...C_m[.]$ ($m \geq 1$). The first case is trivial, while the second follows by induction; in fact, the context $C_0[.][C_1[.]/z]C_2[.]...C_m[.]$ is discriminating M and N and so is **N**-relevant too.

($\Leftarrow$) Let $k = \max\{m_1, ..., m_n\}$; it is easy to see that $M \equiv DD$ and $N \equiv \lambda x_1...x_k z.z$ make $C[.]$ relevant. □

By observing the details of the proof, it is easy to see that actually, for all $M \in \Lambda^0$, if $C[M] \Downarrow_{\mathbf{N}}$ then in the derivation of $C[M] \Downarrow_{\mathbf{N}}$ there are contexts $C_1^i[.], ..., C_{m_i}^i[.]$ ($m_i \in \mathbb{N}$) and there are n subderivations proving $MC_1^i[M]...C_{m_i}^i[M] \Downarrow_{\mathbf{N}}$ ($1 \leq i \leq n$).

The notion of **N**-relevant context is weaker than that of **H**-relevant context. In particular, it does not enjoy a property similar to that proved in Lemma 6.1.7. Let $C[.] \equiv [.](\lambda x.I)(DD)$, so $C[\lambda xy.x(DD)] \Downarrow_{\mathbf{N}} I$ but $\lambda xy.y(DD) \Uparrow_{\mathbf{N}}$; moreover, $C[\lambda yx.x] \Uparrow_{\mathbf{N}}$ while $\lambda yx.x \Downarrow_{\mathbf{N}}$.

So, in order to prove the operational equality between closed unsolvable terms, we will use a less general property of relevant contexts, but sufficient for our purpose.

Lemma 6.2.7. *Let $C[.]$ be **N**-relevant.*
If $M \in \Lambda^0$ and $C[M] \Downarrow_{\mathbf{N}}$ then M is solvable.

Proof. By induction on the derivation proving $C[M] \Downarrow_{\mathbf{N}}$.
If the last applied rule is (var) then either $C[.] \equiv xC_1[.]...C_m[.]$ ($m \in \mathbb{N}$) or $C[.] \equiv [.]C_1[.]...C_m[.]$ ($m \in \mathbb{N}$). In the first case the relevance implies $m \geq 1$, so the proof follows by induction; the second case is not possible, since $M \in \Lambda^0$.
If the last applied rule is (abs) then either $C[.] \equiv [.]$ or $C[.] \equiv \lambda z.C'[.]$. The first case is trivial, since M has Λ-nf; the second case follows by induction.
If the last applied rule is (head) then either $C[.] \equiv [.]C_1[.]...C_m[.]$ ($m \in \mathbb{N}$) or $C[.] \equiv (\lambda z.C_0[.])C_1[.]...C_m[.]$ ($m \geq 1$).
The first case follows by Property 2.1.9.(ii), while the second follows by induction on $C_0[M][C_1[M]/z]C_2[M]...C_m[M]$. □

Also the **N**-operational semantics, like the **H** one, equates all closed Λ-unsolvable terms.

Theorem 6.2.8 (N-Incompleteness).
*The $\lambda\Lambda$-calculus is incomplete with respect to the **N**-operational semantics.*

Proof. Let P and Q be two closed Λ-unsolvable terms such that $P \neq_{\Lambda} Q$. A non-**N**-relevant context cannot discriminate P and Q. Let $C[.]$ be an **N**-relevant context; P, Q do not have hnf (by Theorem 2.1.1), thus both $P \Uparrow_{\mathbf{N}}$ and $Q \Uparrow_{\mathbf{N}}$; so $C[P] \Uparrow_{\mathbf{N}}$ and $C[Q] \Uparrow_{\mathbf{N}}$ by Lemma 6.2.7. Hence, $P \approx_{\mathbf{N}} Q$. □

The following property holds.

Property 6.2.9. $I \approx_{\mathbf{N}} E$.

Proof. By absurdum assume $I \not\approx_{\mathbf{N}} E$. This means that there is a context $C[.]$ discriminating them. Let $C[.]$ be such that $C[I] \Downarrow_{\mathbf{N}}$ while $C[E] \Uparrow_{\mathbf{N}}$.

Let $C[.]$ be a minimal discriminating context for I and E, in the sense that the derivation of $C[I] \Downarrow_{\mathbf{N}}$ has a minimal size between all the proofs of $C'[I] \Downarrow_{\mathbf{N}}$, for every $C'[.]$ such that $C'[I] \Downarrow_{\mathbf{N}}$ and $C'[E] \Uparrow_{\mathbf{N}}$. The proof is done by considering the last applied rule in the derivation proving $C[I] \Downarrow_{\mathbf{N}}$.

If the last applied rule is (var) then $C[.] \equiv xC_1[.]...C_m[.]$ $(m \in \mathbb{N})$, so there is a $C_k[.]$ $(1 \leq k \leq m)$ discriminating I and E with a derivation having smaller size, against the hypothesis that $C[.]$ is minimum.

If the last used rule is (abs) then, either $C[.] \equiv \lambda x.C'[.]$ or $C[.] \equiv [.]$. In the former case, $C'[.]$ would be a discriminating context such that the derivation of $C'[I] \Downarrow_{\mathbf{N}}$ has smaller size than the derivation of $C[I] \Downarrow_{\mathbf{N}}$, against the hypothesis. The latter case is not possible, since clearly $[.]$ is not a discriminating context for I and E.

Let the last used rule be (head), thus either $C[.] \equiv (\lambda x.C_0[.])C_1[.]...C_m[.]$ $(m \geq 1)$ or $C[.] \equiv [.]C_1[.]...C_m[.]$ $(m \in \mathbb{N})$. In the former case, the context $C'[.] \equiv C_0[.][C_1[.]/x]C_2[.]...C_m[.]$ would be a discriminating context, such that the derivation $C'[I]$ has smaller size than $C[.]$, against the hypothesis that $C[.]$ is minimum. The case $C[.] \equiv [.]C_1[.]...C_m[.]$ leads to a similar contradiction. In fact, in this case $C_1[.]...C_m[.]$ would be a discriminating context "smaller" than $C[.]$.

The case $C[I] \Uparrow_{\mathbf{N}}$ and $C[E] \Downarrow_{\mathbf{N}}$ is symmetric. □

Theorem 6.2.10 shows that the **N**-operational semantics is fully extensional (see Sect. 1.3).

Theorem 6.2.10. *The theory* $\approx_{\mathbf{N}}$ *is fully extensional.*

Proof. By Properties 2.1.7 and 6.2.9. □

6.3 L-Operational Semantics

$\mathbf{L} \in \mathcal{E}(\Lambda, \Lambda\text{-LHNF})$ is the evaluation relation studied in this section; it is the universal evaluation relation $\mathcal{U}^{\Lambda}_{\Lambda\text{-LHNF}}$.

The **L**-operational semantics models the so-called lazy evaluation in a call-by-name parameter passing environment. It is characterized by the fact that a Λ-redex is never reduced in case it occurs under the scope of an abstraction. This behaviour is similar to that of the real (call-by-name) programming languages, where the body of a procedure is evaluated only when its parameters are supplied.

Definition 6.3.1 (L-Operational semantics).

(i) $\mathbf{L} \in \mathcal{E}(\Lambda, \Lambda\text{-}LHNF)$ *is the evaluation relation induced by the formal system proving judgments of the shape*

$$M \Downarrow_{\mathbf{L}} N$$

where $M \in \Lambda$ *and* $N \in \Lambda\text{-}LHNF$. *It consists of the following rules:*

$$\frac{m \geq 0}{xM_1 \ldots M_m \Downarrow_{\mathbf{L}} xM_1 \ldots M_m}\ (var)$$

$$\frac{}{\lambda x.M \Downarrow_{\mathbf{L}} \lambda x.M}\ (lazy)$$

$$\frac{P[Q/x]M_1 \ldots M_m \Downarrow_{\mathbf{L}} N}{(\lambda x.P)QM_1 \ldots M_m \Downarrow_{\mathbf{L}} N}\ (head)$$

(ii) $M \preceq_{\mathbf{L}} N$ *if and only if, for all contexts* $C[.]$ *such that* $C[M], C[N] \in \Lambda^0$, $\big(C[M] \Downarrow_{\mathbf{L}}$ *implies* $C[N] \Downarrow_{\mathbf{L}}\big)$.

(iii) $M \approx_{\mathbf{L}} N$ *if and only if* $M \preceq_{\mathbf{L}} N$ *and* $N \preceq_{\mathbf{L}} M$.

The formal system described before, when restricted to closed terms, corresponds to the call-by-name lazy evaluation machine introduced by Plotkin [78]. It is easy to check that $\mathbf{L}$ is deterministic.

Example 6.3.2. $(\lambda xy.x)(DD) \Downarrow_{\mathbf{L}} \lambda y.DD$. In fact, we can build the following derivation:

$$\frac{\dfrac{}{\lambda y.DD \Downarrow_{\mathbf{L}} \lambda y.DD}\ (lazy)}{(\lambda xy.x)(DD) \Downarrow_{\mathbf{L}} \lambda y.DD}\ (head)$$

The following theorem proves that the system $\mathbf{L}$ characterizes completely the class of Λ-lazy head normal forms.

Theorem 6.3.3. (i) $M \Downarrow_{\mathbf{L}} N$ *implies* $M \rightarrow^{*p}_{\Lambda} N$ *and* N *is in* Λ-*lhnf.*
(ii) $M \Downarrow_{\mathbf{L}}$ *if and only if* M *has a* Λ-*lhnf.*

Proof. (i) By induction on the definition of $\Downarrow_{\mathbf{L}}$.
(ii) ($\Rightarrow$) The proof is a consequence of (i).
($\Leftarrow$) M has a Λ-lhnf means that there is $N \in \Lambda$-LHNF such that $M =_\Lambda N$. But Λ-LHNF is a set of output values with respect to Λ, by Property 5.0.3; so there is a reduction sequence $M \to_\Lambda^{*p} M' \in \Lambda$-LHNF. The proof is done by induction on the length of the reduction sequence $M \to_\Lambda^{*p} M'$. Let $M \equiv \lambda x_1 \ldots x_n.\zeta M_1 \ldots M_m$.
If either $n \geq 1$ or ζ is a variable, then a Λ-lhnf of M is M itself, so $M \Downarrow_{\mathbf{L}} M$, by an application of rule (*lazy*) or an application of rule (*var*). If $n = 0$ and $\zeta \equiv (\lambda x.P)Q$ then the Λ-lhnf of M is the Λ-lhnf of $P[Q/x]M_1 \ldots M_m$. By induction $P[Q/x]M_1 \ldots M_m \Downarrow_{\mathbf{L}} N$, for some N, so $M \Downarrow_{\mathbf{L}} N$, by applying the rule (*head*). □

By Theorem 6.3.3, it follows that both $M \Downarrow_{\mathbf{H}}$ and $M \Downarrow_{\mathbf{N}}$ imply $M \Downarrow_{\mathbf{L}}$. The following property will be quite useful in Sect. 15.2.

Property 6.3.4. Let $M, N, T, U \in \Lambda^0$ and $M \Downarrow_{\mathbf{L}} N$.

$$MT \Downarrow_{\mathbf{L}} U \text{ if and only if } NT \Downarrow_{\mathbf{L}} U.$$

Proof. By induction on $M \Downarrow_{\mathbf{L}} N$. The last applied rule cannot be (var), since $M \in \Lambda^0$. If the last applied rule is (lazy) then the proof is trivial. If the last applied rule is (head) then the proof follows by induction. □

Theorem 6.3.5 proves that $\approx_{\mathbf{L}}$ is a Λ-theory.

Theorem 6.3.5 (L-Correctness).
The $\lambda\Lambda$-calculus is correct with respect to the **L***-operational semantics.*

Proof. By definition of correctness, we must prove that $M =_\Lambda N$ implies $M \approx_{\mathbf{L}} N$. Let $M =_\Lambda N$ and let $C[.]$ be a context such that $C[M], C[N] \in \Lambda^0$. By definition of $=_\Lambda$, $C[M] =_\Lambda C[N]$. So the proof follows by Theorem 6.3.3, since by the confluence theorem the property of having Λ-lhnf is closed under $=_\Lambda$. □

We will prove that the $\lambda\Lambda$-calculus is not complete with respect to the **L**-operational semantics by using a syntactical characterization of **L**-*relevant* context.

Lemma 6.3.6 (L-Relevance).
A context $C[.]$ is **L**-relevant *whenever there is a context $C'[.] \equiv [.]C_1[.]...C_m[.]$ ($m \in \mathbb{N}$) such that for all $M \in \Lambda^0$, $C[M] \Downarrow_{\mathbf{L}}$ if and only if $C'[M] \Downarrow_{\mathbf{L}}$.*

Proof. ($\Rightarrow$) Assume that $C[.]$ is **L**-relevant; namely, there are $M, N \in \Lambda^0$ such that $C[M] \Downarrow_{\mathbf{L}}$ and $C[N] \Uparrow_{\mathbf{L}}$. By induction on $C[M] \Downarrow_{\mathbf{L}}$ we will prove that there is a context $C'[.]$ satisfying the statement.
The last applied rule cannot be (var), since $C[.] \equiv xC_1[.]...C_m[.]$ ($m \in \mathbb{N}$) is not relevant. If the last applied rule is (lazy) then either $C[.] \equiv [.]$ or

$C[.] \equiv \lambda z.C''[.]$. The first case is immediate; while the second is not possible, since $\lambda z.C''[.]$ is not relevant.
If the last applied rule is (head) then either $C[.] \equiv [.]C_1[.]...C_m[.]$ $(m \in \mathbb{N})$ or $C[.] \equiv (\lambda z.C_0[.])C_1[.]...C_m[.]$ $(m \geq 1)$. The first case is trivial, while the second follows by induction; in fact, the context $C_0[.][C_1[.]/z]C_2[.]...C_m[.]$ is discriminating M and N and so is **L**-relevant too.

($\Leftarrow$) Let $C'[.]$ be a context satisfying the statement of this lemma, so $C'[M] \Downarrow_{\mathbf{L}}$ if and only if $C[M] \Downarrow_{\mathbf{L}}$, for each $M \in \Lambda$. Thus $M \equiv DD$ and $N \equiv \lambda x_1...x_m z.z$ are witnesses of the **L**-relevance of $C[.]$. □

By observing the details of the proof, it is easy to see that, for all $M \in \Lambda$, if $C[M] \Downarrow_{\mathbf{L}}$ then in the derivation of $C[M] \Downarrow_{\mathbf{L}}$ there are contexts $C_1[.], ..., C_m[.]$ $(m \in \mathbb{N})$ and there is a subderivation proving $MC_1[M]...C_m[M] \Downarrow_{\mathbf{L}}$. Note that the context $\lambda z.[.]$ is **H**-relevant nevertheless it is not **L**-relevant.

Lemma 6.3.7. *Let $C[.]$ be **L**-relevant. If $M \in \Lambda^0$ and $C[M] \Downarrow_{\mathbf{L}}$ then $M \Downarrow_{\mathbf{L}}$.*

Proof. By induction on the derivation proving $C[M] \Downarrow_{\mathbf{L}}$.
The last applied rule cannot be (var), since $C[.] \equiv xC_1[.]...C_m[.]$ $(m \in \mathbb{N})$ is not relevant. If the last applied rule is (abs) then $C[.] \equiv \lambda z.C'[.]$ is not relevant, while the case $C[.] \equiv [.]$ is trivial.
If the last applied rule is (head) then either $C[.] \equiv [.]C_1[.]...C_m[.]$ $(m \in \mathbb{N})$ or $C[.] \equiv (\lambda z.C_0[.])C_1[.]...C_m[.]$ $(m \geq 1)$. The last case follows by induction on $C_0[M][C_1[M]/z]C_2[M]...C_m[M]$; so let $C[.] \equiv [.]C_1[.]...C_m[.]$ $(m \in \mathbb{N})$. If $M \equiv \lambda z.M'$ then immediately $M \Downarrow_{\mathbf{L}}$, so let $M \equiv (\lambda z.P)QM_1...M_n$ $(n \in \mathbb{N})$. $C[M] \Downarrow_{\mathbf{L}}$ implies, by rule (head) that $P[Q/z]M_1...M_nC_1[M]...C_m[M] \Downarrow_{\mathbf{L}}$; since $[.]C_1[M]...C_m[M]$ is a relevant context (it discriminates DD and M), by induction $P[Q/z]M_1...M_n \Downarrow_{\mathbf{L}}$. Thus by applying the rule (head), $M \Downarrow_{\mathbf{L}}$ follows. □

An incompleteness result holds.

Theorem 6.3.8 (L-Incompleteness).
*The $\lambda\Lambda$-calculus is incomplete with respect to the **L**-operational semantics.*

Proof. Let P and Q be two closed Λ-unsolvable terms of order zero such that $P \neq_\Lambda Q$. A non-**L**-relevant context cannot discriminate them. By Definition 2.1.3, if either $P \to_\Lambda^* R$ or $Q \to_\Lambda^* R$, for some R, then R cannot be an abstraction, hence $P \Uparrow_{\mathbf{L}}$ and $Q \Uparrow_{\mathbf{L}}$. Let $C[.]$ be any **L**-relevant context; so, by Lemma 6.3.7, $C[P] \Uparrow_{\mathbf{L}}$ and $C[Q] \Uparrow_{\mathbf{L}}$. Hence, $P \approx_{\mathbf{L}} Q$. □

Property 6.3.9. (i) $m \neq n$ implies $\lambda x_1...x_n.DD \not\approx_{\mathbf{L}} \lambda x_1...x_m.DD$.
(ii) $I \not\approx_{\mathbf{L}} E$.

Proof. (i) It is an exercise, by using the fact that $DD \Uparrow_{\mathbf{L}}$.
(ii) The context $[.](DD)$ discriminates the two given terms. □

From the previous property and by Property 2.1.7, it follows that the operational semantics $\approx_{\mathbf{L}}$, being a Λ-theory, is not fully extensional.

6.3.1 An Example

We will show now that $L_0 \approx_{\mathbf{L}} L_1$, where

$$L_0 \equiv \lambda x.x(x(\lambda x.DD)(DD))(\lambda x.DD)\,,$$
$$L_1 \equiv \lambda x.x(\lambda y.x(\lambda x.DD)(DD)y)(\lambda x.DD).$$

This equivalence was first stated in [2]. The interest of such a result will be clear when we will study the denotational semantics. First, let us prove a general property.

Lemma 6.3.10. *Let $M \to_\eta^* N$. If $N \Downarrow_{\mathbf{L}}$ then $M \Downarrow_{\mathbf{L}}$.*

Proof. Let s be the size of the derivation proving $N \Downarrow_{\mathbf{L}}$, and let l be the length of the reduction sequence from M to N. The proof is given by induction on the pair (s,l); the pairs are ordered according to the lexicographical order.
If $l = 0$ then the proof is immediate, so let $l \geq 1$.
If $N \equiv xN_1...N_n$ ($n \in \mathbb{N}$) then there are three cases.

1. $M \equiv \lambda y.M_0 y$, where $M_0 \to_\eta^* N$ and $y \notin \mathrm{FV}(M_0)$. The proof follows immediately by rule (lazy).
2. $M \equiv (\lambda y.M_0 y)M_1...M_m$ ($1 \leq m \leq n$), where $M_0 M_1...M_m \to_\eta^* N$ and $y \notin \mathrm{FV}(M_0)$. By induction $M_0 M_1...M_m \Downarrow_{\mathbf{L}}$, hence $M \Downarrow_{\mathbf{L}}$ by rule (head).
3. $M \equiv xM_1...M_n$, where $M_i \to_\eta^* N_i$ ($1 \leq i \leq n$). The proof follows by rule (var).

If $N \equiv \lambda x.N_0$ then $M \equiv \lambda y.M_0$ and the proof follows by rule (lazy). If $N \equiv (\lambda x.N_0)N_1...N_n$ ($n \geq 1$), then there are three cases.

1. $M \equiv \lambda y.M_0 y$, where $M_0 \to_\eta^* N$ and $y \notin \mathrm{FV}(M_0)$. The proof follows by rule (lazy).
2. $M \equiv (\lambda y.M_0 y)M_1...M_m$ ($1 \leq m \leq n$), where $M_0 M_1...M_m \to_\eta^* N$ and $y \notin \mathrm{FV}(M_0)$. By induction $M_0 M_1...M_m \Downarrow_{\mathbf{L}}$, hence $M \Downarrow_{\mathbf{L}}$ by rule (head).
3. $M \equiv (\lambda x.M_0)M_1...M_n$, where $M_i \to_\eta^* N_i$ ($1 \leq i \leq n$). It is easy to see that $M_0[M_1/x]M_2...M_n \to_\eta^* N_0[N_1/x]N_2...N_n$. But $N \Downarrow_{\mathbf{L}}$ implies $N_0[N_1/x]N_2...N_n \Downarrow_{\mathbf{L}}$, and there is a derivation having size less than s proving it; hence by induction $M_0[M_1/x]M_2...M_n \Downarrow_{\mathbf{L}}$. Then the proof follows by rule (head). □

Lemma 6.3.10 implies the Corollary 6.3.11.

Corollary 6.3.11. *If $M \to_\eta^* N$ then $N \preceq_{\mathbf{L}} M$.*

Proof. Clearly $C[M] \to_\eta^* C[N]$, for all contexts $C[.]$. Then by Lemma 6.3.10, $C[N] \Downarrow_{\mathbf{L}}$ implies $C[M] \Downarrow_{\mathbf{L}}$. □

In particular, it follows that $L_0 \preceq_{\mathbf{L}} L_1$. The next goal is to prove the reverse relation, namely $L_1 \preceq_{\mathbf{L}} L_0$.

Let $M \in \Lambda^0$; it is easy to check that $M(\lambda y.M(\lambda x.DD)(DD)y)(\lambda x.DD) \Downarrow_{\mathbf{L}}$ if and only if $L_1 M \Downarrow_{\mathbf{L}}$, by rule (head).

Lemma 6.3.12. *Let* $M, N \in \Lambda^0$ *be such that* $N =_\Lambda M$.
If $M(\lambda y.N(\lambda x.DD)(DD)y)(\lambda x.DD) \Downarrow_{\mathbf{L}}$, *then either* $N \to_\Lambda^* \lambda x_0 x_1.x_0$ *or* $N \to_\Lambda^* \lambda x_0 x_1.x_1$ *or* $N \to_\Lambda^* \lambda x_0 x_1 x_2.M'''$, *for some* $M''' \in \Lambda$.

Proof. Let $R \equiv M(\lambda y.N(\lambda x.DD)(DD)y)(\lambda x.DD)$. The proof is given by induction on the size of the derivation proving $R \Downarrow_{\mathbf{L}}$.
The last applied rule cannot be (var), since $R \in \Lambda^0$. The last applied rule cannot be (lazy), since R is not an abstraction. Hence the last applied rule is (head); we consider all the possible shapes of M.

- $M \equiv xM_1...M_m$ ($m \in \mathbb{N}$) is not possible, since $M \in \Lambda^0$.
- $M \equiv (\lambda x_0.M')M_1...M_m$ ($m \geq 1$). The proof follows by induction on

$$M'[M_1/x_0]M_2...M_m(\lambda y.N(\lambda x.DD)(DD)y)(\lambda x.DD) \Downarrow_{\mathbf{L}} .$$

- Let $M \equiv \lambda x_0.M'$, so $\mathrm{FV}(M') \subseteq \{x_0\}$.
  - $M' \equiv x_0 M_1...M_m$ ($m \in \mathbb{N}$) is not possible; in fact, $\approx_{\mathbf{L}}$ is a Λ-theory, so $N(\lambda x.DD)(DD) \approx_{\mathbf{L}} M(\lambda x.DD)(DD) \approx_{\mathbf{L}} (\lambda x.DD)M'_1...M'_m(DD)$, where $M'_i \equiv M_i[\lambda x.DD/x_0]$ ($1 \leq i \leq m$).
    This fact implies $M(\lambda y.N(\lambda x.DD)(DD)y)(\lambda x.DD) \Uparrow_{\mathbf{L}}$.
  - Let $M' \equiv (\lambda x_1.M_0)M_1...M_m$ ($m \geq 1$) and $T \equiv \lambda y.N(\lambda x.DD)(DD)y$, so there is a derivation d and a term R' such that

$$\dfrac{\dfrac{\dfrac{d}{(M_0[M_1/x_1])[T/x_0]M_1[T/x_0]...M_m[T/x_0](\lambda x.DD) \Downarrow_{\mathbf{L}} R'}\,(\cdots)}{(\lambda x_1.M_0[T/x_0])M_1[T/x_0]...M_m[T/x_0](\lambda x.DD) \Downarrow_{\mathbf{L}} R'}\,(\mathit{head})}{(\lambda x_0.(\lambda x_1.M_0)M_1...M_m)(\lambda y.N(\lambda x.DD)(DD)y)(\lambda x.DD) \Downarrow_{\mathbf{L}} R'}\,(\mathit{head})$$

    hence the proof follows by induction on the following derivation

$$\dfrac{\dfrac{d}{(M_0[M_1/x_1])[T/x_0]M_1[T/x_0]...M_m[T/x_0](\lambda x.DD) \Downarrow_{\mathbf{L}} R'}\,(\cdots)}{(\lambda x_0.(M_0[M_1/x_1])M_2...M_m)(\lambda y.N(\lambda x.DD)(DD)y)(\lambda x.DD) \Downarrow_{\mathbf{L}} R'}\,(\mathit{head})$$

  - Let $M' \equiv \lambda x_1.M_0$. Since $M \in \Lambda^0$, there are only three further cases.
    (i) $M \equiv \lambda x_0 x_1.x_0 M_1...M_m$ ($m \in \mathbb{N}$). If $m \geq 1$, then $N(\lambda x.DD)(DD) \approx_{\mathbf{L}} (\lambda x_0 x_1.x_0 M_1...M_m)(\lambda x.DD)(DD) \approx_{\mathbf{L}} (\lambda x.DD)M'_1...M'_m$, where $M'_i \equiv M_i[\lambda x.DD/x_0, DD/x_1]$ ($1 \leq i \leq m$), because $\approx_{\mathbf{L}}$ is a Λ-theory.
    Hence $R \Uparrow_{\mathbf{L}}$ since

$$R \approx_{\mathbf{L}} (\lambda x_0 x_1.x_0 M_1...M_m)(\lambda y.(\lambda x.DD)M'_1...M'_m y)(\lambda x.DD).$$

    So let $m = 0$ and the proof is done.
    (ii) $M \equiv \lambda x_0 x_1.x_1 M_1...M_m$ ($m \in \mathbb{N}$); it is easy to see that $m \geq 1$ implies $R \Uparrow_{\mathbf{L}}$. Thus $m = 0$.
    (iii) The case $M \equiv \lambda x_0 x_1 x_2.M'''$ is immediate. □

By using the previous characterization we can prove that $L_1 \preceq_{\mathbf{L}} L_0$.

Lemma 6.3.13. *Let $C[.]$ be such that $C[L_0], C[L_1] \in \Lambda^0$. If $C[L_1] \Downarrow_{\mathbf{L}}$ then $C[L_0] \Downarrow_{\mathbf{L}}$.*

Proof. The proof is given by induction on the size of $C[L_1] \Downarrow_{\mathbf{L}}$, by considering the last applied rule.

(var) It is not possible, since $C[L_1] \in \Lambda^0$.

(lazy) If $C[.] \equiv [.]$ then the proof is trivial, since $L_0 \Downarrow_{\mathbf{L}}$. If $C[.] \equiv \lambda z.C'[.]$ then the proof is trivial, since $C[.]$ is not relevant.

(head) Either $C[.] \equiv (\lambda z.C_0[.])C_1[.]...C_m[.]$ or $C[.] \equiv [.]C_1[.]...C_m[.]$ $(m \geq 1)$. In the first case, the proof follows by induction on

$$C_0[L_1][C_1[L_1]/z]C_2[L_1]...C_m[L_1].$$

In the second case, the hypothesis that the last applied rule is (*head*) implies $m \geq 1$, and there is a derivation d and a term R such that

$$\cfrac{\cfrac{d}{C_1[L_1](\lambda y.C_1[L_1](\lambda x.DD)(DD)y)(\lambda x.DD)C_2[L_1]...C_m[L_1] \Downarrow_{\mathbf{L}} R}\ (head)}{L_1 C_1[L_1]...C_m[L_1] \Downarrow_{\mathbf{L}} R}\ (head)$$

By inductive hypothesis,

$$C_1[L_0](\lambda y.C_1[L_0](\lambda x.DD)(DD)y)(\lambda x.DD)C_2[L_0]...C_m[L_0] \Downarrow_{\mathbf{L}}$$

so by Lemma 6.3.12, there are three possible cases.

(i) If $C_1[L_0] \to^*_\Lambda \lambda x_0 x_1.x_0$ then it is easy to see that

$$C_1[L_0](\lambda y.C_1[L_0](\lambda x.DD)(DD)y)(\lambda x.DD) =_\Lambda \lambda y.DD;$$

so $m \geq 2$ is not possible. Let $m = 1$; so it is easy to see that

$$\lambda y.C_1[L_0](\lambda x.DD)(DD)y =_\Lambda \lambda x.DD =_\Lambda C_1[L_0](\lambda x.DD)(DD).$$

Hence $C_1[L_0](C_1[L_0](\lambda x.DD)(DD))(\lambda x.DD) \Downarrow_{\mathbf{L}}$, since $\approx_{\mathbf{L}}$ is a Λ-theory. By rule (head), $C[L_0] \Downarrow_{\mathbf{L}}$.

(ii) If $C_1[L_1] \to^*_\Lambda \lambda x_0 x_1.x_1$ then it is easy to see that

$$C_1[L_0](\lambda y.C_1[L_0](\lambda x.DD)(DD)y)(\lambda x.DD) =_\Lambda \lambda x.DD;$$

so $m \geq 2$ is not possible and we can assume $m = 1$.
But $C_1[L_1] \to^*_\Lambda \lambda x_0 x_1.x_1$ implies $C_1[L_0]Q(\lambda x.DD) =_\Lambda \lambda x.DD$, for each $Q \in \Lambda$. Hence $C_1[L_0](C_1[L_0](\lambda x.DD)(DD))(\lambda x.DD) \Downarrow_{\mathbf{L}}$, since $\approx_{\mathbf{L}}$ is a Λ-theory. By rule (head), $C[L_0] \Downarrow_{\mathbf{L}}$.

(iii) If $C_1[L_0] \to_{\Lambda}^{*} \lambda x_0 x_1 x_2.M'''$, for some $M''' \in \Lambda$, then

$$C_1[L_0](\lambda y.C_1[L_0](\lambda x.DD)(DD)y)(\lambda x.DD)C_2[L_0]...C_m[L_0] =_{\Lambda}$$
$$C_1[L_0](\lambda y.M'''[\lambda x.DD/x_0, DD/x_1, y/x_2])(\lambda x.DD)C_2[L_0]...C_m[L_0] =_{\Lambda}$$
$$C_1[L_0](C_1[L_0](\lambda x.DD)(DD))(\lambda x.DD)C_2[L_0]...C_m[L_0]$$

Since $\approx_{\mathbf{L}}$ is a $\lambda\Lambda$-theory,

$$C_1[L_0](C_1[L_0](\lambda x.DD)(DD))(\lambda x.DD)C_2[L_0]...C_m[L_0] \Downarrow_{\mathbf{L}}$$

and the proof follows by applying the rule (head). □

Theorem 6.3.14. $L_1 \approx_{\mathbf{L}} L_0$.

Proof. By Corollary 6.3.11 and Lemma 6.3.13. □

7. Call-by-Value Operational Semantics

7.1 V-Operational Semantics

As proved in Property 5.0.3, the set of Γ-lazy blocked normal forms (Γ-lbnf's), namely Γ-LBNF $= \{\lambda x.M \mid M \in \Lambda\} \cup \{xM_1...M_m \mid M_i \in \Lambda\ ,\ m \in \mathbb{N}\} \cup \{(\lambda x.P)QM_1...M_m \mid P, M_i \in \Lambda\ ,\ Q \notin \Gamma\ ,\ Q \in \Gamma\text{-LBNF}\ ,\ m \in \mathbb{N}\}$, is a set of output values with respect to Γ. Notice that $\Gamma\text{-LBNF}^0 = \Gamma^0$.

$\mathbf{V} \in \mathcal{E}(\Gamma, \Gamma\text{-LBNF})$ is the evaluation relation studied in this section; it is the universal evaluation relation $\mathcal{U}^{\Gamma}_{\Gamma\text{-LBNF}}$. This operational semantics models the call-by-value parameter passing together with lazy evaluation.

Definition 7.1.1 (V-Operational semantics).

(i) $\mathbf{V} \in \mathcal{E}(\Gamma, \Gamma\text{-}LBNF)$ *is the evaluation relation induced by the formal system proving judgments of the shape*

$$M \Downarrow_{\mathbf{V}} N$$

where $M \in \Lambda$ and $N \in \Gamma$-LBNF. It consists of the following rules:

$$\frac{}{xM_1 \ldots M_m \Downarrow_{\mathbf{V}} xM_1 \ldots M_m}\ (var)$$

$$\frac{}{\lambda x.M \Downarrow_{\mathbf{V}} \lambda x.M}\ (lazy)$$

$$\frac{Q \Downarrow_{\mathbf{V}} Q' \quad Q' \in \Gamma \quad P[Q'/x]M_1 \ldots M_m \Downarrow_{\mathbf{V}} N}{(\lambda x.P)QM_1 \ldots M_m \Downarrow_{\mathbf{V}} N}\ (head)$$

$$\frac{Q \Downarrow_{\mathbf{V}} Q' \quad Q' \notin \Gamma}{(\lambda x.P)QM_1 \ldots M_m \Downarrow_{\mathbf{V}} (\lambda x.P)Q'M_1 \ldots M_m}\ (block)$$

(ii) *$M \preceq_{\mathbf{V}} N$ if and only if, for all context $C[.]$ such that $C[M], C[N] \in \Lambda^0$, $(C[M] \Downarrow_{\mathbf{V}}$ implies $C[N] \Downarrow_{\mathbf{V}})$.*

(iii) *$M \approx_{\mathbf{V}} N$ if and only if $M \preceq_{\mathbf{V}} N$ and $N \preceq_{\mathbf{V}} M$.*

The formal system described before, when restricted to closed terms, corresponds to the SECD machine introduced by Landin [63], and further studied by Plotkin [78].

Example 7.1.2. $(\lambda x.yx)(Ky) \Downarrow_{\mathbf{V}} y(\lambda z.y)$. In fact, we can build the following derivation:

$$\dfrac{\dfrac{\dfrac{}{y \Downarrow_{\mathbf{V}} y}\,(var) \quad \dfrac{}{\lambda z.y \Downarrow_{\mathbf{V}} \lambda z.y}\,(lazy)}{Ky \Downarrow_{\mathbf{V}} \lambda z.y}\,(head) \qquad \dfrac{}{y(\lambda z.y) \Downarrow_{\mathbf{V}} y(\lambda z.y)}\,(var)}{(\lambda x.yx)(Ky) \Downarrow_{\mathbf{V}} y(\lambda z.y)}\,(head)$$

$\mathbf{V}$ is deterministic and it characterizes the set Γ-LBNF.

Theorem 7.1.3. (i) *$M \Downarrow_{\mathbf{V}} N$ implies $M \to_{\Gamma}^{*p} N$ and N is in Γ-lbnf.*
(ii) *$M \Downarrow_{\mathbf{V}}$ if and only if M has a Γ-lbnf.*

Proof. (i) By induction on the rules of $\Downarrow_{\mathbf{V}}$.
(ii) ($\Rightarrow$) The proof is a consequence of (i).
($\Leftarrow$) M has a Γ-lbnf means that there is $N \in \Gamma$-LBNF such that $M \to_{\Gamma}^{*} N$. But Γ-LBNF is a set of output values with respect to Γ, by Property 5.0.3; so there is a reduction sequence $M \to_{\Gamma}^{*p} M' \in \Gamma$-LBNF.
Let $M \equiv \lambda x_1...x_n.\zeta M_1...M_m$; we reason by induction on the length of $M \to_{\Gamma}^{*p} M'$. If $n \neq 0$ then the proof follows by rule (lazy), so let $n = 0$. If $\zeta \in \mathrm{Var}$ the proof follows by rule (var). Otherwise $\zeta \equiv (\lambda z.P)Q$; hence, if Q is Γ-valuable then the proof follows by induction and rule (head), otherwise it follows by induction and rule (block). □

Corollary 7.1.4. *Let $M \in \Lambda^0$. Then $M \Downarrow_{\mathbf{V}}$ if and only if M is Γ-valuable.*

Proof. Since $\Gamma^0 = \Gamma$-LBNF^0, the proof follows by Theorem 7.1.3.(ii). □

It follows by Theorem 7.1.5 that $\approx_{\mathbf{V}}$ is a Γ-theory.

Theorem 7.1.5 (V-Correctness).
The $\lambda\Gamma$-calculus is correct with respect to the $\mathbf{V}$-operational semantics.

Proof. By definition of correctness, we must prove that $M =_{\Gamma} N$ implies $M \approx_{\mathbf{V}} N$. Let $M =_{\Gamma} N$ and let $C[.]$ be a context such that $C[M], C[N] \in \Lambda^0$. By definition of $=_{\Gamma}$, $C[M] =_{\Gamma} C[N]$. So the proof follows from Theorem 7.1.3, since by the confluence theorem the property of having Γ-lbnf is closed under $=_{\Gamma}$. □

The **V**-relevant contexts cannot be characterized by using contexts of the shape $[.]C_1[.]...C_m[.]$ ($m \in \mathbb{N}$), as in the call-by-name operational settings; in fact, $[.](DD[.])I$ is not **V**-relevant, while $(\lambda xy.y)([.]I)$ is **V**-relevant. However, the following lemma establishes a negative characterization taking into account not **V**-relevant contexts.

Lemma 7.1.6 (V-Relevance).
*Let $C[.]$ be not **V**-relevant and $M \in \Lambda^0$. If $C[M] \Downarrow_{\mathbf{V}}$ then there is a context $C'[.]$ such that, $\forall P \in \Lambda^0$ $C[P] \Downarrow_{\mathbf{V}} C'[P]$; moreover, $C'[P] \in \Gamma$ if and only if $C'[M] \in \Gamma$.*

Proof. By induction on $C[M] \Downarrow_{\mathbf{V}}$, so by cases on the last applied rule.

(var) If $C[.] \equiv xC_1[.]...C_m[.]$ ($m \in \mathbb{N}$) then the proof is trivial and $C'[.] \equiv C[.]$. The case $C[.] \equiv [.]C_1[.]...C_m[.]$ ($m \in \mathbb{N}$) is not possible, since $M \in \Lambda^0$.

(lazy) The case $C[.] \equiv [.]$ is not possible, since it is trivially **V**-relevant; while in case $C[.] \equiv \lambda z.C''[.]$, the proof is trivial.

(head) $C[.] \equiv [.]C_1[.]...C_m[.]$ ($m \in \mathbb{N}$) is not possible, since $C[M] \Downarrow_{\mathbf{V}}$ while $C[DD] \Uparrow_{\mathbf{V}}$ thus $C[.]$ is **V**-relevant, against the hypothesis; so, let $C[.] \equiv (\lambda z.C_0[.])C_1[.]...C_m[.]$ ($m \geq 1$). If there is $N \in \Lambda^0$ such that $C_1[N] \Uparrow_{\mathbf{V}}$ then $C[N] \Uparrow_{\mathbf{V}}$, by implying that $C[.]$ is **V**-relevant, against the hypothesis. Hence $C_1[.]$ is not relevant and $C_1[M] \Downarrow_{\mathbf{V}}$, thus by induction there is $D_1[.]$ such that $\forall P \in \Lambda^0$ $C_1[P] \Downarrow_{\mathbf{V}} D_1[P]$ and $D_1[P] \in \Gamma$ if and only if $D_1[M] \in \Gamma$. By induction on $C_0[M][D_1[M]/x]C_2[M]...C_m[M]$ there is a context $C'[.]$ such that $C_0[N][D_1[N]/x]C_2[N]...C_m[N] \Downarrow_{\mathbf{V}} C'[N]$ and $C'[N] \in \Gamma$ if and only if $C'[M] \in \Gamma$, for all $N \in \Lambda^0$. The proof follows by rule (head).

(block) The case $C[.] \equiv [.]C_1[.]...C_m[.]$ ($m \in \mathbb{N}$) is not possible, since it is **V**-relevant (as in the previous case); so, let $C[.] \equiv (\lambda z.C_0[.])C_1[.]...C_m[.]$ ($m \geq 1$). Clearly $C_1[.]$ is not relevant and $C_1[M] \Downarrow_{\mathbf{V}}$, so by induction there is $D_1[.]$ such that $\forall P \in \Lambda^0$ $C_1[P] \Downarrow_{\mathbf{V}} D_1[P]$ and $D_1[P] \notin \Gamma$. Let $C'[.] \equiv (\lambda z.C_0[.])D_1[.]C_2[.]...C_m[.]$ and the proof follows easily. □

We will prove that the $\lambda\Gamma$-calculus is not complete with respect to the **V**-operational semantics by using the notion of **V**-relevant context.

Lemma 7.1.7. *Let $C[.]$ be **V**-relevant. If $M \in \Lambda^0$ and $C[M] \Downarrow_{\mathbf{V}}$ then $M \Downarrow_{\mathbf{V}}$.*

Proof. By induction on $C[M] \Downarrow_{\mathbf{V}}$, so by cases on the last applied rule.

(var) $C[.] \equiv xC_1[.]...C_m[.]$ ($m \in \mathbb{N}$), is not **V**-relevant.

(lazy) $C[.] \equiv \lambda z.C'[.]$ is not **V**-relevant, while $C[.] \equiv [.]$ is trivial.

(head) Let $C[.] \equiv [.]C_1[.]...C_m[.]$ ($m \in \mathbb{N}$), so there are two cases.

1. If $M \equiv \lambda z.M'$ then $M \Downarrow_{\mathbf{V}}$.
2. Let $M \equiv (\lambda z.P)QM_1...M_n$ ($n \in \mathbb{N}$). $C[M] \Downarrow_{\mathbf{V}}$ implies, by rule (head) that $Q \Downarrow_{\mathbf{V}} Q'$ and $P[Q'/z]M_1...M_nC_1[M]...C_m[M] \Downarrow_{\mathbf{V}}$; since $[.]C_1[M]...C_m[M]$ is a relevant context (it discriminates M and DD),

by induction $P[Q'/z]M_1...M_n \Downarrow_{\mathbf{V}}$. Thus by rule (head), it follows that $M \Downarrow_{\mathbf{V}}$.

Let $C[.] \equiv (\lambda z.C_0[.])C_1[.]...C_m[.]$ $(m \geq 1)$ so there two cases again.

1. If $C_1[.]$ is not relevant then there exists $C_1'[.]$ satisfying the Lemma 7.1.6, thus by induction on $C_0[M][C_1'[M]/z]C_2[M]...C_m[M] \Downarrow_{\mathbf{L}}$, the proof follows.
2. Otherwise, $C_1[.]$ is relevant and the proof follows by induction on $C_1[M] \Downarrow_{\mathbf{L}}$.

(block) Let $C[.] \equiv [.]C_1[.]...C_m[.]$ $(m \in \mathbb{N})$; if $M \equiv (\lambda x.P)QN_1...N_n$ $(n \in \mathbb{N})$ then $Q \Downarrow_{\mathbf{V}} Q'$, but $Q \in \Lambda^0$ and Q' is a closed Γ-lbnf imply that Q' is an abstraction, against the hypothesis that the last applied rule is (block); thus $M \equiv \lambda x.M'$ and $M \Downarrow_{\mathbf{V}}$ by rule (lazy).

Let $C[.] \equiv (\lambda z.C_0[.])C_1[.]...C_m[.]$ $(m \geq 1)$, so $C_1[.]$ is relevant and the proof follows by induction. □

The following result holds.

Theorem 7.1.8 (V-Incompleteness).
The $\lambda\Gamma$-calculus is incomplete with respect to the **V***-operational semantics.*

Proof. Let P and Q be two closed Γ-unsolvable terms of order zero such that $P \neq_{\Lambda} Q$. A non-**V**-relevant context cannot discriminate them. By Definition 3.1.12, if $P \rightarrow_{\Gamma}^{*} R$ or $Q \rightarrow_{\Gamma}^{*} R$, for some R, then R cannot be an abstraction, hence $P \Uparrow_{\mathbf{V}}$ and $Q \Uparrow_{\mathbf{V}}$. Let $C[.]$ be a **V**-relevant context. So, by Lemma 7.1.7, $C[P] \Uparrow_{\mathbf{V}}$ and $C[Q] \Uparrow_{\mathbf{V}}$. Hence, $P \approx_{\mathbf{V}} Q$. □

As a corollary, we obtain that the **V**-operational equivalence equates all closed Γ-unsolvable terms of the same order.

Corollary 7.1.9. *Let P and Q be closed Γ-unsolvable terms of the same order n. Then $P \approx_{\mathbf{V}} Q$.*

Proof. By induction on n. If $n = 0$, then the proof follows from the proof of the incompleteness result. Otherwise, it follows by induction. □

The next property shows an interesting characterization of the **V**-operational semantics.

Property 7.1.10. Let M and N be such that M is potentially Γ-valuable while N is not potentially Γ-valuable. Then $M \not\approx_{\mathbf{V}} N$.

Proof. According to the definition of potentially Γ-valuable terms, we will consider only substitutions whose codomain is Γ. M potentially Γ-valuable means that there is a substitution $\mathbf{s}$ such that $\mathbf{s}(M) \in \Lambda^0$ is Γ-valuable, while $\mathbf{s}'(N)$ is not Γ-valuable, for each substitution $\mathbf{s}'$ such that $\mathbf{s}'(N) \in \Lambda^0$. So let $\mathbf{s}$ be such that $\mathbf{s}(M)$ is Γ-valuable; we can easily extend $\mathbf{s}$ to a substitution $\mathbf{s}'$ such that $\mathbf{s}'(M) \in \Lambda^0$ is Γ-valuable while $\mathbf{s}'(N) \in \Lambda^0$ is not Γ-valuable. Let $C[.] \equiv (\lambda x_1...x_n.[.])\mathbf{s}'(x_1)...\mathbf{s}'(x_n)$ where $\mathrm{FV}(M) \cup \mathrm{FV}(N) \subseteq \{x_1, ..., x_n\}$; therefore, by Corollary 7.1.4, $C[.]$ is discriminating for M and N. □

The **V**-theory is not fully extensional. In fact, $DD \not\approx_{\mathbf{V}} \lambda x.DDx$.

7.1.1 An Example

We will show now that $V_0 \approx_{\mathbf{V}} V_1$, where

$V_0 \equiv \lambda x.(\lambda x_1 x_2.DD)(x(\lambda x_1.DD)(\lambda x_1.DD))$,
$V_1 \equiv \lambda x.(\lambda x_1 x_2 x_3.DD)(x(\lambda x_1.DD)(\lambda x_1 x_2.DD))(x(\lambda x_1 x_2.DD)(\lambda x_1.DD))$.

This equivalence was first proved in [44]. The interest of such a result will be clear when we will study denotational semantics.

Lemma 7.1.11. *Let* $M \in \Lambda^0$.

(i) *If* $V_0 M \Downarrow_{\mathbf{V}}$ *then* $V_0 M \Downarrow_{\mathbf{V}} \lambda x.DD$.
(ii) *If* $V_1 M \Downarrow_{\mathbf{V}}$ *then* $V_1 M \Downarrow_{\mathbf{V}} \lambda x.DD$.

Proof. Clearly, $P \in \Lambda^0$ and $P \Downarrow_{\mathbf{V}} P'$ imply P' is an abstraction.

(i) Let $D^1 \equiv (\lambda x_1.DD)$ and $D^2 \equiv (\lambda x_1 x_2.DD)$; thus

$$\dfrac{\dfrac{d_0}{M \Downarrow_{\mathbf{V}} M_0}\cdots \qquad \dfrac{\dfrac{d_1}{M_0 D^1 D^1 \Downarrow_{\mathbf{V}} M_1}\cdots \qquad \dfrac{}{\lambda x_2.DD \Downarrow_{\mathbf{V}} \lambda x_2.DD}\,(lazy)}{(\lambda x_1 x_2.DD)(M_0 D^1 D^1) \Downarrow_{\mathbf{V}} \lambda x_2.DD}\,(head)}{(\lambda x.D^2(xD^1D^1))M \Downarrow_{\mathbf{V}} \lambda x_2.DD}\,(head)$$

(ii) Let $D^1 \equiv (\lambda x_1.DD)$, $D^2 \equiv (\lambda x_1 x_2.DD)$ and $D^3 \equiv (\lambda x_1 x_2 x_3.DD)$; thus

$$\dfrac{\dfrac{d_0}{M \Downarrow_{\mathbf{V}} M_0}\cdots \qquad \dfrac{\dfrac{d_{12}}{M_0 D^1 D^2 \Downarrow_{\mathbf{V}} M_{12}}\cdots \qquad \dfrac{\dfrac{d_{21}}{M_0 D^2 D^1 \Downarrow_{\mathbf{V}} M_{21}}\cdots \qquad \dfrac{}{\lambda x.DD \Downarrow_{\mathbf{V}} \lambda x.DD}\,(lazy)}{D^2((M_0 D^2 D^1) \Downarrow_{\mathbf{V}} \lambda x.DD}\,(head)}{D^3(M_0 D^1 D^2)(M_0 D^2 D^1) \Downarrow_{\mathbf{V}} \lambda x.DD}\,(head)}{(\lambda x.D^3(xD^1D^2)(xD^2D^1))M \Downarrow_{\mathbf{V}} \lambda x.DD}\,(head)$$

□

Lemma 7.1.12. *Let* $C[.]$ *be a context such that* $C[V_0], C[V_1] \in \Lambda^0$.

(i) *If* $C[V_0] \Downarrow_{\mathbf{V}}$ *then* $\exists C'[.]$ *such that* $C[V_0] \Downarrow_{\mathbf{V}} C'[V_0]$ *and* $C[V_1] \Downarrow_{\mathbf{V}} C'[V_1]$.
(ii) *If* $C[V_1] \Downarrow_{\mathbf{V}}$ *then* $\exists C'[.]$ *such that* $C[V_0] \Downarrow_{\mathbf{V}} C'[V_0]$ *and* $C[V_1] \Downarrow_{\mathbf{V}} C'[V_1]$.

Proof. (i) By induction on the derivation proving $C[V_0] \Downarrow_{\mathbf{V}}$.

(var) This case is not possible, since by hypothesis $C[V_0] \in \Lambda^0$.

(lazy) $C[.] \equiv \lambda z.C_0[.]$ and $C[.] \equiv [.]$ are both trivial.

(head) Let $C[.] \equiv [.]C_1[.]...C_m[.]$ $(m \geq 1)$; if $m \geq 2$ then by Lemma 7.1.11.(i) it is easy to see that $C[V_0] \Uparrow_{\mathbf{V}}$. In case $m = 1$, the proof follows by Lemma 7.1.11. Let $C[.] \equiv (\lambda z.C_0[.])C_1[.]...C_m[.]$ $(m \geq 1)$, so by induction on $C_1[V_0] \Downarrow_{\mathbf{V}}$ there is $C_1'[.]$ such that $C_1[V_0] \Downarrow_{\mathbf{V}} C_1'[V_0]$ and $C_1[V_1] \Downarrow_{\mathbf{V}} C_1'[V_1]$. The proof follows by induction on $C_0[V_0][C_1'[V_0]/z]C_2[V_0]...C_m[V_0] \Downarrow_{\mathbf{V}}$.

(block) This case is not possible, since by hypothesis $C[V_0] \in \Lambda^0$.

(ii) By induction on the derivation proving $C[V_1] \Downarrow_{\mathbf{V}}$.

(var) This case is not possible, since by hypothesis $C[V_0] \in \Lambda^0$.

(lazy) $C[.] \equiv \lambda z.C_0[.]$ and $C[.] \equiv [.]$ are both trivial.

(head) Let $C[.] \equiv [.]C_1[.]...C_m[.]$ $(m \geq 1)$, if $m \geq 2$ then by Lemma 7.1.11.(ii) it is easy to see that $C[V_0] \Uparrow_{\mathbf{V}}$. In case $m = 1$, the proof follows by Lemma 7.1.11. Let $C[.] \equiv (\lambda z.C_0[.])C_1[.]...C_m[.]$ $(m \geq 1)$ so by induction on $C_1[V_1] \Downarrow_{\mathbf{V}}$ there is $C'_1[.]$ such that $C_1[V_0] \Downarrow_{\mathbf{V}} C'_1[V_0]$ and $C_1[V_1] \Downarrow_{\mathbf{V}} C'_1[V_1]$. The proof follows by induction on $C_0[V_1][C'_1[V_1]/z]C_2[V_1]...C_m[V_1] \Downarrow_{\mathbf{V}}$.

(block) This case is not possible, since by hypothesis $C[V_0] \in \Lambda^0$. □

Theorem 7.1.13. $V_0 \approx_{\mathbf{V}} V_1$.

Proof. The proof follows from Lemma 7.1.12. □

8. Operational Extensionality

8.1 Operational Semantics and Extensionality

In Sect. 1.3, the notion of full extensionality was introduced. A Δ-theory $\mathcal{T}$ is fully extensional when all terms in it can be interpreted as functions, i.e. if and only if the full extensionality principle holds:

$$\text{(EXT)} \quad Mx =_{\mathcal{T}} Nx \Rightarrow M =_{\mathcal{T}} N \qquad x \notin \text{FV}(M) \cup \text{FV}(N).$$

Moreover, we proved that a Δ-theory $\mathcal{T}$ is fully extensional if and only if it is closed under η-equality, which is the congruence relation induced by the η-reduction rule:

$$(\eta) \quad \lambda x.Mx \rightarrow_{\eta} M \text{ if and only if } x \notin \text{FV}(M).$$

Now the notion of extensionality will be considered in the particular setting of the Δ-theories that arise from operational semantics, namely the Δ-*operational theories*. In the rest of this section, we will restrict our discussion to Δ-operational theories induced from a formal system such that the Δ-calculus is correct with respect to them. Sometimes, for sake of simplicity, we will skip the prefix Δ.

Let $\mathbf{O} \in \mathcal{E}(\Delta, \Theta)$; intuitively a term M has a functional behaviour in $\mathbf{O}$, or equivalently it can be interpreted as a function, when $M \approx_{\mathbf{O}} \lambda x.Mx$ ($x \notin \text{FV}(M)$). If all terms have a functional behaviour in $\mathbf{O}$, then it is natural to expect that $\approx_{\mathbf{O}}$ in its turn behaves like the extensional equivalence on functions, i.e. if $\forall P \in \Lambda$, $MP \approx_{\mathbf{O}} NP$ then $M \approx_{\mathbf{O}} N$. But not all operational theories give a functional interpretation to all terms.

Let us consider, for example, the **L**-operational theory: DD and $\lambda x.DD$ have the same applicative behaviour (since, for all $y \in \text{Var}$, both $DDy \Uparrow_{\mathbf{L}}$ and $(\lambda x.DDx)y \Uparrow_{\mathbf{L}}$), nevertheless they cannot be equated, since the context [.] separates them. In some sense DD, in the **L**-operational theory, can be see as a function too, but of arity 0. But the extensionality principle becomes vacuous if extended to 0-arity functions. So we will introduce the notion of *operational extensionality*.

In order to formalize such a notion, let us introduce the key notion of **O**-comparable terms, with respect to an operational theory **O**.

Definition 8.1.1. *Let* $\mathbf{O} \in \mathcal{E}(\Delta, \Theta)$ *be defined by a formal system.*
M *and* N *are* **O**-comparable *(notation* $M \smile_{\mathbf{O}} N$*) when, for each substitution* $\mathbf{s} : \mathrm{Var} \to \Delta$ *such that* $\mathbf{s}(M), \mathbf{s}(N) \in \Lambda^0$

$$\mathbf{s}(M) \Downarrow_{\mathbf{O}} \text{ if and only if } \mathbf{s}(N) \Downarrow_{\mathbf{O}}.$$

Otherwise, M *and* N *are said to be* **O***-incomparable (notation* $M \frown_{\mathbf{O}} N$*).*

We will speak simply about comparable terms if the involved operational semantics is clear by the context.

Note that being **O**-comparable does not imply equivalent. In fact, if $\mathbf{O} \in \{\mathbf{H}, \mathbf{N}, \mathbf{L}, \mathbf{V}\}$ then $\lambda x.xzI \smile_{\mathbf{O}} \lambda x.xzO$ but $\lambda x.xzI \not\approx_{\mathbf{O}} \lambda x.xzO$. Note that $x(DD) \smile_{\mathbf{V}} (\lambda y.xy)(DD)$ are **V**-comparable, although $x(DD) \Downarrow_{\mathbf{V}}$ and $(\lambda y.xy)(DD) \Uparrow_{\mathbf{V}}$.

To be comparable in a given operational semantics is a *necessary* condition for two terms being equivalent.

In the rest of this section, $\lambda^* x.Mx$ will denote the fact that $x \notin \mathrm{FV}(M)$. Now we can state formally the *operational functionality principle.*

(OP-FUN) $\mathbf{O} \in \mathcal{E}(\Delta, \Theta)$ is *op-functional* if and only if, for all $M \in \Lambda$, $M \smile_{\mathbf{O}} \lambda^* x.Mx$ implies $M \approx_{\mathbf{O}} \lambda^* x.Mx$.

It is easy to see that $M \approx_{\mathbf{O}} \lambda^* x.Mx$ implies $M \smile_{\mathbf{O}} \lambda^* x.Mx$.

Informally, an operational theory **O** is *op-extensional* when, for all terms M and N, if they can be interpreted as functions and they have the same applicative behaviour, then $M \approx_{\mathbf{O}} N$.

(OP-EXT) $\mathbf{O} \in \mathcal{E}(\Delta, \Theta)$ is *op-extensional* if and only if, for all $M, N \in \Lambda$, for all $x \notin \mathrm{FV}(M) \cup \mathrm{FV}(N)$, if $M \smile_{\mathbf{O}} \lambda^* x.Mx$, $N \smile_{\mathbf{O}} \lambda^* x.Nx$ and $Mx \approx_{\mathbf{O}} Nx$ then $M \approx_{\mathbf{O}} N$.

It is easy to check that the two principles are equivalent when the operational semantics are correct.

Property 8.1.2. OP-FUN if and only if OP-EXT.

Proof. ($\Rightarrow$) Let $\mathbf{O} \in \mathcal{E}(\Delta, \Theta)$ satisfy OP-FUN. For all $x \notin \mathrm{FV}(M) \cup \mathrm{FV}(N)$, let $M \smile_{\mathbf{O}} \lambda^* x.Mx$, $N \smile_{\mathbf{O}} \lambda^* x.Nx$ and $Mx \approx_{\mathbf{O}} Nx$.
OP-FUN implies both $M \approx_{\mathbf{O}} \lambda^* x.Mx$ and $N \approx_{\mathbf{O}} \lambda^* x.Nx$. Moreover, $\lambda x^*.Mx \approx_{\mathbf{O}} \lambda^* x.Nx$, since $Mx \approx_{\mathbf{O}} Nx$ and $\approx_{\mathbf{O}}$ is a congruence; thus

$$M \approx_{\mathbf{O}} \lambda^* x.Mx \approx_{\mathbf{O}} \lambda^* x.Nx \approx_{\mathbf{O}} N.$$

($\Leftarrow$) Let $\mathbf{O} \in \mathcal{E}(\Delta, \Theta)$ satisfy OP-EXT; let $M \smile_{\mathbf{O}} \lambda^* x.Mx$.
$Mz \approx_{\mathbf{O}} (\lambda^* x.Mx)z$ and $\lambda^* x.Mx \approx_{\mathbf{O}} \lambda^* u.(\lambda^* x.Mx)u$, since $\approx_{\mathbf{O}}$ is a Δ-theory; thus, $\lambda^* x.Mx \smile_{\mathbf{O}} \lambda^* u.(\lambda^* x.Mx)u$ and by OP-EXT, $M \approx_{\mathbf{O}} \lambda^* x.Mx$. □

The notion of op-extensionality can be captured by a suitable reduction rule, parameterized with respect to the considered operational semantics.

Definition 8.1.3. *Let* $\mathbf{O} \in \mathcal{E}(\Delta, \Theta)$ *be an evaluation relation.*

(i) *The* $\mathbf{O}\eta$*-reduction* $\to_{\mathbf{O}\eta}$ *is the contextual closure of the following rule:* $\lambda^* x.Mx \to_{\mathbf{O}\eta} M$ *if and only if* $M \smile_{\mathbf{O}} \lambda^* x.Mx$.
$\lambda^* x.Mx$ *is a* $\mathbf{O}\eta$*-redex and* M *is its contractum.*
(ii) $M \to^*_{\mathbf{O}\eta} N$ *and* $=_{\mathbf{O}\eta}$ *are respectively the reflexive and transitive closure of* $\to_{\mathbf{O}\eta}$ *and the symmetric, reflexive and transitive closure of* $\to_{\mathbf{O}\eta}$.
(iii) $M \to_{\Delta\mathbf{O}\eta} N$ *when either* $M \to_{\mathbf{O}\eta} N$ *or* $M \to_{\Delta} N$.
(iv) $M \to^*_{\Delta\mathbf{O}\eta} N$ *and* $=_{\Delta\mathbf{O}\eta}$ *are respectively the reflexive, symmetric and transitive closure of* $\to_{\Delta\mathbf{O}\eta}$ *and the symmetric, reflexive and transitive closure of* $\to_{\Delta\mathbf{O}\eta}$.

A Δ-theory $\approx_{\mathbf{O}}$ is a $\Delta\mathbf{O}\eta$-theory, when $\approx_{\mathbf{O}}$ is closed under $=_{\mathbf{O}\eta}$, namely $P =_{\mathbf{O}\eta} Q$ implies $P \approx_{\mathbf{O}} Q$.

The relationship between op-extensionality and $\mathbf{O}\eta$-reduction rule is clarified in the next theorem.

Theorem 8.1.4. *Let* $\mathbf{O} \in \mathcal{E}(\Delta, \Theta)$ *be correct with respect to the* $\lambda\Delta$*-calculus.* $\mathbf{O}$ *is op-extensional if and only if* $\approx_{\mathbf{O}}$ *is closed under* $=_{\mathbf{O}\eta}$.

Proof. ($\Rightarrow$) Assume $C[M] =_{\mathbf{O}\eta} C[\lambda^* x.Mx]$, so $M \smile_{\mathbf{O}} \lambda^* x.Mx$ by definition of $\to_{\mathbf{O}\eta}$. Clearly $\lambda^* u.(\lambda^* x.Mx)u =_{\Delta} \lambda^* u.Mu$, so $\lambda^* u.(\lambda^* x.Mx)u \approx_{\mathbf{O}} \lambda^* u.Mu$, since $\mathbf{O}$ is a Δ-theory, thus $\lambda^* x.Mx \smile_{\mathbf{O}} \lambda^* u.(\lambda^* x.Mx)u$ too. $Mz \approx_{\mathbf{O}} (\lambda^* x.Mx)z$, since $\mathbf{O}$ is a Δ-theory. Hence $M \approx_{\mathbf{O}} \lambda^* x.Mx$ by op-extensionality. Thus, $C[M] \approx_{\mathbf{O}} C[\lambda^* x.Mx]$.
($\Leftarrow$) Let $M \smile_{\mathbf{O}} \lambda^* x.Mx$, $N \smile_{\mathbf{O}} \lambda^* x.Nx$ and $Mx \approx_{\mathbf{O}} Nx$, for all $x \notin \mathrm{FV}(M) \cup \mathrm{FV}(N)$. Since $\mathbf{O}$ is a $\Delta\mathbf{O}\eta$-theory, both $M \approx_{\mathbf{O}} \lambda^* x.Mx$ and $N \approx_{\mathbf{O}} \lambda^* x.Nx$. Moreover, $Mx \approx_{\mathbf{O}} Nx$ implies $(\lambda^* x.Mx) \approx_{\mathbf{O}} (\lambda^* x.Nx)$, so the proof follows by transitivity of $\approx_{\mathbf{O}}$. □

We will prove that **H**, **N**, **L** and **V** are operationally extensional.

First we need to characterize the class of terms M such that M and $\lambda^* x.Mx$ are **O**-comparable, when $\mathbf{O} \in \{\mathbf{H}, \mathbf{N}, \mathbf{L}, \mathbf{V}\}$.

Lemma 8.1.5. *Both* $M \smile_{\mathbf{H}} \lambda^* x.Mx$ *and* $M \smile_{\mathbf{N}} \lambda^* x.Mx$, *for all* $M \in \Lambda$.

Proof. This proof is easy. □

In **H** and **N**, the operational extensionality corresponds to full extensionality (see Theorems 6.1.10 and 6.2.10).

Theorem 8.1.6. **H** *and* **N** *are operational extensional* Λ*-theories.*

Proof. Obvious, since full extensionality implies operational extensionality. □

In **L**, it is no longer true that $M \smile_{\mathbf{L}} \lambda^* z.Mz$ holds for all $M \in \Lambda$. In fact, y and $\lambda z.yz$ are not **L**-comparable (take the substitution $\mathbf{s}$ such that $\mathbf{s}(y) = DD$).

Lemma 8.1.7. *$M \smile_{\mathbf{L}} \lambda^* z.Mz$ if and only if $M \to_\Lambda^* \lambda x.N$, for some $N \in \Lambda$.*

Proof. ($\Rightarrow$) Assume M does not reduce to an abstraction.
This means that either $M \to_\Lambda^* x\vec{Q}$ or $M \to_\Lambda^* U$, where U is a Λ-unsolvable term of order 0. By correctness, this implies either $M \approx_{\mathbf{L}} x\vec{Q}$ or $M \approx_{\mathbf{L}} U$. Let $\mathbf{s}$ be a substitution such that $\mathbf{s}(x) = DD$, for all x.
In both cases, $\mathbf{s}(x\vec{Q})$ and $\mathbf{s}(U)$ are Λ-unsolvable of order 0, therefore $\mathbf{s}(x\vec{Q}) \Uparrow_{\mathbf{L}}$ and $\mathbf{s}(U) \Uparrow_{\mathbf{L}}$. On the other hand, $\mathbf{s}(\lambda^* z.Mz) =_\Lambda \lambda^* z.\mathbf{s}(M)z$, so $\mathbf{s}(\lambda^* z.Mz) \Downarrow_{\mathbf{L}}$, against the hypothesis that $M \smile_L \lambda^* z.Mz$.
($\Leftarrow$) Let $M \to_\Lambda^* \lambda x.N$; so $\lambda^* z.Mz \to_\Lambda^* \lambda z.N[z/x] =_\alpha \lambda x.N$, therefore by correctness $M \approx_{\mathbf{L}} \lambda^* z.Mz$, which implies $M \smile_{\mathbf{L}} \lambda^* z.Mz$. □

By the previous lemma, the $\mathbf{L}\eta$-reduction ($\to_{\mathbf{L}\eta}$) can be restated, without any explicit reference to the comparability relation $\mathbf{L}$, as follows:

$$\lambda^* x.Mx \to_{\mathbf{L}\eta} M \text{ if and only if there is } N \in \Lambda \text{ such that } M \to_\Lambda^* \lambda x.N.$$

Theorem 8.1.8. *$\mathbf{L}$ is an operational extensional Λ-theory.*

Proof. By Lemma 8.1.7, if $M =_{\mathbf{L}\eta} N$ then $M =_\Lambda N$ so $M \approx_{\mathbf{L}} N$ by correctness and the proof follows by Theorem 8.1.4. □

Now let us consider the call-by-value operational semantics $\approx_{\mathbf{V}}$.

Lemma 8.1.9. *$M \smile_{\mathbf{V}} \lambda^* z.Mz$ if and only if M is Γ-valuable.*

Proof. ($\Rightarrow$) Assume that $M \to_\Gamma^* N$ implies $N \notin \Gamma$. This means that either $N \equiv x\vec{Q}$ for some sequence $\|\vec{Q}\| > 0$, or $N \equiv (\lambda x.P)Q\vec{R}$ where $Q \to_\Gamma^* Q'$ and $Q' \notin \Gamma$, or $M \to_\Gamma^* U$ where U is a Γ-unsolvable term of order 0. We prove by induction on N that there is a substitution $\mathbf{s}$ such that $\mathbf{s}(N) \Uparrow_{\mathbf{V}}$.
Let $\mathbf{s}$ be a substitution such that $\forall x \in \mathrm{Var}$, $\mathbf{s}(x) = \lambda x.DD \in \Gamma$. The first and the third cases are obvious: both $\mathbf{s}(x\vec{Q})$ and $\mathbf{s}(U)$ are unsolvable of order 0, hence $\mathbf{s}(x\vec{Q}) \Uparrow_{\mathbf{V}}$ and $\mathbf{s}(U) \Uparrow_{\mathbf{V}}$. In the second case, by induction $\mathbf{s}(Q) \Uparrow_{\mathbf{V}}$; so, $\mathbf{s}((\lambda x.P)Q\vec{R}) \Uparrow_{\mathbf{V}}$. In all cases, $\mathbf{s}(M) \Uparrow_{\mathbf{V}}$.
On the other hand, $\mathbf{s}(\lambda^* z.Mz) = \lambda^* z.\mathbf{s}(M)z$, so $\mathbf{s}(\lambda^* z.Mz) \Downarrow_{\mathbf{V}}$ by rule (*lazy*), against the hypothesis that $M \smile_{\mathbf{V}} \lambda^* z.Mz$.
($\Leftarrow$) By definition, $\Gamma = \mathrm{Var} \cup \{\lambda x.M \mid M \in \Lambda\}$. If $M \to_\Gamma^* \lambda x.P'$, for some P', then $M =_\Gamma \lambda^* z.Mz$, so $M \approx_{\mathbf{V}} \lambda^* z.Mz$ and $M \smile_{\mathbf{V}} \lambda^* z.Mz$.
Let $M \to_\Gamma^* x$; for every substitution $\mathbf{s} : \mathrm{Var} \to \Gamma^0$, it is easy to see that $\mathbf{s}(x) \in \Gamma^0$ and $\mathbf{s}(M) \to_\Gamma^* \mathbf{s}(x)$. By correctness, $\mathbf{s}(M) \Downarrow_{\mathbf{V}}$ and $\mathbf{s}(\lambda^* z.Mz) = \lambda^* z.\mathbf{s}(M)z \in \Gamma^0$, so $\mathbf{s}(\lambda^* z.Mz) \Downarrow_{\mathbf{V}}$. This implies, by definition, $M \smile_{\mathbf{V}} \lambda^* z.Mz$. □

By the previous lemma, the $\mathbf{V}\eta$-reduction ($\to_{\mathbf{V}\eta}$) can be restated, without any explicit reference to the evaluation relation $\mathbf{V}$, as follows:

$$\lambda^* x.Mx \to_{\mathbf{V}\eta} M \text{ if and only if } M \text{ is } \Gamma\text{-valuable}.$$

We prove that **V** is an operational extensional Γ-theory by using some denotational tools, in Property 12.1.20.(i). An interesting overview on rewriting and extensionality can be found in [41].

8.1.1 Head-Discriminability

We introduced in Definition 5.0.10, the notion of context discriminating a pair of terms, for a given evaluation relation **O**. We will refine such a notion defining **O** head-discriminable, if the operational difference between two terms can be tested through an head context. Clearly, this notion is in some sense related to extensionality, since filling a head context $C[.]$ by a closed term M corresponds just to applying M to a suitable sequence of arguments.

Definition 8.1.10. *$\mathbf{O} \in \mathcal{E}(\Delta, \Theta)$ is* head discriminable *if and only if $M \not\approx_{\mathbf{O}} N$ implies there is a Δ-valuable head context $C[.]$ such that $C[M], C[N] \in \Lambda^0$ and $C[M] \Downarrow_{\mathbf{O}}$ while $C[N] \Uparrow_{\mathbf{O}}$ (or vice versa).*

Now let us define a particular class of operational semantics.

Definition 8.1.11. *An evaluation relation* **O** *is* uniform *if and only if $\lambda y.M \frown_{\mathbf{O}} \lambda x.N$ implies $M \frown_{\mathbf{O}} N$.*

Informally, the uniformity condition says that a reduction machine either computes under a λ-abstraction or not; in other words, it has either a lazy or a not lazy behaviour, but it cannot mix the two styles of computing. Note that all the semantics we defined in this chapter are uniform. Moreover, we would like to stress that uniformity is quite a natural property to expect for every reasonable operational semantics.

We will prove in the next theorem that, for all the uniform operational semantics, head discriminability implies operational extensionality.

Theorem 8.1.12. *If $\mathbf{O} \in \mathcal{E}(\Delta, \Theta)$ is uniform and head discriminable then it is operationally extensional.*

Proof. Let **O** be head discriminable, $M \smile_{\mathbf{O}} \lambda^* x.Mx$ and $N \smile_{\mathbf{O}} \lambda^* x.Nx$; we will prove that $M \not\approx_{\mathbf{O}} N$ implies $Mx \not\approx_{\mathbf{O}} Nx$, for all $x \notin \mathrm{FV}(M) \cup \mathrm{FV}(N)$. Since $\approx_{\mathbf{O}}$ is head discriminable, $M \not\approx_{\mathbf{O}} N$ implies that there is a Δ-valuable head context $(\lambda\vec{y}.[.])\vec{P}$ such that $(\lambda\vec{y}.M)\vec{P}, (\lambda\vec{y}.N)\vec{P} \in \Lambda^0$, and $(\lambda\vec{y}.M)\vec{P} \Downarrow_{\mathbf{O}}$ and $(\lambda\vec{y}.N)\vec{P} \Uparrow_{\mathbf{O}}$ (or vice versa).

- Let us consider the case $\|\vec{P}\| = \|\vec{y}\|$. Thus $M[\vec{P}/\vec{y}] \Downarrow_{\mathbf{O}}$ and $N[\vec{P}/\vec{y}] \Uparrow_{\mathbf{O}}$, by correctness. Since $M \smile_{\mathbf{O}} \lambda^* x.Mx$ and $N \smile_{\mathbf{O}} \lambda^* x.Nx$, $\lambda^* x.M[\vec{P}/\vec{y}]x \Downarrow_{\mathbf{O}}$ and $\lambda^* x.N[\vec{P}/\vec{y}]x \Uparrow_{\mathbf{O}}$. Thus the context $(\lambda\vec{y}x.[.])\vec{P}$ discriminates Mx and Nx, namely $Mx \not\approx_{\mathbf{O}} Nx$.

- Let $\|\vec{P}\| > \|\vec{y}\|$. Then $\vec{P} \equiv \vec{P_1}\vec{P_2}$, where $\|\vec{P_1}\| = \|\vec{y}\|$, and so $M[\vec{P_1}/\vec{y}]\vec{P_2} \Downarrow_O$ and $N[\vec{P_1}/\vec{y}]\vec{P_2} \Uparrow_O$. Since $x \notin \mathrm{FV}(M) \cup \mathrm{FV}(N)$, $\approx_O$ is closed under $=_\Delta$ and $\|\vec{P_2}\| > 0$, both $(\lambda x.M[\vec{P_1}/\vec{y}]x)\vec{P_2} \Downarrow_O$ and $(\lambda x.N[\vec{P_1}/\vec{y}]x)\vec{P_2} \Uparrow_O$, and consequently, by correctness, $(\lambda \vec{y}x.Mx)\vec{P_1}\vec{P_2} \Downarrow_O$ and $(\lambda \vec{y}x.Nx)\vec{P_1}\vec{P_2} \Uparrow_O$. So the context $(\lambda \vec{y}x.[.])\vec{P_1}\vec{P_2}$ is a head context discriminating Mx and Nx, namely $Mx \not\approx_O Nx$.
- Let $\|\vec{P}\| < \|\vec{y}\|$. Then $\vec{y} \equiv \vec{y_1}\vec{y_2}$, where $\|\vec{P}\| = \|\vec{y_1}\|$, and, by Δ-reduction, $\lambda\vec{y_2}.M[\vec{P}/\vec{y_1}] \Downarrow_O$ and $\lambda\vec{y_2}.N[\vec{P}/\vec{y_1}] \Uparrow_O$. By uniformity, this implies there is a substitution $\mathbf{s}$ such that $\mathbf{s}(M[\vec{P}/\vec{y_1}]) \Downarrow_O$ and $\mathbf{s}(N[\vec{P}/\vec{y_1}]) \Uparrow_O$, and consequently there is a substitution $\mathbf{s}'$ such that $\mathbf{s}'(M) \Downarrow_O$ and $\mathbf{s}'(N) \Uparrow_O$. Since $M \smile_O \lambda^* x.Mx$ and $N \smile_O \lambda^* x.Nx$, $\lambda^* x.\mathbf{s}'(M)x \Downarrow_O$ and $\lambda^* x.\mathbf{s}'(N)x \Uparrow_O$. Let $\mathrm{FV}(M) \cup \mathrm{FV}(N) \subseteq \{z_1, ..., z_k\}$ for some $k \in \mathbb{N}$, and $C'[.] \equiv (\lambda z_1...z_k.[.])\mathbf{s}'(z_1)...\mathbf{s}'(z_k)$. Then $\lambda x.C'[.]$ is the context discriminating Mx and Nx, namely $Mx \not\approx_O Nx$.

□

The previous theorem assures us that the notion of operational extensionality we defined is meaningful under the hypothesis of uniformity. In fact, head discriminability means that terms can be discriminated just observing their applicative behaviour, so by considering them as functions, may be of arity 0. All operational theories we considered are head-discriminable.

9. Further Reading

Operational semantics. An algebraic view of structural operational semantics (SOS) can be found in [94]. A formalization of the SOS. operational semantics based on natural deduction is given in [24]. In [77] the reader can find a presentation of the different approaches to structural operational semantics, concentrating on the advantages and disadvantages of each one for reasoning about operational equivalence of programs.

λ-Reduction machines. Some abstract machines for evaluating λ-terms according to different evaluation strategies have been performed. The CUCH machine, defined in [16] and whose implementation is described in [21], performs the **N**-evaluation. Krivine [60] designed an abstract machine performing a variant of the **H**-evaluation, inducing the same operational semantics. The Bologna Optimal Higher Order Machine (BOHM) machine implements an optimal evaluation relation, in the sense that it optimizes the number of λ-reduction performed in parallel [6]. The optimality is reached through a graph representation of terms based on *linear logic* [48]; a survey on the optimal implementation of $\lambda\Lambda$-calculus can be found in [7].

Part III

Denotational Semantics

10. $\lambda\Delta$-Models

To study the operational behaviour of λ-terms, we will use the *denotational (mathematical)* approach. A denotational semantics for a language is based on the choice of a space of *semantics values,* or *denotations*, where terms are to be interpreted. Choosing a space with nice mathematical properties can help in proving the semantic properties of terms, since to this aim standard mathematical techniques can be used.

In the next definition, we will give the properties that a structure must satisfy in order to be used as denotations space for the $\lambda\Delta$-calculus, or, equivalently, to be a *model* for this calculus.

Definition 10.0.1 ($\lambda\Delta$-Calculus model).
A $\lambda\Delta$-model *is a quadruple* $< \mathbb{D}, \mathbb{I}, \circ, [\![.]\!] >$, *where:*
$\mathbb{D}$ *is a set,* $\circ$ *is a map from* $\mathbb{D}^2$ *in* $\mathbb{D}$ *and* $\mathbb{I} \subseteq \mathbb{D}$. *Moreover, if* $\mathbb{E}$ *is the collection of functions* (environments) *from* Var *to* $\mathbb{I}$, *ranged over by* $\rho, \rho', ..$ *then the* interpretation function $[\![.]\!] : \Lambda \times \mathbb{E} \to \mathbb{D}$ *satisfies the following conditions:*

1. $[\![x]\!]_\rho = \rho(x)$,
2. $[\![MN]\!]_\rho = [\![M]\!]_\rho \circ [\![N]\!]_\rho$,
3. $[\![\lambda x.M]\!]_\rho \circ d = [\![M]\!]_{\rho[d/x]}$ *if* $d \in \mathbb{I}$,
4. *if* $[\![M]\!]_{\rho[d/x]} = [\![M']\!]_{\rho'[d/y]}$ *for each* $d \in \mathbb{I}$, *then* $[\![\lambda x.M]\!]_\rho = [\![\lambda y.M']\!]_{\rho'}$,
5. $M \in \Delta$ *implies* $\forall \rho. [\![M]\!]_\rho \in \mathbb{I}$,

where $\rho[d/x](y) =$ *if* $y \equiv x$ *then* d *else* $\rho(y)$.

This definition ensures that a $\lambda\Delta$-model respects some elementary key properties, namely the interpretation of a term depends only on the behaviour of the environment on the free variables of the term itself, the α-rule is respected, the syntactical substitution is modeled by the environment and the interpretation is contextually closed [59, 50]. Moreover, $\mathbb{I}$ is the semantical counterpart of the set of input values.

Property 10.0.2. Let $< \mathbb{D}, \mathbb{I}, \circ, [\![.]\!] >$ be a $\lambda\Delta$-model.

(i) If $\rho(x) = \rho'(x)$, for all $x \in \mathrm{FV}(M)$, then $[\![M]\!]_\rho = [\![M]\!]_{\rho'}$.
(ii) If $y \notin \mathrm{FV}(M)$ then $[\![M]\!]_{\rho[d/x]} = [\![M[y/x]]\!]_{\rho[d/y]}$, for all $d \in \mathbb{I}$.
(iii) If $y \notin \mathrm{FV}(M)$ then $[\![\lambda x.M]\!]_\rho = [\![\lambda y.M[y/x]]\!]_\rho$.
(iv) If $N \in \Delta$ then $[\![M[N/x]]\!]_\rho = [\![M]\!]_{\rho[[\![N]\!]_\rho/x]}$.
(v) If $[\![M]\!]_\rho = [\![N]\!]_\rho$ then, for every context $C[.]$, $[\![C[M]]\!]_\rho = [\![C[N]]\!]_\rho$.

Proof. (i) By induction on M. If $M \in \text{Var}$ then by Definition 10.0.1.1, the proof is immediate. If $M \equiv PQ$ then the proof follows by induction and Definition 10.0.1.2. If $M \equiv \lambda x.N$ then by induction, $\forall d \in \mathbb{I}$, $[\![N]\!]_{\rho[d/x]} = [\![N]\!]_{\rho'[d/x]}$; so $[\![\lambda x.N]\!]_\rho = [\![\lambda x.N]\!]_{\rho'}$ by Definition 10.0.1.4.

(ii) By induction on M. If $M \in \text{Var}$ then by Definition 10.0.1.1, the proof is immediate. If $M \equiv PQ$ then the proof follows by induction and Definition 10.0.1.2. If $M \equiv \lambda z.N$ then by Definition 10.0.1.3, $\forall d' \in \mathbb{I}$, $[\![N]\!]_{\rho[d/x][d'/z]} = [\![N[y/x]]\!]_{\rho[d/y][d'/z]}$ (clearly $\rho[d_0/x_0][d_1/x_1] = \rho[d_1/x_1][d_0/x_0]$); hence, $[\![\lambda z.N]\!]_{\rho[d/x]} = [\![(\lambda z.N)[y/x]]\!]_{\rho'[d/y]}$ by Definition 10.0.1.4 and the proof is done.

(iii) $\forall d \in \mathbb{I}$, $[\![M]\!]_{\rho[d/x]} = [\![M[y/x]]\!]_{\rho[d/y]}$ by the previous point of this property. The proof follows by Definition 10.0.1.4.

(iv) By induction on M. If $M \in \text{Var}$ then the proof is immediate. If $M \equiv PQ$ then the proof follows by induction.
If $M \equiv \lambda z.P$ then $\forall d \in \mathbb{I}$, $[\![P[N/x]]\!]_{\rho[d/z]} = [\![P]\!]_{\rho[d/z][[\![N]\!]_\rho/x]}$, by Definition 10.0.1.3. Hence, $[\![M[N/x]]\!]_\rho = [\![M]\!]_{\rho[[\![N]\!]_\rho/x]}$, by Definition 10.0.1.4.

(v) By induction on the context $C[.]$.
If $C[.]$ does not contains holes or $C[.] \equiv [.]$, then the proof is obvious.
If $C[.]$ is $C_1[.]C_2[.]$ then the proof follows immediately by induction.
Let $C[.]$ be $\lambda x.C'[.]$; thus, $\forall d \in \mathbb{I}$, $[\![C'[M]]\!]_{\rho[d/x]} = [\![C'[N]]\!]_{\rho[d/x]}$ by Definition 10.0.1.3. The proof follows by Definition 10.0.1.4. □

The previous property implies that condition 3 of Definition 10.0.1 is the semantics counterpart of the Δ-reduction rule. It says that the interpretation of a term is closed under $=_\Delta$, as proved in the following.

Corollary 10.0.3. *Let* $< \mathbb{D}, \mathbb{I}, \circ, [\![.]\!] >$ *be a* $\lambda\Delta$*-model.*
If $M =_\Delta N$ *then* $[\![M]\!]_\rho = [\![N]\!]_\rho$*, for all* ρ*.*

Proof. It is sufficient to prove that if $M \to_\Delta N$ then $[\![M]\!]_\rho = [\![N]\!]_\rho$, for all ρ. Let $Q \in \Delta$; so $[\![(\lambda z.P)Q]\!]_\rho = [\![\lambda z.P]\!]_\rho \circ [\![Q]\!]_\rho = [\![P]\!]_{\rho[[\![Q]\!]_\rho/z]} = [\![P[Q/z]]\!]_\rho$ by the definition of the model and by Property 10.0.2.(iv). The proof follows by Property 10.0.2.(v). □

Given a $\lambda\Delta$-model $\mathcal{M}$, the interpretation function $[\![.]\!]^{\mathcal{M}}$ induces a denotational semantics on Λ. Namely, two terms M and N are denotationally equivalent in $\mathcal{M}$ (and we write $M \sim_{\mathcal{M}} N$) if and only if:

$$[\![M]\!]^{\mathcal{M}}_\rho = [\![N]\!]^{\mathcal{M}}_\rho, \text{ for all environments } \rho.$$

Corollary 10.0.3 ensure us that $\sim_{\mathcal{M}}$ is a Δ-theory; moreover, it implies that if $M =_\Delta N \in \Delta$ then $\forall \rho.[\![M]\!]_\rho \in \mathbb{I}$.

The denotational semantics induced by a model $\mathcal{M}$ is *correct* with respect to an operational equivalence $\approx_{\mathbf{O}}$ if:

$$M \sim_{\mathcal{M}} N \text{ implies } M \approx_{\mathbf{O}} N, \text{ for all } M \text{ and } N;$$

while it is *complete* if:

$$M \approx_{\mathbf{O}} N \text{ implies } M \sim_{\mathcal{M}} N, \text{ for all } M \text{ and } N.$$

A model is called *fully abstract* [68] with respect to an operational equivalence if the induced denotational semantics is both correct and complete with respect to it.

As we will see in the rest of this section, if our aim is to study an operational equivalence then the correctness is the key point. The next lemma gives us a useful tool for testing the correctness of a model.

Lemma 10.0.4. *Let $\mathcal{M}$ be a $\lambda\Delta$-model such that:*

$$M \Downarrow_{\mathbf{O}} \text{ and } N \Uparrow_{\mathbf{O}} \text{ imply } M \not\sim_{\mathcal{M}} N, \text{ for all } M, N \in \Lambda.$$

Then $\mathcal{M}$ is correct with respect to the operational equivalence $\approx_{\mathbf{O}}$.

Proof. Let $M \sim_{\mathcal{M}} N$, so by Property 10.0.2.(v), for each context $C[.]$, $C[M] \sim_{\mathcal{M}} C[N]$; hence, by hypothesis, $C[M]$ and $C[N]$ either both $\Downarrow_{\mathbf{O}}$ or both $\Uparrow_{\mathbf{O}}$. Since this is true in particular when $C[M], C[N] \in \Lambda^0$, it follows that $M \approx_O N$. □

The simplest denotational model is the so-called *term model*. Let $\mid M \mid$ be the Δ-equivalence class of M, i.e. $\mid M \mid = \{N \mid N =_\Delta M\}$; let $\mid \Lambda \mid$ be the set all the equivalence classes of Λ with respect to $=_\Delta$, while $\mid \Delta \mid \subseteq \mid \Lambda \mid$ is the set of equivalence classes containing at least one input value. The term model $\mathcal{TM}(\Delta)$ is the quadruple $< \mid \Lambda \mid, \mid \Delta \mid, \circ, [\![.]\!]^{\mathcal{TM}(\Delta)} >$, where $\circ$ is defined as $\mid M \mid \circ \mid N \mid = \mid MN \mid$. The interpretation of a term M in $\mathcal{TM}(\Delta)$, with $\mathrm{FV}(M) = \{x_1, \ldots, x_m\}$, is given by $[\![M]\!]_\rho = \mid M[N_1/x_1]..[N_m/x_m] \mid$, where $N_i \in \rho(x_i)$ $(1 \leq i \leq m)$. It is easy to verify that $\mathcal{TM}(\Delta)$ satisfies the conditions of Definition 10.0.1.

Theorem 10.0.5. *Let $\mathbf{O}$ be an evaluation relation. If the $\lambda\Delta$-calculus is correct with respect to $\approx_{\mathbf{O}}$ then $\mathcal{TM}(\Delta)$ is correct for $\approx_{\mathbf{O}}$.*

Proof. Since the $\lambda\Delta$-calculus is correct with respect to $\approx_{\mathbf{O}}$, $=_\Delta$ implies $\approx_{\mathbf{O}}$. Since $\sim_{\mathcal{TM}(\Delta)}$ coincides with $=_\Delta$, the result follows. □

It is easy to check that $\mathcal{TM}(\Lambda)$ is not complete with respect to the operational semantics **H**, **N** and **L**; while $\mathcal{TM}(\Gamma)$ is not complete with respect to the operational semantics **V**. Just take two unsolvable terms of order 0, e.g. DD and $(\lambda x.xxx)(\lambda x.xxx)$. They are equated in all the operational semantics above, while they are different in both $\mathcal{TM}(\Lambda)$ and $\mathcal{TM}(\Gamma)$.

Remark 10.0.6. In case $\Delta = \Lambda$ and $\mathbb{D} = \mathbb{I}$, our definition of the $\lambda\Delta$-calculus model becomes the well-known definition of a λ-calculus model. But it looks different from the original one, given by Hindley and Longo in [50]. In fact, they ask the interpretation function to satisfy the following six conditions:

1. $[\![x]\!]_\rho = \rho(x)$;
2. $[\![MN]\!]_\rho = [\![M]\!]_\rho \circ [\![N]\!]_\rho$;
3. $[\![\lambda x.M]\!]_\rho \circ d = [\![M]\!]_{\rho[d/x]}$;
4. $\rho(x) = \rho'(x)$ for all $x \in \mathrm{FV}(M) \Rightarrow [\![M]\!]_\rho = [\![M]\!]_{\rho'}$;
5. $y \notin \mathrm{FV}(M) \Rightarrow [\![\lambda x.M]\!]_\rho = [\![\lambda y.M[y/x]]\!]_\rho$;
6. if $[\![M]\!]_{\rho[d/x]} = [\![M']\!]_{\rho[d/x]}$ for each $d \in \mathbb{I}$, then $[\![\lambda x.M]\!]_\rho = [\![\lambda x.M']\!]_\rho$.

Conditions from 1 to 3 occur identical in our definition. Condition 4 is more restrictive than the corresponding one in Definition 10.0.1, while conditions 5 and 6 ask that the interpretation be closed by α-equivalence and λ-abstraction, respectively. Our definition is shorter, and the strengthening of condition 4 allows one to obtain, as side effect, both the α-equality and the contextual equality (see Property 10.0.2).

It is an useful exercise for the reader to prove that the two definitions are equivalent.

10.1 Filter $\lambda\Delta$-Models

The idea of *filter $\lambda\Delta$-model* is based on the notions of *type* and of *type assignment system*. Types represent properties of terms, and they are expressed through the language of the implicative and conjunctive fragment of intuitionistic logic, i.e. the predicate logic with just two connectives, the implication ($\rightarrow$) and the conjunction ($\wedge$), and the constant *true* (ω). For historical reasons, the conjunction will be called the *intersection*. Intersection types were first introduced in [27].

Very informally, a term M has the property $\sigma \rightarrow \tau$ if its application to every term N having the property σ has the property τ, and M has the property $\sigma \wedge \tau$ if and only if it has both property σ and property τ. The constant ω represents the property of being a term, so it holds for all terms. A formal description of types is sketched in Sect. 13.1 where types are interpreted as compact elements of suitable domains.

A type assignment system is a set of rules assigning types to terms, starting from a basis, i.e. a function assigning types to variables. A type assignment system can induce a $\lambda\Delta$-model, where the interpretation of a term is given by the set of types that can be assigned to it.

This kind of model is particularly interesting, since the type assignment system is not only a support for defining the model itself, but it is a tool for reasoning, in a finitary way, on the interpretations of terms in it. So a filter model, if it is correct with respect to an operational semantics **O**, gives standard and powerful techniques for studying the **O**-operational behaviour of terms.

Definition 10.1.1 (The intersection type assignment system).

(i) *Let C be a non empty countable set of* type constants, *containing at least the constant ω (the universal type).*
The set $T(C)$ of types *is inductively defined as follows:*

$$\begin{array}{lcl} \sigma \in C & \Rightarrow & \sigma \in T(C), \\ \sigma, \tau \in T(C) & \Rightarrow & (\sigma \to \tau) \in T(C), \\ \sigma, \tau \in T(C) & \Rightarrow & (\sigma \wedge \tau) \in T(C). \end{array}$$

(ii) *An* intersection relation $\leq$ *is a preorder relation on $T(C)$, closed under the following rules:*

$$\frac{}{\sigma \leq \omega}\,(a) \qquad \frac{}{\sigma \leq \sigma \wedge \sigma}\,(b) \qquad \frac{}{\sigma \wedge \tau \leq \sigma}\,(c) \qquad \frac{}{\sigma \wedge \tau \leq \tau}\,(c')$$

$$\frac{}{(\sigma \to \tau) \wedge (\sigma \to \pi) \leq \sigma \to (\tau \wedge \pi)}\,(d) \qquad \frac{\sigma \leq \sigma',\ \tau \leq \tau'}{\sigma \wedge \tau \leq \sigma' \wedge \tau'}\,(e)$$

$$\frac{\sigma' \leq \sigma,\ \tau \leq \tau'}{\sigma \to \tau \leq \sigma' \to \tau'}\,(f) \qquad \frac{}{\sigma \to \omega \leq \omega \to \omega}\,(g) \qquad \frac{}{\sigma \leq \sigma}\,(r) \qquad \frac{\sigma \leq \rho,\ \rho \leq \tau}{\sigma \leq \tau}\,(t)$$

(iii) *Let $\leq$ be a intersection relation on $T(C)$.*
$\leq$ induce a type theory $\simeq$

$$\sigma \simeq \tau \textit{ if and only if } \sigma \leq \tau \textit{ and } \tau \leq \sigma.$$

(iv) *A* type system ∇ *is a triple $< C, \leq_\nabla, I(C) >$, where C is a set of type constants, $\leq_\nabla$ is an intersection relation on $T(C)$ and $I(C) \subseteq T(C)$ is a* set of input types *with respect to $\leq_\nabla$; namely, it is not empty and it is closed under the following conditions:*
1. *$\sigma \in I(C)$ and $\sigma \simeq_\nabla \tau$ imply $\tau \in I(C)$,*
2. *$\sigma \in I(C)$ and $\tau \notin I(C)$ imply $\sigma \leq_\nabla \tau$.*

(v) *Given a type system ∇, the corresponding* type assignment system $\vdash_\nabla$ *is a formal system proving statements of the shape:*

$$B \vdash_\nabla M : \sigma$$

where M is a term, $\sigma \in T(C)$ and B is a basis, i.e a function from Var *to $I(C)$.*
$B[\sigma/x]$ denotes the basis such that $B[\sigma/x](y) =$ if $y \equiv x$ then σ else $B(y)$.

The type assignment system consists of the following rules:

$$\frac{}{B[\sigma/x] \vdash_{\nabla} x : \sigma}\ (var) \qquad \frac{}{B \vdash_{\nabla} M : \omega}\ (\omega)$$

$$\frac{B[\sigma/x] \vdash_{\nabla} M : \tau}{B \vdash_{\nabla} \lambda x.M : \sigma \to \tau}\ (\to I) \qquad \frac{\sigma \in I(C) \quad B \vdash_{\nabla} M : \sigma \to \tau \quad B \vdash_{\nabla} N : \sigma}{B \vdash_{\nabla} MN : \tau}\ (\to E)$$

$$\frac{B \vdash_{\nabla} M : \sigma \quad B \vdash_{\nabla} M : \tau}{B \vdash_{\nabla} M : \sigma \wedge \tau}\ (\wedge I) \qquad \frac{B \vdash_{\nabla} M : \sigma \quad \sigma \leq_{\nabla} \tau}{B \vdash_{\nabla} M : \tau}\ (\leq_{\nabla})$$

$$\frac{B \vdash_{\nabla} M : \sigma \wedge \tau}{B \vdash_{\nabla} M : \sigma}\ (\wedge E_l) \qquad \frac{B \vdash_{\nabla} M : \sigma \wedge \tau}{B \vdash_{\nabla} M : \tau}\ (\wedge E_r)$$

Note that rules $(\wedge E_l)$ *and* $(\wedge E_r)$ *are redundant, since the rule* $(\leq_{\nabla})$.

In the definition of intersection relation, note the controvariance of $\leq_{\nabla}$ with respect to the left-hand argument of $\to$ in rule (f). Moreover, note that the rule $(\to E)$ imposes a condition on the type of the argument of application, namely that it belongs to the set of input types.

A *derivation* is a tree whose nodes are instances of rules, such that the premises of a rule are the consequences of its son nodes. If the root of a derivation d has as consequence the statement $B \vdash_{\nabla} M : \sigma$, we will say that d proves $B \vdash_{\nabla} M : \sigma$. The notion of *subderivation* then corresponds to the notion of subtree. We write $B \vdash_{\nabla} M : \sigma$ when *there is a derivation proving* $B \vdash_{\nabla} M : \sigma$; moreover, $B \vdash_{\nabla} M : \sigma$ is called a *typing*. If $B \vdash_{\nabla} M : \sigma$ and $B \vdash_{\nabla} N : \sigma$, then we will say that the terms M and N have a typing in common. $B \not\vdash_{\nabla} M : \sigma$ will denote that there are not derivations proving $B \vdash_{\nabla} M : \sigma$.

Example 10.1.2. Let $\mathfrak{U} =< \{\omega\}, \leq_{\mathfrak{U}}, I_{\mathfrak{U}} >$ where $I_{\mathfrak{U}} = T(\{\omega\})$ and $\leq_{\mathfrak{U}}$ is the least intersection relation such that $\omega \leq_{\mathfrak{U}} \sigma$, for all $\sigma \in T(\{\omega\})$.
It is easy to check that $\sigma \simeq_{\mathfrak{U}} \omega$, for all $\sigma \in T(\{\omega\})$; in particular, $\omega \simeq_{\mathfrak{U}} \omega \to \omega$. Let B_ω be the basis such that $B_\omega(x) = \omega$, for all $x \in \mathrm{Var}$; hence, the following derivation proves $B_\omega \vdash_{\mathfrak{U}} (\lambda x.xx)(\lambda x.xx) : \omega \to \omega$:

$$\frac{\dfrac{}{B_\omega \vdash_{\mathfrak{U}} DD : \omega}\ (\omega) \qquad \omega \leq_{\mathfrak{U}} \omega \to \omega}{B_\omega \vdash_{\mathfrak{U}} DD : \omega \to \omega}\ (\leq_{\mathfrak{U}})$$

It is easy to check that, for every set C of type-constants, it is correct to choose $I(C) = T(C)$.

Lemma 10.1.3. *Let* ∇ *be the type system* $< C, \leq_{\nabla}, I(C) >$.

(i) *If* $\pi \in I(C)$ *and* $\sigma \leq_{\nabla} \pi$ *then* $\sigma \in I(C)$.
(ii) *If* $\sigma \in I(C)$ *then* $\sigma \wedge \tau \in I(C)$, *for all* $\tau \in T(C)$.
(iii) *If* $\pi \notin I(C)$ *and* $\pi \leq_{\nabla} \sigma$ *then* $\sigma \notin I(C)$.

Proof. (i) Assume $\pi \in I(C)$ and $\sigma \leq_\nabla \pi$ and $\sigma \notin I(C)$; so $\pi \leq_\nabla \sigma$ by condition 2 on the set of input types. Hence $\sigma \simeq_\nabla \pi$; by condition 1 on the set of input types, this is absurd.

(ii) $\sigma \wedge \tau \leq_\nabla \sigma$, by rule (c) of intersection relations. The proof follows by the result proved in the previous point.

(iii) Similar to the proof of point (i). □

Note that if $I(C) \neq T(C)$ then $\omega \notin I(C)$, by the previous lemma.

In the next lemma some useful equivalences between types, true in all type theories, are proved.

Lemma 10.1.4. *Let* ∇ *be the type system* $< C, \leq_\nabla, I(C) >$.

(i) $\sigma \simeq_\nabla \sigma \wedge \sigma$;
(ii) $\omega \wedge \sigma \simeq_\nabla \sigma$;
(iii) $\sigma \rightarrow (\tau \wedge \pi) \simeq_\nabla (\sigma \rightarrow \tau) \wedge (\sigma \rightarrow \pi)$;
(iv) $\sigma \rightarrow \omega \simeq_\nabla \omega \rightarrow \omega$;
(v) $(\sigma \wedge \tau) \wedge \pi \simeq_\nabla \sigma \wedge (\tau \wedge \pi)$;
(vi) $\sigma \wedge \tau \simeq_\nabla \tau \wedge \sigma$;
(vii) $\pi \leq_\nabla \sigma$ *and* $\pi \leq_\nabla \tau$ *if and only if* $\pi \leq_\nabla \sigma \wedge \tau$.

Proof. (i) By rules (b) and (c) of the definition of intersection relations.

(ii) By the rule (c) of definition of intersection relations, $\omega \wedge \sigma \leq_\nabla \sigma$. On the other side, $\sigma \leq_\nabla \sigma \wedge \sigma \leq_\nabla \omega \wedge \sigma$, by rules (a), (b) and (e) and by the reflexivity of $\leq_\nabla$. The proof follows by (t).

(iii) Since rule (d), we need just to prove $\sigma \rightarrow (\tau \wedge \pi) \leq_\nabla (\sigma \rightarrow \tau) \wedge (\sigma \rightarrow \pi)$. $\sigma \rightarrow (\tau \wedge \pi) \leq_\nabla \sigma \rightarrow \tau$ and $\sigma \rightarrow (\tau \wedge \pi) \leq_\nabla \sigma \rightarrow \pi$ by rules (f), (c) and (c'); thus, $\sigma \rightarrow (\tau \wedge \pi) \leq_\nabla (\sigma \rightarrow (\tau \wedge \pi)) \wedge (\sigma \rightarrow (\tau \wedge \pi)) \leq_\nabla (\sigma \rightarrow \tau) \wedge (\sigma \rightarrow \pi)$ by rules (b) and (e). The proof follows by (t).

(iv) Since rule (g), we need just to prove $\omega \rightarrow \omega \leq_\nabla \sigma \rightarrow \omega$, which follows by rules (a) and (f), and by the reflexivity of $\leq_\nabla$.

(v) Let $\mu_0 \equiv (\sigma \wedge \tau) \wedge \pi$ and $\mu_1 \equiv \sigma \wedge (\tau \wedge \pi)$.
$\mu_0, \mu_1 \leq_\nabla \sigma, \tau, \pi$ by rules (c), (c') and (t), hence $\mu_0 \wedge (\mu_0 \wedge \mu_0) \leq_\nabla \mu_1$ by rule (e). Thus $\mu_0 \leq_\nabla \mu_0 \wedge (\mu_0 \wedge \mu_0)$ by rule (b) and then $\mu_0 \leq_\nabla \mu_1$ by rule (t). The reverse relation can be proved in a symmetric way.

(vi) Both $\sigma \wedge \tau \leq_\nabla \sigma, \tau$ and $\tau \wedge \sigma \leq_\nabla \sigma, \tau$ by rules (c) and (c'); hence $(\sigma \wedge \tau) \wedge (\sigma \wedge \tau) \leq_\nabla \tau \wedge \sigma$, and then $\sigma \wedge \tau \leq_\nabla \tau \wedge \sigma$ by rules (b) and (e). The reverse relation can be proved in a symmetric way.

(vii) By rules (b), (c), (c') and (e). □

In order to decrease the number of parenthesis in types, we will use the following precedence rules between connectives: $\wedge$ binds stronger than $\rightarrow$, moreover $\rightarrow$ associates to the right. For example, $\sigma \rightarrow \tau \wedge \rho$, $\sigma \rightarrow \tau \rightarrow \rho$ and $\sigma \wedge \tau \rightarrow \rho$ stand respectively for $\sigma \rightarrow (\tau \wedge \rho)$, $\sigma \rightarrow (\tau \rightarrow \rho)$ and $(\sigma \wedge \tau) \rightarrow \rho$.

Since the result of the previous lemma, at point (v), when no ambiguity can arise, we will use $\sigma \wedge \tau \wedge \rho$ for denoting both $\sigma \wedge (\tau \wedge \rho)$ and $\sigma \wedge (\tau \wedge \rho)$.

The notion of *legal type theory*, given in the next definition, is a key one, since we will prove that to be legal is a necessary condition for a type theory to induce a $\lambda\Delta$-model.

Definition 10.1.5. *Let ∇ be the type system $< C, \leq_\nabla, I(C) >$.*
∇ is legal *if and only if for all $\sigma \in I(C)$ and $\tau \not\simeq_\nabla \omega$:*

$$(\sigma_1 \to \tau_1) \wedge ... \wedge (\sigma_n \to \tau_n) \leq_\nabla \sigma \to \tau \ (1 \leq n) \text{ implies}$$
$$\exists\{i_1, ..., i_k\} \subseteq \{1, ..., n\} \text{ such that } (\sigma_{i_1} \wedge ... \wedge \sigma_{i_k}) \geq_\nabla \sigma$$
$$\text{and } (\tau_{i_1} \wedge ... \wedge \tau_{i_k}) \leq_\nabla \tau.$$

Let ∇ be a type system $< C, \leq_\nabla, I(C) >$ such that $I(C) = T(C)$ and $\leq_\nabla$ is the least inclusion relation: ∇ is legal.

In case of a legal type theory, rule (f) of the intersection relation defined in Definition 10.1.1 becomes a double implication. This will be useful in the following for proving properties of $\lambda\Delta$-models.

Property 10.1.6. Let $\nabla =< C, \leq_\nabla, I(C) >$ be legal.
If $\tau \not\simeq_\nabla \omega$ and $\sigma \in I(C)$ then

$$\sigma' \to \tau' \leq_\nabla \sigma \to \tau \text{ if and only if } \sigma \leq_\nabla \sigma' \text{ and } \tau' \leq_\nabla \tau.$$

Proof. By Definition 10.1.5 and by rule (f) of Definition 10.1.1.(ii). □

In order to show that a legal type theory induces a $\lambda\Delta$-model, first some syntactical properties of a type assignment system induced by a legal type theory must be proved.

Lemma 10.1.7. *Let $\nabla =< C, \leq_\nabla, I(C) >$ be legal.*

(i) *$B \vdash_\nabla M : \sigma$ and $x \notin \mathrm{FV}(M)$ imply $\forall\tau \in I(C)$, $B[\tau/x] \vdash_\nabla M : \sigma$.*
(ii) *$B \vdash_\nabla x : \sigma$ if and only if $B(x) \leq_\nabla \sigma$.*
(iii) *$B[\tau/x] \vdash_\nabla M : \sigma$ and $\pi \leq_\nabla \tau$ imply $B[\pi/x] \vdash_\nabla M : \sigma$.*
(iv) *If $B \vdash_\nabla \lambda x.M : \sigma$ then either $\sigma \simeq_\nabla \omega$ or $\sigma \geq_\nabla \pi_1 \wedge ... \wedge \pi_n$ $(n \geq 1)$, where $\pi_i \simeq_\nabla \mu_i \to \nu_i$, $\mu_i \in I(C)$ and $B[\mu_i/x] \vdash_\nabla M : \nu_i$ $(1 \leq i \leq n)$.*
(v) *If $B \vdash_\nabla MN : \sigma$ then either $\sigma \simeq_\nabla \omega$ or $\sigma \geq_\nabla \pi_1 \wedge ... \wedge \pi_n$ $(n \geq 1)$ such that $B \vdash_\nabla M : \tau_i \to \pi_i$ and $B \vdash_\nabla N : \tau_i$, for some $\tau_1, ..., \tau_n \in I(C)$ $(1 \leq i \leq n)$.*
(vi) *Let $\tau \not\simeq_\nabla \omega$ and $\sigma \in I(C)$.*
$B \vdash_\nabla \lambda x.M : \sigma \to \tau$ if and only if $B[\sigma/x] \vdash_\nabla M : \tau$.
(vii) *Let $\sigma \not\simeq_\nabla \omega$. $B \vdash_\nabla MN : \sigma$ if and only if $B \vdash_\nabla M : \tau \to \sigma$ and $B \vdash_\nabla N : \tau$, for some $\tau \in I(C)$.*

Proof. (i) Immediate, from the definition of the type assignment system.
(ii) $(\Leftarrow)$ By rules (var) and $(\leq_\nabla)$.
$(\Rightarrow)$ By induction on the derivation. Note that $(\to I)$ and $(\to E)$ cannot be used.
(iii) By induction on the derivation. Note that $\pi \in I(C)$ by Lemma 10.1.3.(i).

(iv) By induction on the derivation proving $B \vdash_\nabla \lambda x.M : \sigma$. If the last applied rule is either (ω) or $(\rightarrow I)$, then the proof is trivial. In case the last applied rule is $(\leq_\nabla)$, the proof follows from the inductive hypothesis and the transitivity of $\leq_\nabla$. In case the last applied rule is $(\wedge I)$, the proof follows from the inductive hypothesis. The case of rule $(\rightarrow E)$ is not possible.

(v) By induction on the derivation proving $B \vdash_\nabla MN : \sigma$. If the last applied rule is (ω) or $(\rightarrow E)$, then the proof is trivial. In case the last applied rule is $(\leq_\nabla)$, the proof follows from the inductive hypothesis and the transitivity of $\leq_\nabla$. In case the last applied rule is $(\wedge I)$, the proof follows from the inductive hypothesis. $(\rightarrow I)$ cannot be the last applied rule.

(vi) $(\Leftarrow)$ By rule $(\rightarrow I)$.
$(\Rightarrow)$ $B \vdash_\nabla \lambda x.M : \sigma \rightarrow \tau$ implies $(\mu_1 \rightarrow \nu_1) \wedge ... \wedge (\mu_m \rightarrow \nu_m) \leq_\nabla \sigma \rightarrow \tau$ for some $m \geq 1$, where $\mu_i \in I(C)$ and $B[\mu_i/x] \vdash_\nabla M : \nu_i$ $(1 \leq i \leq m)$, by point (iv). Since the type theory is legal, $\exists\{i_1, ..., i_k\} \subseteq \{1, ..., m\}$ such that $\mu_{i_1} \wedge ... \wedge \mu_{i_k} \geq_\nabla \sigma$ and $\nu_{i_1} \wedge ... \wedge \nu_{i_k} \leq_\nabla \tau$. By Lemma 10.1.3.(ii) $\mu_{i_1} \wedge ... \wedge \mu_{i_k} \in I(C)$, so $B[\mu_{i_j}/x] \vdash_\nabla M : \nu_{i_j}$ $(1 \leq j \leq k)$ imply $B[\mu_{i_1} \wedge ... \wedge \mu_{i_k}/x] \vdash_\nabla M : \nu_{i_1} \wedge ... \wedge \nu_{i_k}$, by rule $(\wedge I)$ and by point (iii) of this lemma. Again by point (iii) of this lemma, $B[\sigma/x] \vdash_\nabla M : \nu_{i_1} \wedge ... \wedge \nu_{i_k}$ and by rule $(\leq_\nabla)$ we can conclude $B[\sigma/x] \vdash_\nabla M : \tau$.

(vii) $(\Leftarrow)$ By rule $(\rightarrow E)$. Note that $\tau \in I(C)$ is a necessary hypothesis.
$(\Rightarrow)$ By point (v), if $B \vdash_\nabla MN : \sigma$ and $\sigma \not\leq_\nabla \omega$ then $\sigma \geq_\nabla \mu_1 \wedge ... \wedge \mu_m$ for some $m \geq 1$; moreover, $B \vdash_\nabla M : \sigma_i \rightarrow \mu_i$ and $B \vdash_\nabla N : \sigma_i$, for some $\sigma_i \in I(C)$ $(1 \leq i \leq m)$. Hence $\sigma_1 \wedge ... \wedge \sigma_m \in I(C)$ by Lemma 10.1.3.(ii). By rule $(\wedge I)$, $B \vdash_\nabla M : (\sigma_1 \rightarrow \mu_1) \wedge ... \wedge (\sigma_m \rightarrow \mu_m)$. Note that

$$\begin{aligned}(\sigma_1 \rightarrow \mu_1) \wedge ... \wedge (\sigma_m \rightarrow \mu_m) &\leq_\nabla (\sigma_1 \wedge ... \wedge \sigma_m \rightarrow \mu_1) \wedge ... \wedge (\sigma_1 \wedge ... \wedge \sigma_m \rightarrow \mu_m)\\ &\leq_\nabla \sigma_1 \wedge ... \wedge \sigma_m \rightarrow \mu_1 \wedge ... \wedge \mu_m\end{aligned}$$

since $\sigma_1 \wedge ... \wedge \sigma_m \leq_\nabla \sigma_i$ $(1 \leq i \leq m)$ and by Lemma 10.1.4.(iii).
So $B \vdash_\nabla M : \sigma_1 \wedge ... \wedge \sigma_m \rightarrow \mu_1 \wedge ... \wedge \mu_m$ by rule $(\leq_\nabla)$; moreover, $B \vdash_\nabla N : \sigma_1 \wedge ... \wedge \sigma_m$, by rule $(\wedge I)$.
Since $B \vdash_\nabla M : (\sigma_1 \wedge ... \wedge \sigma_m) \rightarrow \sigma$ by rule $(\leq_\nabla)$, the proof is done. □

Note that the notion of legality is essential in the proof of point (vi) of Lemma 10.1.7.

Remark 10.1.8. By Lemma 10.1.7.(i), it follows that the derivation of a type for a closed term is independent from the basis, i.e. if M is a closed term then $B \vdash_\nabla M : \sigma$ implies that $B' \vdash_\nabla M : \sigma$, for all B'.

Now we are ready to introduce the basic ingredients for defining a filter model.

Definition 10.1.9. *Let ∇ be the type system $< C, \leq_\nabla, I(C) >$.*

(i) *A filter f on ∇ is any set containing ω and closed under $\wedge$ and $\leq_\nabla$, namely:*

- *$\mu, \nu \in f$ implies $\mu \wedge \nu \in f$;*
- *$\mu \in f$ and $\mu \leq_\nabla \tau$ imply $\tau \in f$.*

Let $\mathcal{F}(\nabla)$ be the set of all filters on ∇, and let $\mathcal{I}(\nabla)$ be the set of filters containing at least one type belonging to $I(C)$.

(ii) *Let S be a set of types; $\uparrow S$ is the filter obtained from S by closing it under $\wedge$ and $\leq_\nabla$, i.e the least filter containing S.*

(iii) *Let $\circ_\nabla$ be the binary operation defined on $\mathcal{F}(\nabla)$ in the following way:*

$$f_1 \circ_\nabla f_2 = \uparrow \{\omega\} \cup \{\tau \mid \sigma \rightarrow \tau \in f_1 \text{ and } \sigma \in f_2 \text{ and } \sigma \in I(C)\}.$$

Note that $f \in \mathcal{I}(\nabla)$ and $\sigma \notin I(C)$ imply $\sigma \in f$ by the conditions on the set of input types.

The following lemma shows that the definition of $\circ_\nabla$ is correct.

Lemma 10.1.10. *Let $\nabla =< C, \leq_\nabla, I(C) >$.*
If $f_1, f_2 \in \mathcal{F}(\nabla)$ then $f_1 \circ_\nabla f_2 \in \mathcal{F}(\nabla)$.

Proof. Clearly $\omega \in \mathcal{F}(\nabla)$.
Let us prove that $f_1 \circ_\nabla f_2$ is closed under intersection.
Let $\alpha, \beta \in f_1 \circ_\nabla f_2$; so, there are $\sigma, \tau \in f_2 \cap I(C)$ such that $\sigma \rightarrow \alpha, \tau \rightarrow \beta \in f_1$. Both f_1 and f_2 are filters, thus $\sigma \wedge \tau \in f_2 \cap I(C)$ and $\sigma \wedge \tau \rightarrow \alpha, \sigma \wedge \tau \rightarrow \beta \in f_1$ (since $\sigma \rightarrow \alpha \leq_\nabla \sigma \wedge \tau \rightarrow \alpha$ and $\tau \rightarrow \beta \leq_\nabla \sigma \wedge \tau \rightarrow \beta$). Hence $(\sigma \wedge \tau \rightarrow \alpha) \wedge (\sigma \wedge \tau \rightarrow \beta) \in f_1$ and $\sigma \wedge \tau \rightarrow \alpha \wedge \beta \in f_1$, by Lemma 10.1.4.(iii). Thus $\sigma \wedge \tau \in f_2 \cap I(C)$ implies $\alpha \wedge \beta \in f_1 \circ_\nabla f_2$, by definition of $\circ_\nabla$.
Now let us prove that $\alpha \in f_1 \circ_\nabla f_2$ implies $\tau \in f_1 \circ_\nabla f_2$, for every $\tau \geq_\nabla \alpha$. In fact, $\alpha \in f_1 \circ_\nabla f_2$ implies that there is $\sigma \in f_2 \cap I(C)$ such that $\sigma \rightarrow \alpha \in f_1$. $\sigma \rightarrow \alpha \leq_\nabla \sigma \rightarrow \tau$ and f_1 is a filter imply that $\sigma \rightarrow \tau \in f_1$, thus $\tau \in f_1 \circ_\nabla f_2$. □

The interpretation function will associate to every term all the types that can be assigned to it.

Let $\nabla =< C, \leq_\nabla, I(C) >$.
$[\![.]\!]^{\mathcal{F}(\nabla)} : \Lambda \times (\text{Var} \rightarrow \mathcal{I}(\nabla)) \rightarrow \mathcal{F}(\nabla)$ is the *interpretation function*, defined as follows:

$$[\![M]\!]_\rho^{\mathcal{F}(\nabla)} = \{\sigma \in T(C) \mid \exists B \propto \rho \text{ such that } B \vdash_\nabla M : \sigma\},$$

where $B \propto \rho$ means that $\forall z \in \text{Var} \;\; B(z) \in \rho(z)$, and it can be read as "B agree with ρ".

Remember that a basis B is a function from Var to $I(C)$.

Theorem 10.1.11 (Filter $\lambda\Delta$-models).
Let $\nabla =< C, \leq_\nabla, I(C) >$ be a legal type system, and let $M \in \Delta$ implies $[\![M]\!]_\rho \in \mathcal{I}(\nabla)$, for all environments ρ.
Then $< \mathcal{F}(\nabla), \mathcal{I}(\nabla), \circ_\nabla, [\![.]\!]^{\mathcal{F}(\nabla)} >$ is a $\lambda\Delta$-model.

Proof. It is easy to see that $[\![M]\!]_\rho^{\mathcal{F}(\nabla)}$ is a filter, for all terms M. The proof is carried out by verifying the conditions of Definition 10.0.1.

1. We will prove $\rho(x) = \{\sigma \in T(C) \mid \exists B \propto \rho \text{ such that } B \vdash_\nabla x : \sigma\}$.
   Let $\sigma \in I(C)$; so, $\sigma \in \rho(x)$ if and only if $\exists B \propto \rho$ such that $B(x) = \sigma$, which means $B \vdash_\nabla x : \sigma$ by rule (var). Note that $\rho(x) \in \mathcal{I}(\nabla)$ implies that there is $\mu \in I(C)$ such that $\mu \in \rho(x)$; if $\sigma \notin I(C)$ then $\sigma \in \rho(x)$ since $\mu \leq_\nabla \sigma$ by conditions on the input types, hence $B \vdash_\nabla x : \mu$ implies $B \vdash_\nabla x : \sigma$ by rule $(\leq_\nabla)$.
2. Let $B_0 \cup B_1$ be the basis such that $(B_0 \cup B_1)(x) = B_0(x) \wedge B_1(x)$.
   By using Lemmas 10.1.7.(iii) and 10.1.7.(vii):

$$[\![MN]\!]_\rho^{\mathcal{F}(\nabla)} = \{\sigma \in T(C) \mid \exists B \propto \rho \text{ such that } B \vdash_\nabla MN : \sigma\} =$$

$$\uparrow\{\omega\} \cup \left\{\sigma \not\leq_\nabla \omega \;\middle|\; \begin{array}{l}\exists B \propto \rho \\ \exists \tau \in I(C)\end{array} \text{ such that both } \begin{array}{l} B \vdash_\nabla N : \tau \quad \text{and} \\ B \vdash_\nabla M : \tau \to \sigma \end{array}\right\} =$$

$$\uparrow\{\omega\} \cup \left\{\sigma \not\leq_\nabla \omega \;\middle|\; \begin{array}{l}\exists B_0, B_1 \propto \rho \\ \exists \tau \in I(C)\end{array} \text{ such that } \begin{array}{l} B_0 \cup B_1 \propto \rho \quad \text{and} \\ B_0 \vdash_\nabla N : \tau \quad \text{and} \\ B_1 \vdash_\nabla M : \tau \to \sigma \end{array}\right\} =$$

$$\uparrow\{\omega\} \cup \left\{\sigma \not\leq_\nabla \omega \;\middle|\; \begin{array}{l}\exists B_0, B_1 \propto \rho \\ \exists \tau \in I(C)\end{array} \text{ such that } \begin{array}{l} B_0 \vdash_\nabla N : \tau \quad \text{and} \\ B_1 \vdash_\nabla M : \tau \to \sigma \end{array}\right\} =$$

$$\uparrow\{\omega\} \cup \left\{\sigma \not\leq_\nabla \omega \mid \exists \tau \in I(C) \text{ such that } \begin{array}{l} \tau \in [\![N]\!]_\rho^{\mathcal{F}(\nabla)} \quad \text{and} \\ \tau \to \sigma \in [\![M]\!]_\rho^{\mathcal{F}(\nabla)} \end{array}\right\} =$$

$$[\![M]\!]_\rho^{\mathcal{F}(\nabla)} \circ_\nabla [\![N]\!]_\rho^{\mathcal{F}(\nabla)}$$

3. Let $f \in \mathcal{I}(\nabla)$; so by Lemma 10.1.7.(vi)

$$\begin{array}{l}
[\![\lambda x.M]\!]_\rho^{\mathcal{F}(\nabla)} \circ_\nabla f = \\
\uparrow\{\omega\} \cup \{\tau \not\leq_\nabla \omega \mid \sigma \in f \text{ and } \sigma \in I(C) \text{ and } \sigma \to \tau \in [\![\lambda x.M]\!]_\rho^{\mathcal{F}(\nabla)}\} = \\
\uparrow\{\omega\} \cup \{\tau \not\leq_\nabla \omega \mid \sigma \in f \cap I(C) \text{ and } \exists B \propto \rho \text{ such that } B \vdash_\nabla \lambda x.M : \sigma \to \tau\} = \\
\uparrow\{\omega\} \cup \{\tau \not\leq_\nabla \omega \mid \exists B \propto \rho \text{ such that } B[\sigma/x] \vdash_\nabla M : \tau \text{ and } \sigma \in I(C) \text{ and } \sigma \in f\}
\end{array}$$

   Thus the proof is done, since

$$[\![M]\!]_{\rho[f/x]}^{\mathcal{F}(\nabla)} = \{\tau \in T(C) \mid \exists B' \propto \rho[f/x] \text{ such that } B' \vdash_\nabla M : \tau\}.$$

4. We assume that $[\![M]\!]_{\rho[d/x]}^{\mathcal{F}(\nabla)} = [\![M']\!]_{\rho'[d/y]}^{\mathcal{F}(\nabla)}$, for all $d \in \mathbb{I}$; namely $\forall d \in \mathbb{I}$

$$\{\sigma \mid \exists B \propto \rho[d/x] \text{ s.t. } B \vdash_\nabla M : \sigma\} = \{\sigma \mid \exists B' \propto \rho'[d/y] \text{ s.t. } B' \vdash_\nabla M' : \sigma\},$$

   in particular, for all $\tau \in I(C)$

$$\{\sigma \mid \exists B \propto \rho[\uparrow \tau/x] \text{ s.t. } B \vdash_\nabla M : \sigma\} = \{\sigma \mid \exists B' \propto \rho'[\uparrow \tau/y] \text{ s.t. } B' \vdash_\nabla M' : \sigma\}.$$

   Hence, it is easy to see that for all $\tau \in I(C)$:

$\exists B \propto \rho$ s.t. $B[\tau/x] \vdash_\nabla M : \sigma$ if and only if $\exists B' \propto \rho'$ s.t. $B'[\tau/y] \vdash_\nabla M' : \sigma$.

Let us assume $\pi \in [\![\lambda x.M]\!]_\rho^{\mathcal{F}(\nabla)}$, so $\exists B \propto \rho$ such that $B \vdash_\nabla \lambda x.M : \pi$ by the definition of the interpretation function. By Lemma 10.1.7.(iv), $\pi \geq_\nabla (\mu_1 \to \nu_1) \wedge ... \wedge (\mu_m \to \nu_m)$ for some $m \geq 1$, where $\mu_i \in I(C)$ and $B[\mu_i/x] \vdash_\nabla M : \nu_i$ $(1 \leq i \leq m)$. But $B[\mu_i/x] \vdash_\nabla M : \nu_i$ if and only if $\exists B' \propto \rho'$ such that $B'[\mu_i/y] \vdash_\nabla M' : \nu_i$. So by rules $(\to I)$, $(\wedge I)$ and $(\leq_\nabla)$ $\exists B' \propto \rho'$ such that $B' \vdash_\nabla \lambda y.M' : \pi$. Hence, $[\![\lambda x.M]\!]_\rho^{\mathcal{F}(\nabla)} = [\![\lambda y.M']\!]_{\rho'}^{\mathcal{F}(\nabla)}$ by the definition of the interpretation function.
5. By hypothesis. □

Now we can define the notion of filter $\lambda\Delta$-model (or briefly filter model, when Δ is either not instantiated or clear from the context).

Definition 10.1.12. (i) *A* filter $\lambda\Delta$-model *is any quadruple*

$$< \mathcal{F}(\nabla), \mathcal{I}(\nabla), \circ_\nabla, [\![.]\!]^{\mathcal{F}(\nabla)} >$$

such that $\nabla =< C, \leq_\nabla, I(C) >$ *is a legal type system, and* $M \in \Delta$ *implies* $[\![M]\!]_\rho \in \mathcal{I}(\nabla)$ *for all environments* ρ.

(ii) *The partial order between terms induced by a filter* $\lambda\Delta$*-model* $\mathcal{F}$ *is defined as follows:*

$$M \sqsubseteq_\mathcal{F} N \quad \textit{if and only if} \quad \forall \rho,\ [\![M]\!]_\rho^\mathcal{F} \subseteq [\![N]\!]_\rho^\mathcal{F}\ ,$$

i.e.

$$\{\sigma | \exists B \propto \rho \textit{ such that } B \vdash_\nabla M : \sigma\} \subseteq \{\sigma | \exists B \propto \rho \textit{ such that } B \vdash_\nabla N : \sigma\}.$$

$M \sqsubset_\mathcal{F} N$ *will denote the proper inclusion.*

The two properties proved in the next lemma are quite useful for proving operational properties of terms.

Lemma 10.1.13. *Let* $\mathcal{F}$ *be a* $\lambda\Delta$*-model.*

(i) *If* $M \sqsubseteq_\mathcal{F} N$ *then* $C[M] \sqsubseteq_\mathcal{F} C[N]$*, for all context* $C[.]$.
(ii) *If* $M \sqsubseteq_\mathcal{F} N$ *and* $B \vdash_\nabla M : \sigma$ *then* $B \vdash_\nabla N : \sigma$*, for all bases* B.

Proof. (i) By induction on the context $C[.]$, similarly to the proof of Property 10.0.2.(v).
(ii) $B \vdash_\nabla M : \sigma$ implies $\sigma \in [\![M]\!]_{\rho_B}^\mathcal{F}$, where $\rho_B(x) = \uparrow B(x)$, for each $x \in$ Var. But $M \sqsubseteq_\mathcal{F} N$ implies $\sigma \in [\![N]\!]_{\rho_B}^\mathcal{F}$, so $\exists B' \propto \rho_B$ such that $B' \vdash_\nabla N : \sigma$. It is easy to see that $B(x) \leq_\nabla \tau \in \rho_B(x)$, so $B(x) \leq_\nabla B'(x)$, for each $x \in$ Var and, by Lemma 10.1.7.(iii), $B \vdash_\nabla N : \sigma$. □

We can refine both the notion of correctness and completeness of a model with respect to a given operational semantics by taking into account the preorder relation instead of the equivalence one.

Definition 10.1.14. *Let $\mathcal{F}$ be a filter model. $\mathcal{F}$ is* correct *with respect to the* **O***-operational semantics if and only if $M \sqsubseteq_{\mathcal{F}} N$ implies $M \preceq_{\mathbf{O}} N$, for all $M, N \in \Lambda$. $\mathcal{F}$ is* complete *with respect to the* **O***-operational semantics if and only if the inverse implication holds.*
Moreover, $\mathcal{F}$ is fully abstract *with respect to* **O***, in case it is both correct and complete with respect to* **O**.

The next property will be very useful in what follows in order to prove the correctness of some filter models.

Property 10.1.15. Let $\mathcal{F}$ be a filter model such that

$$M \Downarrow_{\mathbf{O}} \text{ and } N \Uparrow_{\mathbf{O}} \text{ implies } N \sqsubset_{\mathcal{F}} M, \text{ for all } M, N \in \Lambda.$$

Then $\mathcal{F}$ is correct with respect to **O**.

Proof. We will prove that the given hypothesis implies the following implication: $Q \not\preceq_{\mathbf{O}} P$ implies $Q \not\sqsubseteq_{\mathcal{F}} P$. If $Q \not\preceq_{\mathbf{O}} P$ then there is a context $C[.]$ such that $C[Q] \Downarrow_{\mathbf{O}}$ while $C[P] \Uparrow_{\mathbf{O}}$, which implies, by hypothesis, $C[P] \sqsubset_{\mathcal{F}} C[Q]$, which in turn implies $C[Q] \not\sqsubseteq_{\mathcal{F}} C[P]$. So $Q \not\sqsubseteq_{\mathcal{F}} P$ by Property 10.1.13.(i). □

The first filter model for the $\lambda\Lambda$-calculus was built in [10]. A presentation of a class of filter models, which includes models of both $\lambda\Lambda$-calculus and $\lambda\Gamma$-calculus, can be found in [38]; it is less general the that one given in this book, and it is based on the notion of partial intersection type assignment system.

11. Call-by-Name Denotational Semantics

To model a $\lambda\Lambda$-theory, it is necessary to reflect in the model the fact that the set of input values coincides with the whole set Λ. So it is natural to ask that in every model the set $\mathbb{D}$ and the set of semantic input values $\mathbb{I}$ must coincide. In a filter model, it must be $\mathcal{F}(C,\nabla) = \mathcal{I}(C,\nabla)$, so $T(C) = I(C)$.

The characterization of a model of $\lambda\Lambda$-calculus follows quite easily, remembering the notion of a legal type system.

Theorem 11.0.1. *Let* $\nabla =< C, \leq_{\nabla}, T(C) >$.
If ∇ *is legal then* $< \mathcal{F}(\nabla), \mathcal{F}(\nabla), \circ_{\nabla}, [\![.]\!]^{\mathcal{F}(\nabla)} >$ *is a* $\lambda\Lambda$*-model.*

Proof. The proof follows from Theorem 10.1.11, since $\Delta = \Lambda$. □

11.1 The Model $\mathcal{H}$

In this section, we will introduce a filter model that is fully abstract with respect to the **H**-operational semantics. By keeping in mind Property 10.1.15 and by observing that if N is Λ-solvable and M Λ-unsolvable then $M \prec_{\mathbf{H}} N$, in order to define a model that is correct with respect to the **H**-operational semantics, it is sufficient to ask for the property:

$$\forall M, N \text{ if } M \text{ is } \Lambda\text{-solvable and } N \text{ is } \Lambda\text{-unsolvable then } N \sqsubset_{\mathcal{H}} M.$$

So we must define a legal type system ∇ based on a set of constants C such that, for every Λ-solvable term M and Λ-unsolvable term N, there is a basis B and at least one type σ such that $B \vdash_{\nabla} M : \sigma$ while not $B \vdash_{\nabla} N : \sigma$. It is not possible to find a type σ having this property, because solvable terms are not closed under application (as we already observed at the end of Sect. 1.2), while a type must have a uniform functional behaviour.

In order to characterize the operational semantics **H**, we will introduce the class of openly solvable terms, which is a proper subclass of Λ-solvable terms closed under application. We will prove that it is possible to characterize this class in the sense as before, and moreover the same model characterizes, in a weaker sense, the class of all Λ-solvable terms.

Definition 11.1.1. *A term M is openly Λ-solvable if and only if $M \to_\Lambda^* \lambda x_1...x_n.zM_1...M_m$ where $z \in \mathrm{FV}(M)$.*

By abuse of notation, we will speak about the head variable of a term for denoting the head variable of its Λ-hnf.

Example 11.1.2. $I(Ix)$ and $I(\lambda y.xx)$ are openly Λ-solvable, while $I(\lambda x.xy)$ is Λ-solvable but non-openly Λ-solvable.

It is immediate to verify that the property of being openly Λ-solvable is closed under application; namely if M is openly solvable then $M\vec{N}$ is openly solvable too, for all $\vec{N}$. This behaviour suggests that a type σ characterizing the openly solvable terms must satisfy the equation $\sigma \simeq \omega \to \sigma$ (remember that the constant ω can be assigned to every term).

Definition 11.1.3. *Let $C_\infty = \{\phi, \omega\}$ and $I(C_\infty) = T(C_\infty)$.*
∞ is the type system $< C_\infty, \leq_\infty, I(C_\infty) >$, where $\leq_\infty$ is the least intersection relation induced by the rules in Fig. 11.1.

$$\frac{}{\sigma \leq_\infty \omega}\ (a) \qquad \frac{}{\sigma \leq_\infty \sigma \wedge \sigma}\ (b) \qquad \frac{}{\sigma \wedge \tau \leq_\infty \sigma}\ (c) \qquad \frac{}{\sigma \wedge \tau \leq_\infty \tau}\ (c')$$

$$\frac{}{(\sigma \to \tau) \wedge (\sigma \to \pi) \leq_\infty \sigma \to (\tau \wedge \pi)}\ (d) \qquad \frac{}{\sigma \to \omega \leq_\infty \omega \to \omega}\ (g)$$

$$\frac{\sigma \leq_\infty \sigma', \tau \leq_\infty \tau'}{\sigma \wedge \tau \leq_\infty \sigma' \wedge \tau'}\ (e) \qquad \frac{\sigma' \leq_\infty \sigma, \tau \leq_\infty \tau'}{\sigma \to \tau \leq_\infty \sigma' \to \tau'}\ (f) \qquad \frac{}{\sigma \leq_\infty \sigma}\ (r) \qquad \frac{\sigma \leq_\infty \rho \quad \rho \leq_\infty \tau}{\sigma \leq_\infty \tau}\ (t)$$

$$\frac{}{\phi \leq_\infty \omega \to \phi}\ (h1) \qquad \frac{}{\omega \to \phi \leq_\infty \phi}\ (h2) \qquad \frac{}{\omega \leq_\infty \omega \to \omega}\ (h3)$$

Fig. 11.1. ∞-intersection relation

Example 11.1.4. It is easy to check that $\phi \simeq_\infty \omega \to \phi$, $\omega \simeq_\infty \omega \to \omega$ and $\sigma \to \omega \simeq_\infty \omega$, for all σ. Moreover $B \vdash_\infty D : \phi \to \phi$, for every basis B. In fact, we can build the following derivation:

$$\dfrac{\dfrac{\dfrac{\dfrac{}{B[\phi/x] \vdash_\infty x : \phi}\ (var)}{B[\phi/x] \vdash_\infty x : \omega \to \phi}\ (\leq_\infty) \qquad \dfrac{}{B[\phi/x] \vdash_\infty x : \omega}\ (\omega)}{B[\phi/x] \vdash_\infty xx : \phi}\ (\to E)}{B \vdash_\infty D : \phi \to \phi}\ (\to I)$$

In Sect. 11.1.1, the intersection relation $\leq_\infty$ is extensively studied.

Theorem 11.1.5. *The type system ∞ is legal.*

Proof. The proof is in Sect. 11.1.1. □

Hence the following definition is well posed, by Theorem 11.0.1.

Definition 11.1.6. *$\mathcal{H}$ is the $\lambda\Lambda$-model:* $< \mathcal{F}(\infty), \mathcal{F}(\infty), \circ_\infty, [\![\cdot]\!]^{\mathcal{F}(\infty)} >$.

We will prove that $\mathcal{H}$ is fully abstract with respect to **H**.

Property 11.1.7.

(i) Let M be openly Λ-solvable and let z be its head variable. If B is the basis such that $B(z) = \phi$, then $B \vdash_\infty M : \phi$.

(ii) Let $M \equiv \lambda x_1...x_n.zM_1...M_m$ $(m, n \in \mathbb{N})$. If $B \vdash_\infty M : \phi$ then M is openly solvable.

(iii) If $M =_\Lambda \lambda x_1 \ldots x_n.x_k M_1 \ldots M_m$ $(1 \leq k \leq n)$, then for all bases B:

- $B \vdash_\infty M : \underbrace{\phi \to ... \to \phi}_{p}$, for every p such that $k+1 \leq p \leq n+1$;
- $B \not\vdash_\infty M : \underbrace{\phi \to ... \to \phi}_{p}$, for any $p \leq k$.

Proof. (i) Let $M =_\Lambda \lambda x_1...x_n.zM_1...M_m$, where $z \in FV(M)$. Clearly

$$B[\omega/x_1, ..., \omega/x_n] \vdash_\infty z : \phi \text{ and } B[\omega/x_1, ..., \omega/x_n] \vdash_\infty z : \underbrace{\omega \to ... \to \omega}_{m} \to \phi,$$

by rule $(\leq_\infty)$. Since $B[\omega/x_1, ..., \omega/x_n] \vdash_\infty M_i : \omega$ $(1 \leq i \leq m)$ by rule (ω), by m applications of rule $(\to E)$ we can build a derivation proving $B[\omega/x_1, ..., \omega/x_n] \vdash_\infty zM_1...M_m : \phi$. Hence, by n applications of rule $(\to I)$, we have $B \vdash_\infty \lambda x_1...x_n.zM_1...M_m : \underbrace{\omega \to ... \to \omega}_{n} \to \phi$, which implies $B \vdash_\infty \lambda x_1...x_n.zM_1...M_m : \phi$, by rule $(\leq_\infty)$. Thus $B \vdash_\infty M : \phi$, by Corollary 10.0.3.

(ii) By induction on n. If $n = 0$ then $M \equiv zM_1...M_m$ and the proof is trivial; so let $n \geq 1$ and $B \vdash_\infty \lambda x_1...x_n.zM_1...M_m : \phi$.
So $B \vdash_\infty M : \omega \to \phi$ by rule $(\leq_\infty)$; so $B[\omega/x_1] \vdash_\infty \lambda x_2...x_n.zM_1...M_m : \phi$, by Lemma 10.1.7.(vi). By inductive hypothesis $z \notin \{x_2, ..., x_n\}$. Moreover, $B(z) \not\leq_\infty \omega$; otherwise it would be $\phi \simeq_\infty \omega$ by Lemma 10.1.7.(ii), against the fact that $\phi \not\leq_\infty \omega$, proved in Sect. 11.1.1. Then $z \not\equiv x_1$ since $B(x_1) = \omega$, so M is openly solvable.

(iii) Let $M =_\Lambda \lambda x_1...x_n.x_k M_1...M_m$, and let $p \geq k$. As for the point (i), we can build a derivation proving

$$B[\phi/x_1, ..., \phi/x_p, \omega/x_{p+1}, ..., \omega/x_n] \vdash_\infty x_k M_1...M_m : \phi,$$

and so, by rules $(\to I)$ and $(\leq_\infty)$,

$$B[\phi/x_1, ..., \phi/x_p] \vdash_\infty \lambda x_{p+1}...x_n.x_k M_1...M_m : \phi.$$

Hence, by rule $(\to I)$, $B \vdash_\infty \lambda x_1...x_n.x_k M_1...M_m : \underbrace{\phi \to ... \to \phi}_{p} \to \phi$.

A derivation assigning to M a type of the shape $\underbrace{\phi \to ... \to \phi}_{p} \to \phi$, with $p < k$ cannot exist, since the only possibility to derive it would be to assign to x_k a type equivalent to ω, and in this case only a type equivalent to ω can be assigned to M. □

By using the type assignment system, we can easily prove that the Λ-theory induced by the model $\mathcal{H}$ is a $\Lambda\eta$-theory.

Property 11.1.8. $\sim_\mathcal{H}$ is a $\Lambda\eta$-theory.

Proof. The theory induced by $\mathcal{H}$ is a Λ-theory by Corollary 10.0.3, since $\mathcal{H}$ is a $\lambda\Lambda$-model by Theorem 11.0.1. In order to prove that it is also a $\Lambda\eta$-theory, by Property 2.1.7, it is sufficient to prove that $I \sim_\mathcal{H} E$. We will prove both $I \sqsubseteq_\mathcal{H} E$ and $E \sqsubseteq_\mathcal{H} I$.

($I \sqsubseteq_\mathcal{H} E$) We will prove that $B \vdash_\infty I : \sigma$ implies $B \vdash_\infty E : \sigma$.
By Property 11.1.36.(i), either $\sigma \simeq_\infty \omega$ or $\sigma \simeq_\infty \sigma_0 \wedge ... \wedge \sigma_n$ $(n \geq 0)$ such that $\forall i \leq n,\ \sigma_i \simeq_\infty \tau_1^i \to ... \to \tau_{m_i}^i \to \phi$ for some $m_i \in \mathbb{N}$. The case $\sigma \simeq_\infty \omega$ is trivial, by rule (ω); so the proof follows by induction on n.
Let $n = 0$, thus we can assume $\sigma \simeq_\infty \mu \to \nu \to \tau$, by Property 11.1.36.(ii). Note that $B \vdash_\infty I : \sigma$ implies $B \vdash_\infty I : \mu \to \nu \to \tau$, by rule $(\leq_\infty)$; thus $B[\mu/x] \vdash_\infty x : \nu \to \tau$ by Lemma 10.1.7.(vi), and $\mu \leq_\infty \nu \to \tau$ by Lemma 10.1.7.(ii). We can build the following derivation:

$$\dfrac{\dfrac{\dfrac{\dfrac{\dfrac{\dfrac{}{B[\mu/x][\nu/y] \vdash_\infty x : \mu}(var) \quad \mu \leq_\infty \nu \to \tau}{B[\mu/x][\nu/y] \vdash_\infty x : \nu \to \tau}(\leq_\infty) \quad \dfrac{}{B[\mu/x][\nu/y] \vdash_\infty y : \nu}(var)}{B[\mu/x][\nu/y] \vdash_\infty xy : \tau}(\to E)}{B[\mu/x] \vdash_\infty \lambda y.xy : \nu \to \tau}(\to I)}{B \vdash_\infty \lambda xy.xy : \mu \to \nu \to \tau}(\to I) \quad \mu \to \nu \to \tau \leq_\infty \sigma}{B \vdash_\infty \lambda xy.xy : \sigma}(\leq_\infty)$$

If $n \geq 1$ then the proof follows by inductive hypothesis.

($E \sqsubseteq_\mathcal{H} I$) We will prove that $B \vdash_\infty E : \sigma$ implies $B \vdash_\infty I : \sigma$.
By Property 11.1.36.(i), either $\sigma \simeq_\infty \omega$ or $\sigma \simeq_\infty \sigma_0 \wedge ... \wedge \sigma_n$ $(n \geq 0)$ such that $\forall i \leq n,\ \sigma_i \simeq_\infty \tau_1^i \to ... \to \tau_{m_i}^i \to \phi$ and $m_i \in \mathbb{N}$. The case $\sigma \simeq_\infty \omega$ is trivial, by rule (ω); so the proof follows by induction on n.
Let $n = 0$, thus we can assume $\sigma \simeq_\infty \mu \to \nu \to \tau$, by Property 11.1.36.(ii). Note that $B \vdash_\infty \lambda xy.xy : \sigma$ implies $B \vdash_\infty \lambda xy.xy :$

$\mu \to \nu \to \tau$, by rule $(\leq_\infty)$; thus, by Lemma 10.1.7.(vi), this implies $B[\mu/x][\nu/y] \vdash_\infty xy : \tau$, which in turn implies $B[\mu/x][\nu/y] \vdash_\infty x : \theta \to \tau$ and $B[\mu/x][\nu/y] \vdash_\infty y : \theta$, for some θ, by Lemma 10.1.7.(vii). Hence, by Lemma 10.1.7.(ii), $\mu \leq_\infty \theta \to \tau$ and $\nu \leq_\infty \theta$ (so $\mu \leq_\infty \nu \to \tau$). Then we can build the following derivation:

$$\dfrac{\dfrac{\dfrac{\dfrac{}{B[\mu/x] \vdash_\infty x : \mu}(var) \qquad \mu \leq_\infty \nu \to \tau}{B[\mu/x] \vdash_\infty x : \nu \to \tau}(\leq_\infty)}{B \vdash_\infty \lambda x.x : \mu \to \nu \to \tau}(\to I) \qquad \mu \to \nu \to \tau \leq_\infty \sigma}{B \vdash_\infty \lambda x.x : \sigma}(\leq_\infty)$$

If $n \geq 1$ then the proof follows by inductive hypothesis. □

In order to prove that $\mathcal{H}$ is correct with respect to the **H**-operational semantics, it is necessary to prove that an unsolvable term cannot be assigned a type of the shape $\underbrace{\phi \to \ldots \to \phi}_{n}$, for every n. To this aim we will introduce a general tool for reasoning about interpretations of terms in a model, namely an approximation theorem. More precisely, we will prove that the meaning of a term (i.e. the set of types that can be derived for it) is the collection of the meanings of a set of normal forms of an extended language. So it will be possible to reason on the denotational semantics of terms simply by induction on normal forms.

First, we extend the language by adding a constant Ω to the formation rules of terms, then we define two new reduction rules on the so-obtained language.

Definition 11.1.9. (i) *$\Lambda\Omega$ is the language obtained by adding to Λ the constant Ω, namely the language inductively defined as follows:*
- $\Omega \in \Lambda\Omega$,
- $x \in \mathrm{Var}$ *implies* $x \in \Lambda\Omega$,
- $M \in \Lambda\Omega$ *and* $x \in \mathrm{Var}$ *implies* $(\lambda x.M) \in \Lambda\Omega$,
- $M \in \Lambda\Omega$ *and* $N \in \Lambda\Omega$ *implies* $(MN) \in \Lambda\Omega$.

(ii) $\to_\Omega$ *is the reduction defined as the contextual closure of the following rules:*

$$\Omega M \to \Omega, \qquad \lambda x.\Omega \to \Omega.$$

(iii) *The $\Lambda\Omega$-*reduction *$(\to_{\Lambda\Omega})$ is the contextual closure of the following rules:*

$$(\lambda x.M)N \to M[N/x], \qquad \Omega M \to \Omega, \qquad \lambda x.\Omega \to \Omega.$$

$\to^*_{\Lambda\Omega}$ *is the reflexive and transitive closure of* $\to_{\Lambda\Omega}$.
The η-reduction $(\to_\eta)$ can be directly applied to the language $\Lambda\Omega$ (see Definition 1.3.7).

$M \in \Lambda\Omega$ is in $\Lambda\Omega$-normal form ($\Lambda\Omega$-nf) if and only if it does not contain $\Lambda\Omega$-redexes.

As usual, terms of $\Lambda\Omega$ will be considered modulo α-equality. The type assignment system of Definition 10.1.1 can be applied to $\Lambda\Omega$ without modifications. It is easy to see that to the term Ω only the type ω can be assigned, by rule (ω).

Property 11.1.10. The type assignment system $\vdash_\infty$ is closed under $=_{\Lambda\Omega}$.

Proof. Easy. □

The intuitive interpretation of the constant Ω is that it represents a term with an unknown behaviour. The interpretation function is therefore naturally extended to $\Lambda\Omega$, i.e. the interpretation of a term of $\Lambda\Omega$ is the set of all types that can be assigned to it.

Definition 11.1.11. *The set of* approximants *of a term M is defined as follows:*

$$\mathcal{A}(M) = \left\{ A \;\middle|\; \begin{array}{r} \textit{there is } M' \textit{ such that } M =_\Lambda M' \textit{ and } A \textit{ is a } \Lambda\Omega\textit{-normal form} \\ \textit{obtained from } M' \textit{ by replacing some subterms with } \Omega. \end{array} \right\}$$

Example 11.1.12. Some sets of approximants are shown.

$$\begin{array}{ll} \mathcal{A}(I) = \{\Omega, \lambda x.x\}, & \mathcal{A}(D) = \{\Omega, \lambda x.xx, \lambda x.x\Omega\}, \\ \mathcal{A}(DD) = \{\Omega\}, & \mathcal{A}\big(K(\lambda x.x(II))(DD)\big) = \{\Omega, \lambda x.x\Omega, \lambda x.xI\}. \end{array}$$

Approximants can be defined inductively as follows.

Definition 11.1.13. *The set $\mathcal{A}$ of $\Lambda\Omega$-normal forms can be inductively defined as follows:*

- *$\Omega \in \mathcal{A}$,*
- *$A_j \in \mathcal{A}$ imply $\lambda x_1 ... x_n.xA_1 ... A_m \in \mathcal{A}$ $(1 \le j \le m)$.*

$\Lambda\Omega$-normal forms will be ranged over by A, A', possibly indexed.

Clearly every Λ-normal form is an approximant of some term. So, we will simply call approximant a $\Lambda\Omega$-normal form.

Property 11.1.14. (i) For every A, there is a term M such that $A \in \mathcal{A}(M)$.
(ii) If $M =_\Lambda N$ then $\mathcal{A}(M) = \mathcal{A}(N)$.

Proof. (i) Easy.
(ii) By definition of approximant. □

An approximant A is openly solvable if and only if $A \equiv \lambda x_1...x_n.zA_1...A_m$ and $z \in \mathrm{FV}(A)$. It is easy to check that Property 11.1.7 holds for approximants too.

A key property of the approximants is that the $\sqsubseteq_{\mathcal{H}}$ order relation between them can be syntactically axiomatized.

Definition 11.1.15. *Let A and A' be two approximants.*
$A \ll A'$ if and only if one of the two following cases arises:

(i) $A \equiv \Omega$;
(ii) $A =_\eta \lambda x_1...x_n.zA_1...A_m$, *and* $A' =_\eta \lambda x_1...x_n.zA'_1...A'_m$, *and* $A_i \ll A'_i$ *for each* i $(1 \leq i \leq m;\ n, m \in \mathbb{N})$.

Although the η-reduction $(\rightarrow_\eta)$ can be directly applied to the language $\Lambda\Omega$, often we are only interested in its use on approximants. Note that $\Omega =_\eta \lambda z.\Omega z$ and $\lambda x.x =_\eta \lambda y.(\lambda x.x)y$, but both $\lambda z.\Omega z \notin \mathcal{A}$ and $\lambda y.(\lambda x.x)y \notin \mathcal{A}$, since they are not $\Lambda\Omega$-normal forms. However, $\lambda z.\Omega z \rightarrow_{\Lambda\Omega} \Omega \in \mathcal{A}$, and $\lambda y.(\lambda x.x)y \rightarrow_{\Lambda\Omega} \lambda x.x \in \mathcal{A}$.

Example 11.1.16. $\mathcal{A}(\lambda x.xII) = \{\Omega, \lambda x.x\Omega\Omega, \lambda x.x\Omega I, \lambda x.xI\Omega, \lambda x.xII\}$. It is easy to see that $\lambda x.x\Omega I \not\ll \lambda x.xI\Omega$, while $\lambda x.x\Omega\Omega \ll \lambda x.xII$.

Clearly $\ll$ is a preorder relation on $\mathcal{A}$ and a partial order on the set of approximants considered up to η-equivalence.

The next theorem proves that two approximants A and A' such that $A \not\ll A'$ can be semiseparated, in the sense that there is a context that $\Lambda\Omega$-reduces either to I or to Ω where it is filled respectively by A and A'. The fact that the model $\mathcal{H}$ is fully abstract with respect to the **H**-operational semantics is based essentially on this property.

Theorem 11.1.17 (Semiseparability).
Let $A \not\ll A'$. Then there is a context $C[.]$ such that $C[A] \rightarrow^*_{\Lambda\Omega} I$ while $C[A'] \rightarrow^*_{\Lambda\Omega} \Omega$.

Proof. The proof is in Sect. 11.1.3. □

Hence, $\ll$ is a syntactical axiomatization of the $\sqsubseteq_{\mathcal{H}}$ order relation between approximants.

Lemma 11.1.18. *$A \sqsubseteq_{\mathcal{H}} A'$ if and only if $A \ll A'$.*

Proof. $(\Leftarrow)$ We will prove that, if $A \ll A'$, then $B \vdash_\infty A : \sigma$ implies $B \vdash_\infty A' : \sigma$, by induction on the definition of $\ll$. If $A \equiv \Omega$, then $\sigma \equiv \omega$ and the proof is trivial, by rule (ω). Otherwise, let $A =_\eta \lambda x_1...x_n.zA_1...A_m$ and $A' =_\eta \lambda x_1...x_n.zA'_1...A'_m$ and $A_i \ll A'_i$ $(1 \leq i \leq m)$. By Property 11.1.36.(i), either $\sigma \simeq_\infty \omega$ or $\sigma \simeq_\infty \sigma_0 \wedge ... \wedge \sigma_k$ $(k \geq 1)$, where $\sigma_i \simeq_\infty \tau^i_1 \rightarrow ... \rightarrow \tau^i_{h_i} \rightarrow \phi$ $(h_i \in \mathbb{N}, i \leq k)$. The case $\sigma \simeq_\infty \omega$ is trivial; so let $\sigma \simeq_\infty \sigma_0 \wedge ... \wedge \sigma_k$ $(k \geq 1)$.

$B \vdash_\infty A : \sigma$ implies $B \vdash_\infty A : \sigma_i$ $(i \leq k)$, by rule $(\leq_\infty)$. As shown in Property 11.1.8, the type assignment system is closed under η-equality, thus $B \vdash_\infty \lambda x_1...x_n.zA_1...A_m : \tau_1^i \to ... \to \tau_{h_i}^i \to \phi$.
Without loss of generality, we can assume $h_i \geq n$ $(\forall i \leq k)$ by Property 11.1.36.(ii), thus $B[\tau_1^i/x_1, ..., \tau_n^i/x_n] \vdash_\infty zA_1...A_m : \tau_{n+1}^i \to ... \to \tau_{h_i}^i \to \phi$, by Lemma 10.1.7.(vi). Hence, by Lemma 10.1.7.(vii) there are $\mu_1^i, ..., \mu_{m_i}^i \in T(C_\infty)$ such that

$$B[\tau_1^i/x_1, ..., \tau_n^i/x_n] \vdash_\infty z : \mu_1^i \to ... \to \mu_m^i \to \tau_{n+1}^i \to ... \to \tau_{h_i}^i \to \phi,$$

and $B[\tau_1^i/x_1, ..., \tau_n^i/x_n] \vdash_\infty A_j : \mu_j^i$ $(j \leq m,\ i \leq k)$.
By induction $B[\tau_1^i/x_1, ..., \tau_n^i/x_n] \vdash_\infty A'_j : \mu_j^i$ $(j \leq m,\ i \leq k)$; so by m applications of rule $(\to E)$ and n applications of rule $(\to I)$, it follows that $B \vdash_\infty A' : \sigma_i$, for all $i \leq k$; the proof follows by rule $(\wedge I)$.

$(\Rightarrow)$ By the semiseparability theorem, $A \not\leq A'$ implies there is a context $C[.]$ such that $C[A] \to^*_{\Lambda\Omega} I$ while $C[A'] \to^*_{\Lambda\Omega} \Omega$. By Property 11.1.10, this implies $C[A] \not\sqsubseteq_{\mathcal{H}} C[A']$, which in turn implies $A \not\sqsubseteq_{\mathcal{H}} A'$. □

The following theorem states that the interpretation of a term is the collection of the interpretations of its approximants.

Theorem 11.1.19 ($\mathcal{H}$-Approximation).
$B \vdash_\infty M : \sigma$ if and only if $B \vdash_\infty A : \sigma$, for some $A \in \mathcal{A}(M)$.

Proof. The proof is in Sect. 11.1.2. □

The syntactic shape of the approximants of a term having Λ-hnf is related to the syntactic shape of its Λ-hnf's, as showed in the following property.

Property 11.1.20. (i) M is Λ-unsolvable if and only if $\mathcal{A}(M) = \{\Omega\}$.
(ii) Let $A \in \mathcal{A}(M)$ and $A \not\equiv \Omega$.
$M \to^*_\Lambda \lambda x_1...x_n.zM_1...M_m$ if and only if $A \equiv \lambda x_1 ... x_n.zA_1 ... A_m$, for some $A_i \in \mathcal{A}(M_i)$.
(iii) Let $A \in \mathcal{A}(M)$ and $A \not\equiv \Omega$. $M =_{\Lambda\eta} \lambda x_1...x_n.zM_1...M_m$ if and only if $A =_\eta \lambda x_1 ... x_n.zA_1 ... A_m$, for some $A_i \in \mathcal{A}(M_i)$ $(1 \leq i \leq m)$.

Proof. (i) If M is a Λ-unsolvable and $M =_\Lambda N \equiv \lambda x_1 \dots x_n.RN_1 \dots N_m$, then R is a Λ-redex. So the only $\Lambda\Omega$-normal form that can be obtained from N by replacing some of its subterms by Ω is Ω itself.
(ii) By definition of approximants and Property 2.1.2.
(iii) By definition of approximants and by definition of of $\Lambda\eta$.

□

Corollary 11.1.21. *The theory induced by $\mathcal{H}$ is sensible.*

Proof. By Property 11.1.20.(i), remembering Definition 1.3.4. □

So the correctness of $\mathcal{H}$ with respect to the **H**-operational semantics follows easily.

Theorem 11.1.22 ($\mathcal{H}$-Correctness).
The model $\mathcal{H}$ is correct with respect to the **H**-*operational semantics.*

Proof. By Property 10.1.15 it is sufficient to prove that, for all M, N, if M is Λ-solvable and N is Λ-unsolvable then $N \sqsubset_{\mathcal{H}} M$.
By Property 11.1.7 a solvable term can be assigned at least one type $\not\simeq_\infty \omega$, while only type $\simeq_\infty \omega$ can be assigned to an unsolvable term, since it has Ω as the only approximant. □

The correctness of the model $\mathcal{H}$ with respect to the operational semantics **H** allows us to transfer some results from the denotational world to the operational one. So some interesting properties of **H** can be proved. The reader can see that the approximation theorem is a quite useful tool.

Corollary 11.1.23. *The* **H**-*operational semantics is fully extensional.*

Proof. By correctness and Property 11.1.8. □

In Sect. 2.1 the notion of call-by-name fixed-point operator was introduced, and a term with this behaviour was shown in the proof of Theorem 2.1.8. An example of another fixed-point operator will be show in the last chapter. It is an easy exercise to prove that it is possible to build an infinite number of such operators (see [83]). The next theorem proves that all are equated in the **H**-operational semantics.

Theorem 11.1.24. *All call-by-name fixed-point operators are* **H**-*equivalent.*

Proof. Let P be any call-by-name fixed-point operator, so $PM =_\Lambda M(PM)$, for all $M \in \Lambda$. Then $P =_\eta \lambda z.Pz =_\Lambda \lambda z.z(Pz)$, where $z \notin FV(P)$; by iterating this reasoning, by the confluence theorem and by Properties 11.1.8 and 11.1.20, every approximant of P is either Ω or of the shape $\lambda z.\underbrace{z(z...(z}_{n}\,\Omega)...)(n \geq 1)$. Hence the result follows, by the approximation theorem. □

The next theorem proves a key result, which implies directly the full-abstraction of the model $\mathcal{H}$ with respect to the **H**-operational semantics. It proves that, if $M \not\sqsubseteq_{\mathcal{H}} N$, then there is a context **H**-discriminating them. Its proof uses the semiseparability property between approximants, namely if $M \not\sqsubseteq_{\mathcal{H}} N$ then it is possible to find two approximants $A \in \mathcal{A}(M)$ and $A' \in \mathcal{A}(N)$, such that $A \not\preccurlyeq A'$ and the context discriminating M and N is a minor modification of the context semiseparating A and A'.

Theorem 11.1.25 ($\mathcal{H}$-Discriminability).
If $M \not\sqsubseteq_{\mathcal{H}} N$ then there is a context $C[.]$ such that $C[M], C[N] \in \Lambda^0$ and $C[M] \Downarrow_{\mathbf{H}}$ while $C[N] \Uparrow_{\mathbf{H}}$.

Proof. The proof is in Sect. 11.1.3. □

Theorem 11.1.26 ($\mathcal{H}$-Completeness).
The model $\mathcal{H}$ is complete with respect to the **H**-*operational semantics.*

Proof. We must prove that $M \preceq_{\mathbf{H}} N$ implies $M \sqsubseteq_{\mathcal{H}} N$. Let $M \not\sqsubseteq_{\mathcal{H}} N$. By the $\mathcal{H}$-discriminability Theorem, there is a context $C[.]$ such that both $C[M]$ and $C[N]$ are closed and $C[M] \Downarrow_{\mathbf{H}}$ while $C[N] \Uparrow_{\mathbf{H}}$, hence $M \not\preceq_{\mathbf{H}} N$. □

Corollary 11.1.27 ($\mathcal{H}$-Full abstraction).
The model $\mathcal{H}$ is fully abstract with respect to the **H**-*operational semantics.*

Proof. By Theorems 11.1.22 and 11.1.26. □

The $\mathcal{H}$-discriminability Theorem allows for a finite axiomatization of the preorder $\sqsubseteq_{\mathcal{H}}$ between terms.

Definition 11.1.28. (i) *The relation $\sqsubseteq_k \subseteq \Lambda \times \Lambda$ $(k \in \mathbb{N})$ is defined, by induction on k, as follows:*
- *$M \sqsubseteq_k N$, for all k, if M is Λ-unsolvable;*
- *$M \sqsubseteq_0 N$ if and only if $M =_{\Lambda\eta} \lambda x_1...x_n.xM_1...M_m$ and $N =_{\Lambda\eta} \lambda x_1...x_n.xN_1...N_m$;*
- *$M \sqsubseteq_{k+1} N$ if and only if $M =_{\Lambda\eta} \lambda x_1...x_n.xM_1...M_m$ and $N =_{\Lambda\eta} \lambda x_1...x_n.xN_1...N_m$, where $M_i \sqsubseteq_k N_i$, for all i $(1 \leq i \leq m)$.*

(ii) *$M \sim_k N$ if both $M \sqsubseteq_k N$ and $N \sqsubseteq_k M$.*

Example 11.1.29. $\lambda x.x(DD)I \sqsubseteq_0 \lambda x.xIK$ while $\lambda x.x(DD)I \not\sqsubseteq_1 \lambda x.xIK$.

Roughly speaking, two terms M and N are in the relation $\sqsubseteq_k$ if there are two terms $M' =_{\Lambda\eta} M$ and $N' =_{\Lambda\eta} N$, having the same structure "up to level k". So, if $\forall k \geq 0, M \sim_k N$, any structural difference between them is pushed to the infinite.

Theorem 11.1.30 ($\mathcal{H}$-Characterization).
$M \sqsubseteq_{\mathcal{H}} N$ if and only if $\forall k \geq 0, M \sqsubseteq_k N$.

Proof. The proof is in Sect. 11.1.3. □

The next theorem proves an unexpected result. It has been proved that the **H**-operational semantics is fully-extensional, i.e. it is closed under η-equality. We will prove now that it also equates terms being differents for an infinite number of η-reductions.

Let $E_\infty \equiv Y(\lambda xyz.y(xz))$ where Y is a call-by-name fixed point operator. Observe that, if $M \in \Lambda$ then $E_\infty M =_\Lambda \lambda z.M(E_\infty z) =_\Lambda \lambda z.M(\lambda z_1.z(E_\infty z_1))$ $=_\Lambda \lambda z.M(\lambda z_1.z(\lambda z_2.z_1(E_\infty z_2)))$, and so on. So z can be viewed as obtained from $E_\infty z$ by means of an infinite number of applications of the η -reduction rule. We will prove that $E_\infty M \approx_{\mathbf{H}} M$.

Theorem 11.1.31. $E_\infty M \approx_{\mathbf{H}} M$.

Proof. We prove $I \sim_{\mathcal{H}} E_\infty$; the result follows from Λ-reduction and correctness. Clearly $\mathcal{A}(I) = \{\Omega, \lambda x.x\}$. Moreover, it is straightforward to check that $\mathcal{A}(E_\infty) = \{\Omega, \lambda z_0 z_1.z_0\Omega, \lambda z_0 z_1.z_0(\lambda z_2.z_1(...(\lambda z_{n+1}.z_n\Omega)...)) \mid n \geq 0\}$.

($E_\infty \sqsubseteq_{\mathcal{H}} I$) It is immediate to check that every $A \in \mathcal{A}(E_\infty)$ is such that $A \ll I$. Thus the proof follows from Lemma 11.1.18 and from the $\mathcal{H}$-approximation theorem.

($I \sqsubseteq_{\mathcal{H}} E_\infty$) It is easy to see that $I \sqsubseteq_k E_\infty$, for all k. Then the proof follows from the $\mathcal{H}$-characterization theorem. □

$E_\infty M$ has no normal form, also when M has a normal form. So a term with normal form is always **H**-operationally equivalent to a term without normal form. A syntactical proof of the **H**-equivalence between the two terms I and E_∞ was done in [36].

11.1.1 The $\leq_\infty$-Intersection Relation

$\leq_\infty$ is a preorder relation on $T(C_\infty)$; so it induces a partial order on $T(C_\infty)$ considered up to $\simeq_\infty$.

Note that $\sigma \in T(C_\infty)$ implies that $\sigma \equiv \sigma_0 \wedge ..... \wedge \sigma_n$ ($n \in \mathbb{N}$) where, $\sigma_i \in T(C_\infty)$, and either $\sigma_i \equiv \omega$ or $\sigma_i \equiv \phi$ or $\sigma_i \equiv \pi_i \to \tau_i$ ($i \leq n$).

Some key properties of ∞-type theory will be shown. The next theorem characterizes the set of types that are ∞ equivalent to ω.

Theorem 11.1.32. *$\sigma \simeq_\infty \omega$ if and only if $\sigma \in \Omega^{\mathcal{H}}$, where*

$$\Omega^{\mathcal{H}} = \left\{ \sigma \in T(C_\infty) \;\middle|\; \begin{array}{l} \sigma \equiv \sigma_0 \wedge ..... \wedge \sigma_n \ (n \geq 0) \text{ such that} \\ \forall i \leq n \text{ either } \sigma_i \equiv \omega, \text{ or } \sigma_i \equiv \pi_i \to \tau_i \text{ and } \tau_i \simeq_\infty \omega \end{array} \right\}.$$

Proof. Note that $\omega \simeq_\infty \sigma$ if and only if $\omega \leq_\infty \sigma$, by the rule ($a$).

$\Leftarrow$ We will prove that $\sigma \in \Omega^{\mathcal{H}}$ implies $\omega \leq_\infty \sigma$, by induction on n. Let $n = 0$. If $\sigma_0 \equiv \omega$ then the proof is trivial; otherwise, $\sigma_0 \equiv \pi \to \tau$ and $\tau \simeq_\infty \omega$. Hence $\omega \leq_\infty \omega \to \omega \leq_\infty \pi \to \tau$, respectively, by rules ($h3$), (f) and (a); by rule (t), the proof follows.
Let $n \geq 1$ and $\sigma \equiv \sigma_0 \wedge ..... \wedge \sigma_n$; so $\forall i \leq n \;\; \omega \leq_\infty \sigma_i$ by induction. So

$$\omega \leq_\infty \underbrace{\omega \wedge ..... \wedge \omega}_{n} \leq_\infty \sigma_0 \wedge ..... \wedge \sigma_n$$

by rules (b), (e) and (t).

$\Rightarrow$ Let us first prove that, if $\sigma \in \Omega^{\mathcal{H}}$ and $\sigma \leq_\infty \tau$ then $\tau \in \Omega^{\mathcal{H}}$, by induction on the rules of $\leq_\infty$.
(a) Trivial, since $\omega \in \Omega^{\mathcal{H}}$.
(b), (c), (c') By definition of $\Omega^{\mathcal{H}}$.

(d) $(\mu \to \rho) \wedge (\mu \to \pi) \in \Omega^{\mathcal{H}}$ implies $\rho, \pi \simeq_\infty \omega$ by definition of $\Omega^{\mathcal{H}}$, so $\omega \leq_\infty \omega \wedge \omega \leq_\infty \rho \wedge \pi$, by rules (b) and (e); hence $\rho \wedge \pi \simeq_\infty \omega$ and $\mu \to (\rho \wedge \pi) \in \Omega^{\mathcal{H}}$.
(e) Let $\mu \leq_\infty \mu'$ and $\pi \leq_\infty \pi'$; $\mu \wedge \pi \in \Omega^{\mathcal{H}}$ implies $\mu, \pi \in \Omega^{\mathcal{H}}$, so $\mu', \pi' \in \Omega^{\mathcal{H}}$ by induction; hence $\mu' \wedge \pi' \in \Omega^{\mathcal{H}}$.
(f) Let $\mu' \leq_\infty \mu$ and $\pi \leq_\infty \pi'$; $\mu \to \pi \in \Omega^{\mathcal{H}}$ implies $\pi \simeq_\infty \omega$, hence $\omega \leq_\infty \pi \leq_\infty \pi'$ implies $\pi' \simeq_\infty \omega$; thus, $\mu' \to \tau' \in \Omega^{\mathcal{H}}$ by definition of $\Omega^{\mathcal{H}}$.
(g) trivial, since $\omega \to \omega \in \Omega^{\mathcal{H}}$.
(r), (t), (h3) Easy.
(h1), (h2) Not possible.
Then the proof is done, since $\sigma \simeq_\infty \omega$ implies $\omega \leq_\infty \sigma$, and therefore $\omega \in \Omega^{\mathcal{H}}$ implies $\sigma \in \Omega^{\mathcal{H}}$. □

As a corollary, we can prove that $\leq_\infty$ is well posed, in the sense that it does not equate different type constants in C_∞.

Corollary 11.1.33. (i) $\phi \not\simeq_\infty \omega$.
(ii) $\sigma_1 \to ... \to \sigma_n \to \omega \simeq_\infty \omega$, *for all* $\sigma_1, ..., \sigma_n$ $(n \in \mathbb{N})$.

Proof. (i) Since $\phi \notin \Omega^{\mathcal{H}}$.
(ii) Easy, by Theorem 11.1.32. □

ϕ is the "minimum" type with respect to $\leq_\infty$.

Lemma 11.1.34. *If* $\sigma \in T(C_\infty)$ *then* $\phi \leq_\infty \sigma$.

Proof. By induction on σ. If $\sigma \equiv \omega$ then the proof is trivial, by rule (a). If $\sigma \equiv \phi$ then the proof is trivial, by rule (r). If $\sigma \equiv \sigma_0 \wedge \sigma_1$ then $\phi \leq_\infty \sigma_i$ $(i \leq 1)$ by induction, thus $\phi \leq_\infty \phi \wedge \phi \leq_\infty \sigma_0 \wedge \sigma_1$, by rules (b) and (e). If $\sigma \equiv \sigma_0 \to \sigma_1$ then $\phi \leq_\infty \sigma_1$ by induction, and $\sigma_0 \leq_\infty \omega$ by rule (a); thus $\phi \leq_\infty \omega \to \phi \leq_\infty \sigma_0 \to \sigma_1$ by rules (h1) and (f). □

The class of types equivalent to ϕ is characterized in the following theorem.

Theorem 11.1.35. *Let* $\Phi^{\mathcal{H}} = \bigcup_{m \geq 0} \Phi^{\mathcal{H}}_m$, *where* $\Phi^{\mathcal{H}}_0 = \{\phi\}$ *and, for* $m \in \mathbb{N}$,

$$\Phi^{\mathcal{H}}_{m+1} = \left\{ \sigma \in T(C_\infty) \left| \begin{array}{l} \sigma \equiv \sigma_0 \wedge ..... \wedge \sigma_n \ (n \in \mathbb{N}) \textit{ and } \exists k \leq n \textit{ such that,} \\ \textit{either } \sigma_k \in \Phi^{\mathcal{H}}_m \\ \quad \textit{or } \sigma_k \equiv \pi_k \to \tau_k \textit{ with } \pi_k \simeq_\infty \omega \textit{ and } \tau_k \in \Phi^{\mathcal{H}}_m \end{array} \right. \right\}.$$

$\sigma \simeq_\infty \phi$ *if and only if* $\sigma \in \Phi^{\mathcal{H}}$, *for all* $\sigma \in T(C_\infty)$.

Proof. Note that $\sigma \simeq_\infty \phi$ if and only if $\sigma \leq_\infty \phi$, by Lemma 11.1.34. Moreover, it is easy to check that

$$\Phi^{\mathcal{H}} = \left\{ \sigma \in T(C_\infty) \left| \begin{array}{l} \sigma \equiv \sigma_0 \wedge ..... \wedge \sigma_n \ (n \in \mathbb{N}) \text{ and } \exists k \leq n, \text{ either } \sigma_k \equiv \phi \\ \text{or } \sigma_k \equiv \pi_k \to \tau_k \text{ with } \pi_k \simeq_\infty \omega \text{ and } \tau_k \simeq_\infty \phi \end{array} \right. \right\}.$$

$\Leftarrow$ By induction on m. If $m = 0$ then the proof is trivial; so let $m \geq 1$, $\sigma \equiv \sigma_0 \wedge ..... \wedge \sigma_n$ $(n \in \mathbb{N})$ and either $\sigma_k \in \Phi^{\mathcal{H}}_{m-1}$ or $\sigma_k \equiv \pi_k \rightarrow \tau_k$, $\pi_k \simeq_\infty \omega$ and $\tau_k \in \Phi^{\mathcal{H}}_{m-1}$. If $\sigma_k \in \Phi^{\mathcal{H}}_{m-1}$ then $\sigma_k \simeq_\infty \phi$ by induction; otherwise, $\sigma_k \leq_\infty \omega \rightarrow \phi \leq_\infty \phi$ by rules $(h1)$ and (f) and by induction. Hence, by rules (e), (c), (r) and (t):

$$\sigma_0 \wedge ... \wedge \sigma_k ... \wedge \sigma_n \leq_\infty \sigma_0 \wedge ... \wedge \phi \wedge ... \wedge \sigma_n \leq_\infty \phi.$$

$\Rightarrow$ Let us first prove that, if $\sigma \in \Phi^{\mathcal{H}}$ and $\tau \leq_\infty \sigma$ then $\tau \in \Phi^{\mathcal{H}}$, by induction on the rules of $\leq_\infty$.
(a) Not possible, since $\omega \notin \Phi^{\mathcal{H}}$.
$(b), (c), (c')$ By definition of $\Phi^{\mathcal{H}}$.
(d) $\mu \rightarrow (\rho \wedge \pi) \in \Phi^{\mathcal{H}}$ implies $\mu \simeq_\infty \omega$ and $\rho \wedge \pi \in \Phi^{\mathcal{H}}$, so it is easy to see that either $\rho \in \Phi^{\mathcal{H}}$ or $\pi \in \Phi^{\mathcal{H}}$; thus either $\mu \rightarrow \rho \in \Phi^{\mathcal{H}}$ or $\mu \rightarrow \pi \in \Phi^{\mathcal{H}}$; hence $(\mu \rightarrow \rho) \wedge (\mu \rightarrow \pi) \in \Phi^{\mathcal{H}}$.
(e) Let $\mu \leq_\infty \mu'$ and $\pi \leq_\infty \pi'$; so $\mu' \wedge \pi' \in \Phi^{\mathcal{H}}$ implies either $\mu' \in \Phi^{\mathcal{H}}$ or $\pi' \in \Phi^{\mathcal{H}}$, thus $\mu \in \Phi^{\mathcal{H}}$ or $\pi \in \Phi^{\mathcal{H}}$ by induction; hence $\mu \wedge \pi \in \Phi^{\mathcal{H}}$.
(f) Let $\mu' \leq_\infty \mu$ and $\pi \leq_\infty \pi'$; $\mu' \rightarrow \pi' \in \Phi^{\mathcal{H}}$ implies $\mu' \simeq_\infty \omega$ and $\pi' \in \Phi^{\mathcal{H}}$; hence, $\pi \in \Phi^{\mathcal{H}}$ by induction and $\omega \leq_\infty \mu$ by rule (t), so $\mu \rightarrow \pi \in \Phi^{\mathcal{H}}$.
$(g), (h3)$ Not possible, since $\omega \rightarrow \omega \notin \Phi^{\mathcal{H}}$.
$(r), (t), (h1), (h2)$ Easy.
Then the proof is done, since $\sigma \simeq_\infty \phi$ implies $\sigma \leq_\infty \phi$, and therefore $\phi \in \Phi^{\mathcal{H}}$ implies $\sigma \in \Phi^{\mathcal{H}}$. □

The next property shows some type characterization.

Property 11.1.36. (i) If $\sigma \not\simeq_\infty \omega$ then $\sigma \simeq_\infty \sigma_0 \wedge ... \wedge \sigma_n$ $(n \geq 0)$ such that $\forall i \leq n,\ \sigma_i \simeq_\infty \tau^i_1 \rightarrow ... \rightarrow \tau^i_{m_i} \rightarrow \phi$ for some $m_i \in \mathbb{N}$.
(ii) If $\sigma \simeq_\infty \tau_1 \rightarrow ... \rightarrow \tau_n \rightarrow \phi$ $(n \in \mathbb{N})$ then, for all $p \in \mathbb{N}$,

$$\sigma \simeq_\infty \tau_1 \rightarrow ... \rightarrow \tau_n \rightarrow \underbrace{\omega \rightarrow ... \rightarrow \omega}_{p} \rightarrow \phi.$$

Proof. (i) By induction on σ. If either $\sigma \equiv \omega$ or $\sigma \equiv \phi$ then the proof is trivial. Let $\sigma \equiv \mu \rightarrow \pi$; so, by induction $\pi \simeq_\infty \pi_0 \wedge ... \wedge \pi_h$ $(h \in \mathbb{N})$ such that $\forall i \leq h,\ \pi_i \simeq_\infty \nu^i_1 \rightarrow ... \rightarrow \nu^i_{m_i} \rightarrow \phi$. By Lemma 10.1.4.(iii), $\sigma \simeq_\infty \sigma_0 \wedge ... \wedge \sigma_h$, where $\sigma_i \simeq_\infty \mu \rightarrow \nu^i_1 \rightarrow ... \rightarrow \nu^i_{m_i} \rightarrow \phi$.
The case $\sigma \equiv \mu \wedge \pi$ follows by induction.
(ii) By rules $(h1)$, $(h2)$ and (f) it is easy to see that:

$$\tau_1 \rightarrow ... \rightarrow \tau_n \rightarrow \phi \simeq_\infty \tau_1 \rightarrow ... \rightarrow \tau_n \rightarrow \omega \rightarrow \phi$$

so the proof follows by induction on p. □

The next property implies the legality of ∞. Let us notice that every type in $T(C_\infty)$ has the following syntactical shape:

$$(\sigma_1 \to \tau_1) \wedge ... \wedge (\sigma_n \to \tau_n) \wedge c_1 \wedge ... \wedge c_m \quad \text{where} \quad m, n \geq 0, m+n \geq 1,$$

where $c_i \in \{\omega, \phi\}$ $(1 \leq i \leq m)$. Moreover remember that every constant in C_∞ is equivalent to an arrow type; indeed, $\omega \simeq_\infty \omega \to \omega$ and $\phi \simeq_\infty \omega \to \phi$.

Property 11.1.37. Let $m, n, p, q \geq 0$, $m+n \geq 1$, $p+q \geq 1$, and

$$(\sigma_1 \to \tau_1) \wedge ... \wedge (\sigma_n \to \tau_n) \wedge c_1 \wedge ... \wedge c_m \leq_\infty (\sigma'_1 \to \tau'_1) \wedge ... \wedge (\sigma'_p \to \tau'_p) \wedge d_1 \wedge ... \wedge d_q$$

where $c_i, d_j \in \{\omega, \phi\}$, $c_i \simeq_\infty \sigma_{n+i} \to \tau_{n+i}$ and $d_j \simeq_\infty \sigma'_{p+j} \to \tau'_{p+j}$ $(\tau_{n+i}, \tau'_{p+j} \in \{\omega, \phi\}, 1 \leq i \leq m, 1 \leq j \leq q)$. If $\tau'_h \not\simeq \omega$ $(1 \leq h \leq p+q)$, then there is $\{i_1, ..., i_k\} \subseteq \{1, ..., n+m\}$ such that $\sigma_{i_1} \wedge ... \wedge \sigma_{i_k} \geq_\infty \sigma'_h$, $\tau_{i_1} \wedge ... \wedge \tau_{i_k} \leq_\infty \tau'_h$ and $\tau_j \not\simeq \omega$, for each $j \in \{i_1, ..., i_k\}$.

Proof. We reason by induction on the last rule of the derivation proving

$$(\sigma_1 \to \tau_1) \wedge ... \wedge (\sigma_n \to \tau_n) \wedge c_1 \wedge ... \wedge c_m \leq_\infty (\sigma'_1 \to \tau'_1) \wedge ... \wedge (\sigma'_p \to \tau'_p) \wedge d_1 \wedge ... \wedge d_q.$$

(a) Trivial, since $\omega \simeq_\infty \omega \to \omega$ makes the implication empty.
$(g), (h3)$ Trivial, since $\omega \to \omega$ makes the implication empty.
$(b), (c), (c'), (d), (r), (h1), (h2)$ Immediate.
$(e), (f)$ By induction.
(t) Let $\rho \in T(C_\infty)$ be such that $(\sigma_1 \to \tau_1) \wedge ... \wedge (\sigma_n \to \tau_n) \wedge c_1 \wedge ... \wedge c_m \leq_\infty \rho$ and $\rho \leq_\infty (\sigma'_1 \to \tau'_1) \wedge ... \wedge (\sigma'_p \to \tau'_p) \wedge d_1 \wedge ... \wedge d_q$. ρ must be of the shape $(\mu_1 \to \nu_1) \wedge ... \wedge (\mu_t \to \nu_t) \wedge e_1 \wedge ... \wedge e_s$, where $e_i \in \{\omega, \phi\}$ and $e_i \simeq_\infty \mu_{t+i} \to \nu_{t+i}$ $(\nu_{t+i} \in \{\omega, \phi\}, 1 \leq i \leq s)$. By induction, for every h such that $\tau'_h \not\simeq_\infty \omega$, there is $\{i^h_1, ..., i^h_{k_h}\} \subseteq \{1, ..., t+s\}$ such that $\mu_{i^h_1} \wedge ... \wedge \mu_{i^h_{k_h}} \geq_\infty \sigma'_h$, $\nu_{i^h_1} \wedge ... \wedge \nu_{i^h_{k_h}} \leq_\infty \tau'_h$ and $\nu_{i^h_j} \not\simeq \omega$, for all $1 \leq j \leq k_h$. By induction, for every $\nu_{i^h_j}$, there is $\{r^{h,j}_1, ..., r^{h,j}_{w_{h,j}}\} \subseteq \{1, 2, ..., n+m\}$ such that $\sigma_{r^{h,j}_1} \wedge ... \wedge \sigma_{r^{h,j}_{w_{h,j}}} \geq_\infty \mu_{i^h_j}$ and $\tau_{r^{h,j}_1} \wedge ... \wedge \tau_{r^{h,j}_{w_{h,j}}} \leq_\infty \nu_{i^h_j}$ and $\tau_{r^{h,j}_u} \not\simeq \omega$, for all $1 \leq u \leq w_{h,j}$. So the proof follows by rule (e). □

▼ **Proof of Theorem 11.1.5** (pag. 121).

The legality of the type system ∞ is a particular case of Property 11.1.37. ■

11.1.2 Proof of the $\mathcal{H}$-Approximation Theorem

The most difficult part of the proof is to prove that $B \vdash_\infty M : \sigma$ implies $B \vdash_\infty A : \sigma$, for some $A \in \mathcal{A}(M)$.
The technique used here is a variant of Tait's notion of computability, first defined for intersection types in [30].

Let us sketch the general idea of the proof.

Definition 11.1.38. *$App_{\mathcal{H}}$ is the predicate defined as follows: $App_{\mathcal{H}}(B, \sigma, M)$ if and only if there is $A \in \mathcal{A}(M)$ such that $B \vdash_{\infty} A : \sigma$.*

Our aim is to prove the following implication:

$$B \vdash_{\infty} M : \sigma \text{ implies } App_{\mathcal{H}}(B, \sigma, M).$$

We will build the proof in two steps. First, it will be proved that:

$$B \vdash_{\infty} M : \sigma \text{ implies } Comp_{\mathcal{H}}(B, \sigma, M), \tag{11.1}$$

and then

$$Comp_{\mathcal{H}}(B, \sigma, M) \text{ implies } App_{\mathcal{H}}(B, \sigma, M), \tag{11.2}$$

where $Comp_{\mathcal{H}}(B, \sigma, M)$ (read: the term M is *computable* of type σ with respect to a basis B) is a property of the triple $< B, \sigma, M >$. We will prove Eq. (11.1) by induction on terms and Eq. (11.2) by induction on types.

A basis B is *finite* if and only if $B(y) \simeq_{\infty} \omega$ except in a finite number of variables. We will use $[\sigma_1/x_1, ..., \sigma_n/x_n]$ to denote a finite basis. By Lemma 10.1.7.(i), in this section we limit ourselves to consider only such a kind of basis.

Let B and B' be two basis. $B \cup B'$ denotes the basis such that, for every x, $B \cup B'(x) = B(x) \wedge B'(x)$ (remember that, for every type σ, $\sigma \wedge \omega \simeq \sigma$).

Definition 11.1.39. *The predicate $Comp_{\mathcal{H}}$ is defined by induction on types as follows:*

- *$Comp_{\mathcal{H}}(B, \omega, M)$ is true;*
- *$Comp_{\mathcal{H}}(B, \phi, M)$ if and only if $\forall \vec{N}$, $App_{\mathcal{H}}(B, \phi, M\vec{N})$;*
- *$Comp_{\mathcal{H}}(B, \sigma \to \tau, M)$ if and only if*
  $\forall N \in \Lambda$, $Comp_{\mathcal{H}}(B', \sigma, N)$ implies $Comp_{\mathcal{H}}(B \cup B', \tau, MN)$;
- *$Comp_{\mathcal{H}}(B, \sigma \wedge \tau, M)$ if and only if*
  both $Comp_{\mathcal{H}}(B, \sigma, M)$ and $Comp_{\mathcal{H}}(B, \tau, M)$).

Note that $Comp_{\mathcal{H}}(B, \phi, M)$ implies $App_{\mathcal{H}}(B, \phi, M)$, as a particular case.

Lemma 11.1.40. *$Comp_{\mathcal{H}}(B, \sigma, M)$ and $M =_{\Lambda} M'$ imply $Comp_{\mathcal{H}}(B, \sigma, M')$, i.e. $Comp_{\mathcal{H}}$ is defined modulo $=_{\Lambda}$ on terms.*

Proof. The proof is given by induction on σ.
The case $\sigma \equiv \omega$ is obvious. Let $\sigma \equiv \phi$ and $M =_{\Lambda} M'$; so $Comp_{\mathcal{H}}(B, \phi, M)$ if and only if $\forall \vec{N}$, $App_{\mathcal{H}}(B, \phi, M\vec{N})$, thus $\forall \vec{N}$, $\exists A \in \mathcal{A}(M\vec{N})$, $B \vdash_{\infty} A : \phi$, which imply $\forall \vec{N}$, $\exists A' \in \mathcal{A}(M'\vec{N})$, $B \vdash_{\infty} A' : \phi$ (since approximants are defined up to $=_{\Lambda}$) if and only if $\forall \vec{N}$, $App_{\mathcal{H}}(B, \phi, M'\vec{N})$ (by definition of $App_{\mathcal{H}}$) if and only if $Comp_{\mathcal{H}}(B, \phi, M')$ (by definition of $Comp_{\mathcal{H}}$). The cases $\sigma \equiv \mu \to \nu$ and $\sigma \equiv \mu \wedge \nu$ follow immediately from the inductive hypothesis. □

Point *(ii)* of the following lemma proves the implication of Eq. (11.2).

Lemma 11.1.41. (i) *$App_{\mathcal{H}}(B, \sigma, x\vec{M})$ implies $Comp_{\mathcal{H}}(B, \sigma, x\vec{M})$.*
(ii) *$Comp_{\mathcal{H}}(B, \sigma, M)$ implies $App_{\mathcal{H}}(B, \sigma, M)$.*

Proof. The proof is done by mutual induction on σ. The only not obvious case is when $\sigma \equiv \tau \to \rho$ and $\rho \not\simeq \omega$.

(i) By induction on (ii), $Comp_{\mathcal{H}}(B', \tau, N)$ implies $App_{\mathcal{H}}(B', \tau, N)$. Thus, if $App_{\mathcal{H}}(B, \tau \to \rho, x\vec{M})$ then $App_{\mathcal{H}}(B \cup B', \rho, x\vec{M}N)$ by rule $(\to E)$, since if $x\vec{A} \in \mathcal{A}(\vec{M})$ and $A' \in \mathcal{A}(N)$ then $x\vec{A}A' \in \mathcal{A}(\vec{M}N)$. Hence, by induction, $Comp_{\mathcal{H}}(B \cup B', \rho, x\vec{M}N)$, so $Comp_{\mathcal{H}}(B, \tau \to \rho, x\vec{M})$ by definition of $Comp_{\mathcal{H}}$.

(ii) Let $z \notin FV(M)$ be such that $B(z) \simeq_\infty \omega$. Note that there exists a such z, from the hypothesis that B is a finite basis. Clearly $z \in \mathcal{A}$ and $[\tau/z] \vdash_\infty z : \tau$, so by induction on (i), $App_{\mathcal{H}}([\tau/z], \tau, z)$ implies $Comp_{\mathcal{H}}([\tau/z], \tau, z)$. Thus $Comp_{\mathcal{H}}(B, \tau \to \rho, M)$ and $Comp_{\mathcal{H}}([\tau/z], \tau, z)$ imply $Comp_{\mathcal{H}}(B[\tau/z], \rho, Mz)$, and this implies $App_{\mathcal{H}}(B[\tau/z], \rho, Mz)$, by induction; which means there is $A \in \mathcal{A}(Mz)$ such that $B[\tau/z] \vdash_\infty A : \rho$. Note that $A \equiv \Omega$ is not possible, since by hypothesis $\rho \not\simeq \omega$. By rule $(\to I)$, $B \vdash_\infty \lambda z.A : \tau \to \rho$. By definition of approximant, $A \in \mathcal{A}(Mz)$ implies $\lambda z.A \in \mathcal{A}(\lambda z.Mz)$. Now there are two cases.
   1. $M \to_\Lambda^* xM_1...M_m$, thus A is of the shape $xA_1...A_m z$.
      So $B \vdash_\infty xA_1...A_m : \tau \to \rho$, by Property 11.1.8. The proof is given, since $xA_1...A_m \in \mathcal{A}(M)$.
   2. Otherwise $M =_\Lambda \lambda y.M'$, so $\lambda z.Mz =_\Lambda \lambda z.M'[z/y] =_\alpha \lambda y.M'$, which implies $\lambda z.A \in \mathcal{A}(M)$, and the proof is given. □

$Comp_{\mathcal{H}}$ is closed under $\leq_\infty$.

Lemma 11.1.42. (i) *$Comp_{\mathcal{H}}(B, \phi, M)$ if and only if $Comp_{\mathcal{H}}(B, \omega \to \phi, M)$.*
(ii) *If $Comp_{\mathcal{H}}(B, \sigma, M)$ and $\sigma \leq_\infty \tau$ then $Comp_{\mathcal{H}}(B, \tau, M)$.*

Proof. (i) $(\Rightarrow)$ $Comp_{\mathcal{H}}(B, \phi, M)$ implies $\forall N$, $\forall \vec{P}$, $App_{\mathcal{H}}(B, \phi, MN\vec{P})$ (by definition of $Comp_{\mathcal{H}}$), which imply $\forall N$, $Comp_{\mathcal{H}}(B, \phi, MN)$ (by definition of $Comp_{\mathcal{H}}$). Hence, $Comp_{\mathcal{H}}(B, \omega \to \phi, M)$ again by definition of $Comp_{\mathcal{H}}$.
$(\Leftarrow)$ We prove that $\forall \vec{N}$, $App_{\mathcal{H}}(B, \phi, M\vec{N})$, by induction on $\|\vec{N}\|$. Then the proof follows, by definition of $Comp_{\mathcal{H}}$. Clearly, $Comp_{\mathcal{H}}(B, \omega \to \phi, M)$ implies $App_{\mathcal{H}}(B, \omega \to \phi, M)$ by Lemma 11.1.41.(ii), and this implies $App_{\mathcal{H}}(B, \phi, M)$ by rule $(\leq_\infty)$.
In case $\|\vec{N}\| \geq 1$, let $\vec{N} \equiv \vec{N'}P$. Therefore $App_{\mathcal{H}}(B, \phi, M\vec{N'})$ by induction. Clearly $App_{\mathcal{H}}(B, \omega, P)$, hence the proof follows by rules $(\leq_\infty)$ and $(\to E)$.
(ii) By induction on the rules of $\leq_\infty$.
$(a), (b), (c), (c'), (e), (r), (t)$ Trivial.

(d) $Comp_{\mathcal{H}}(B,(\sigma_0 \to \tau_0) \wedge (\sigma_0 \to \pi_0), M)$ implies, by definition of $Comp_{\mathcal{H}}$, $Comp_{\mathcal{H}}(B, \sigma_0 \to \tau_0, M)$ and $Comp_{\mathcal{H}}(B, \sigma_0 \to \pi_0, M)$.
If $Comp_{\mathcal{H}}(B', \sigma_0, N)$ then $Comp_{\mathcal{H}}(B \cup B', \tau_0, MN)$ and $Comp_{\mathcal{H}}(B \cup B', \pi_0, MN)$; therefore, both $Comp_{\mathcal{H}}(B \cup B', \tau_0 \wedge \pi_0, MN)$ and $Comp_{\mathcal{H}}(B, \sigma_0 \to (\tau_0 \wedge \pi_0), M)$, by definition of $Comp_{\mathcal{H}}$.

(f) Let $\sigma'_0 \leq_\infty \sigma_0$, $\tau_0 \leq_\infty \tau'_0$ and $Comp_{\mathcal{H}}(B, \sigma_0 \to \tau_0, M)$.
If $Comp_{\mathcal{H}}(B', \sigma'_0, N)$ then $Comp_{\mathcal{H}}(B', \sigma_0, N)$ by induction; hence $Comp_{\mathcal{H}}(B \cup B', \tau_0, MN)$ by definition of $Comp_{\mathcal{H}}$. Again, by induction, $Comp_{\mathcal{H}}(B \cup B', \tau'_0, MN)$, so the proof is done.

(g), (h3) Easy, since $Comp_{\mathcal{H}}(B, \omega, MN)$ is always true.

(h1), (h2) By point (i) of this lemma. □

The following lemma allows us to prove the implication of Eq. (11.1).

Lemma 11.1.43. *Let* $\mathrm{FV}(M) \subseteq \{x_1, ..., x_n\}$ *and* $B = [\sigma_1/x_1, ..., \sigma_n/x_n]$. *If* $Comp_{\mathcal{H}}(B_i, \sigma_i, N_i)$ $(1 \leq i \leq n)$ *and* $B \vdash_\infty M : \tau$ *then*

$$Comp_{\mathcal{H}}(B_1 \cup ... \cup B_n, \tau, M[N_1/x_1, ..., N_n/x_n]).$$

Proof. By induction on the derivation of $B \vdash_\infty M : \tau$. The most interesting case is when the last applied rule is $(\to I)$; so $M \equiv \lambda x.M'$, $\tau \equiv \mu \to \nu$ and the derivation is:

$$\dfrac{\dfrac{\vdots}{B[\mu/x] \vdash_\infty M' : \nu}\,\cdots}{B \vdash_\infty \lambda x.M' : \mu \to \nu}\,(\to I)$$

If $Comp_{\mathcal{H}}(B', \mu, N)$ then $Comp_{\mathcal{H}}(B' \cup_{1 \leq j \leq n} B_j, \nu, M'[N_1/x_1, ..., N_n/x_n, N/x])$, by induction; hence $Comp_{\mathcal{H}}(B' \cup_{1 \leq j \leq n} B_j, \nu, (\lambda x.M'[N_1/x_1, ..., N_n/x_n])N)$ by Lemma 11.1.40, thus $Comp_{\mathcal{H}}(B_1 \cup ... \cup B_n, \tau, M[N_1/x_1, ..., N_n/x_n])$ by definition of $Comp_{\mathcal{H}}$.
All other cases follow directly from the inductive hypothesis. □

▼ **Proof of the $\mathcal{H}$-Approximation Theorem** (Theorem 11.1.19 pag. 126).

(⇒) Let us prove that $B \vdash_\infty M : \sigma$ implies $B \vdash_\infty A : \sigma$, for some $A \in \mathcal{A}(M)$.
Clearly $Comp_{\mathcal{H}}([\tau/x], \tau, x)$, by Lemma 11.1.41.(i).
Let $\mathrm{FV}(M) \subseteq \{x_1, ..., x_n\}$, so we can assume $B = [\sigma_1/x_1, ..., \sigma_n/x_n]$ without loss of generality, by Lemma 10.1.7.(i). Hence $B \vdash_\infty M : \sigma$ and $Comp_{\mathcal{H}}([\sigma_i/x_i], \sigma_i, x_i)$ $(1 \leq i \leq n)$ imply $Comp_{\mathcal{H}}(B, \sigma, M)$ by Lemma 11.1.43. Thus by Lemma 11.1.41.(ii) and the definition of approximant, the proof is done.

(⇐) We must prove that $B \vdash_\infty A : \sigma$ for some $A \in \mathcal{A}(M)$ implies $B \vdash_\infty M : \sigma$. By definition, there is M' such that $M =_\Lambda M'$ and A matches M' except at occurrences of Ω. A derivation of $B \vdash_\infty A : \sigma$ can be transformed into a derivation of $B \vdash_\infty M' : \sigma$, simply by replacing every subderivation

$$\frac{}{B \vdash_\infty \Omega : \omega}(\omega) \quad \text{by} \quad \frac{}{B \vdash_\infty N : \omega}(\omega),$$

where N is the subterm replaced by Ω in M'. $B \vdash_\infty M' : \sigma$ implies $B \vdash_\infty M : \sigma$, since the type assignment system is closed under $=_\Lambda$ on terms, as a consequence of the fact that it induces a $\lambda\Lambda$-model, and the proof is given. ■

11.1.3 Proof of Semiseparability, $\mathcal{H}$-Discriminability and $\mathcal{H}$-Characterization Theorems

In order to prove the three theorems, we need a deeper investigation on the preorder relation induced by the model $\mathcal{H}$ on $\mathcal{A}$. First of all, let us formalize a stratified version of the relation $\ll$ and of its negation $\not\ll$.

Definition 11.1.44. *Let a* path c *be a finite sequence of natural numbers greater than* 0 *(ϵ is the empty sequence).*

(i) $A \ll_c A'$ *if and only if one of the following cases arises:*
  (a) $A \equiv \Omega$;
  (b) $c \equiv \epsilon$, $A =_\eta \lambda x_1 ... x_n . x A_1 ... A_m$ *and* $A' =_\eta \lambda x_1 ... x_n . x A'_1 ... A'_m$ *for some* $n, m \in \mathbb{N}$;
  (c) $c \equiv j, c'$, $A =_\eta \lambda x_1 ... x_n . x A_1 ... A_m$, $A' =_\eta \lambda x_1 ... x_n . x A'_1 ... A'_m$, *and* $A_j \ll_{c'} A'_j$ $(1 \leq j \leq m)$.

(ii) $A \not\ll_c A'$ *if and only if one of the following cases arises:*
  (a) $A \not\equiv \Omega$ *and* $A' \equiv \Omega$;
  (b) $A =_\eta \lambda x_1 ... x_n . x A_1 ... A_m$, $A' =_\eta \lambda x_1 ... x_p . y A'_1 ... A'_q$ *and, either* $x \not\equiv y$ *or* $| n - m | \neq | p - q |$;
  (c) $c \equiv j, c'$, $A =_\eta \lambda x_1 ... x_n . x A_1 ... A_m$, $A' =_\eta \lambda x_1 ... x_n . x A'_1 ... A'_m$ *and* $A_j \not\ll_{c'} A'_j$ $(1 \leq j \leq m)$.

Two approximants A and A' are structurally different in a path c *if* $A \not\ll_c A'$ *is proved without using rule* (a), *they are* structurally similar in a path c *if and only if* $A \not\ll_c A'$ *is proved using rule* (a).

Example 11.1.45. $x \ll_{3,1} x$; in fact, $x =_\eta \lambda x_1 x_2 x_3 . x x_1 x_2 x_3$ and $x_3 \ll_1 x_3$, since $x_3 =_\eta \lambda y_1 . x_3 y_1$ and $y_1 \ll_\epsilon y_1$.
Moreover, $\lambda x . x O \Omega \ll_{2,3} \lambda x . x K I$ since $\Omega \ll_3 I$, but $\lambda x . x O \Omega \not\ll_1 \lambda x . x K I$ since $O \not\ll_\epsilon K$; in particular, $\lambda x . x O \Omega$ and $\lambda x . x K I$ are structurally similar along path 2, but they are structurally different along the path 1.

Property 11.1.46. Let c be a path.

(i) $A \ll A'$ if and only if $A \ll_c A'$, for all path c.
(ii) $A \not\ll A'$ if and only if $A \not\ll_c A'$, for some path c.

Proof. (i) ($\Rightarrow$) Easy, by Definition 11.1.15 (pag. 125).
($\Leftarrow$) By induction on A.

(ii) We show that either $A \ll_c A'$ or $A \not\ll_c A'$ in an exclusive sense, for all paths c. In case $A \equiv \Omega$ then $A \ll_c A$ by Definition 11.1.44.(i).(a); while A, A' are not in relation $\not\ll_c$, for all paths c.
Let $A \not\equiv \Omega$ and $A' \equiv \Omega$ then $A \not\ll_c A'$, for all paths c; while A, A' are not in relation $\ll_c$, for all path c.
Otherwise, $A \equiv \lambda x_1...x_n.xA_1...A_m$, $A' \equiv \lambda x_1...x_p.yA'_1...A'_q$, and the proof follows by taking into account all possible cases.

Thus by the previous point, the proof is done. □

The proof of the semiseparability theorem will be done by showing a semiseparability algorithm, which is in some sense an extension to approximants of the Λ-separability algorithm. The main difference between the two algorithms is that the semiseparability one is defined depending on a particular path. Namely, given two approximants A and A' and a path c such that $A \not\ll_c A'$, it gives as output a context $C[.]$ such that $C[A] \to^*_{\Lambda\Omega} I$ while $C[A'] \to^*_{\Lambda\Omega} \Omega$. The path c is explicitly used by the algorithm.

The terms B^n, O^n and U^i_m are defined as in Sect. 2.1.2. Since approximants are $\Lambda\Omega$-normal forms, the notion of args (see Definition 2.1.12 pag. 29) is naturally extended to them, by $\mathsf{args}(\Omega) = 0$. Moreover, if $P \in \Lambda\Omega$ has a $\Lambda\Omega$-normal form, then $\mathsf{nf}_{\Lambda\Omega}(P)$ will denote the $\Lambda\Omega$-normal form of P.

The algorithm is defined as a formal system, proving statements of the shape:

$$A, A' \Rrightarrow^c_D C[.],$$

where $A \not\ll_c A'$ and $C[.]$ is a context.

The design of the algorithm follows the same pattern as the separability algorithm, but for some rules dealing with Ω; it is presented in Fig. 11.2 (pag.138). For the sake of simplicity, we assume that all bound and free variables have different names.

Lemma 11.1.47. *Let A, A' be two approximants, $r \geq \max\{\mathsf{args}(A), \mathsf{args}(A')\}$ and $C^r_x[.] \equiv (\lambda x.[.])B^r$.*

(i) *There is $\bar{A} \in \mathcal{A}$ such that $C^r_x[A] \to^*_{\Lambda\Omega} \bar{A}$ and $r \geq \mathsf{args}(\bar{A})$.*
(ii) *If $A \not\ll_c A'$ then $\mathsf{nf}_{\Lambda\Omega}(C^r_x[A]) \not\ll_c \mathsf{nf}_{\Lambda\Omega}(C^r_x[A'])$.*

Proof. The proof is quite similar to the proof of Lemma 2.1.15, by replacing Λ-NF by $\mathcal{A}$ and $\not\sim_c$ by $\not\ll_c$.

(i) By induction on A.
(ii) By induction on c. □

Now we can prove that the algorithm is correct and complete.

Let A, A' be approximants such that $A \not\lessapprox_c A'$ and $r \geq \max\{\mathsf{args}(A), \mathsf{args}(A')\}$.

The rules of the system proving statements $A, A' \Rrightarrow^c_D C[.]$, are the following:

$$\frac{}{xA_1...A_m, \Omega \Rrightarrow^c_D (\lambda x.[.])O^m} \ (D1)$$

$$\frac{xA_1...A_m, \Omega \Rrightarrow^c_D C[.]}{\lambda x_1...x_n.xA_1...A_m, \Omega \Rrightarrow^c_D C[[.]x_1...x_n]} \ (D2)$$

$$\frac{p \leq q \qquad xA_1...A_m x_{p+1}...x_q, yA'_1...A'_n \Rrightarrow^c_D C[.]}{\lambda x_1...x_p.xA_1...A_m, \lambda x_1...x_q.yA'_1...A'_n \Rrightarrow^c_D C[[.]x_1...x_q]} \ (D3)$$

$$\frac{q < p \qquad xA_1...A_m, yA'_1...A'_n x_{q+1}...x_p \Rrightarrow^c_D C[.]}{\lambda x_1...x_p.xA_1...A_m, \lambda x_1...x_q.yA'_1...A'_n \Rrightarrow^c_D C[[.]x_1...x_p]} \ (D4)$$

$$\frac{n < m}{xA_1...A_m, xA'_1...A_n \Rrightarrow^c_D (\lambda x.[.])O^m \underbrace{I.....I}_{m-n-2} KI\Omega} \ (D5)$$

$$\frac{m < n}{xA_1...A_m, xA'_1...A'_n \Rrightarrow^c_D (\lambda x.[.])O^n \underbrace{I.....I}_{n-m-2} K\Omega I} \ (D6)$$

$$\frac{x \not\equiv y}{xA_1...A_m, yA'_1...A'_n \Rrightarrow^c_D (\lambda xy.[.])(\lambda x_1...x_m.I)(\lambda x_1...x_n.\Omega)} \ (D7)$$

$$\frac{x \notin \mathrm{FV}(A_k) \cup \mathrm{FV}(A'_k) \qquad A_k \not\lessapprox A'_k \qquad A_k, A'_k \Rrightarrow^c_D C[.]}{xA_1...A_m, xA'_1...A'_m \Rrightarrow^{k,c}_D C[(\lambda x.[.])U^k_m]} \ (D8)$$

$$\frac{x \in \mathrm{FV}(A_k) \cup \mathrm{FV}(A'_k) \qquad A_k \not\lessapprox A'_k \qquad C^r_x[.] \equiv (\lambda x.[.])B^r \qquad \mathsf{nf}_{\Lambda\Omega}(C^r_x[A_k]), \mathsf{nf}_{\Lambda\Omega}(C^r_x[A'_k]) \Rrightarrow^c_D C[.]}{xA_1...A_m, xA'_1...A'_m \Rrightarrow^{k,c}_D C[C^r_x[.] \underbrace{I.....I}_{r-m} U^k_r]} \ (D9)$$

Fig. 11.2. Semiseparability algorithm

Lemma 11.1.48 (Termination).
If $A \not\ll_c A'$ then $A, A' \Rrightarrow^c_D C[.]$.

Proof. The proof can be done by induction on c. The proof follows essentially the same pattern as the proof of termination of the Λ-separability algorithm (see Lemma 2.1.17). The only different cases are rules ($D1$) and ($D2$), and in both cases the proof is immediate. □

Lemma 11.1.49 (Correctness).
Let $A \not\ll_c A'$, for some path c.
If $A, A' \Rrightarrow^c_D C[.]$ then $C[A] \to^*_{\Lambda\Omega} I$, while $C[A'] \to^*_{\Lambda\Omega} \Omega$.

Proof. By induction on the derivation of $A, A' \Rrightarrow^c_\Lambda C[.]$.

(D1),(D2) Obvious.
(D3),(D4),(D5),(D6),(D7),(D8) Respectively similar to case (Λ1), (Λ2), (Λ3), (Λ4), (Λ5) and (Λ6) in the proof of correctness of the Λ-separability algorithm (see Lemma 2.1.18).
(D9) The proof is similar to the proof of rule (Λ7) of the proof of correctness of the Λ-Separability algorithm, using Lemma 11.1.47 instead of Lemma 2.1.15. □

▼ **Proof of the Semiseparability Theorem** (Theorem 11.1.17 pag. 125).
By Lemmas 11.1.48 and 11.1.49. ■

The proof of the $\mathcal{H}$-discriminability theorem is based on the semiseparability property. In fact, we will prove that, given two terms M and N such that $M \not\sqsubseteq_{\mathcal{H}} N$, it is always possible to find two approximants $A \in \mathcal{A}(M)$ and $A' \in \mathcal{A}(N)$ and a path c such that $A \not\ll_c A'$, and the context $C[.]$ such that $A, A' \Rrightarrow^c_D C[.]$ can be easily transformed in a context discriminating M and N. In order to choose the correct approximants, we need a lemma, based on the following formalization of a "strict" version of $\ll_c$.

Definition 11.1.50. *$A \ll^\circ_c A'$ if and only if one of the following cases arises:*

(a) *$A \equiv \Omega$ and $A' \not\equiv \Omega$;*
(b) *$c \equiv j, c'$, $A =_\eta \lambda x_1 ... x_n.xA_1 ... A_m$, $A' =_\eta \lambda x_1 ... x_n.xA'_1 ... A'_m$ where $j \leq m$, and $A_j \ll^\circ_{c'} A'_j$.*

Moreover, $A \in \mathcal{A}(M)$ is maximal along c *in $\mathcal{A}(M)$ if and only if there is no $A' \in \mathcal{A}(M)$ such that $A \ll^\circ_c A'$.*

It is easy to check that $A \ll^\circ_c A'$ implies $A \ll_c A'$, while the opposite does not hold; in particular, $A \not\ll^\circ_c A$, for all $A \in \mathcal{A}$.

Example 11.1.51. Let E_∞ be defined as in Sect. 11.1, before Theorem 11.1.31. $\lambda x.xI\Omega$ is maximal along 1 in $\mathcal{A}(\lambda x.xIE_\infty)$, although $\lambda x.xI\Omega$ is not maximal along 2 in $\mathcal{A}(\lambda x.xE_\infty I)$. In particular, there is no maximal approximant along paths starting with 2 in $\mathcal{A}(\lambda x.xIE_\infty)$.

Property 11.1.52. Let $A \in \mathcal{A}$ be such that $B \vdash_\infty A : \sigma$. Let N be such that, for each $A' \in \mathcal{A}(N)$:

$$A \not\ll_c A' \text{ implies both } \begin{cases} (a) \ A, A' \text{ are structurally similar in } c, \\ (b) \ A' \text{ is not maximal along } c \text{ in } \mathcal{A}(N). \end{cases}$$

Then $B \vdash_\infty N : \sigma$.

Proof. Let σ be a type generated from the following grammar:

$$\sigma ::= \omega \mid \sigma_1 \wedge ... \wedge \sigma_n \mid \sigma_1 \to ... \to \sigma_n \to \phi.$$

If $B \vdash_\infty A : \tau$ where $\tau \in T(C_\infty)$ then by Property 11.1.36.(i), there is a σ produced from the previous grammar such that $\tau \simeq_\infty \sigma$; hence there is no loss of generality by considering only this kind of types.

The proof is given by induction on the pair (A, σ) endowed by the lexicographic order, where σ is a type of the considered grammar. If $\sigma \equiv \omega$ then the proof is trivial. If $\sigma \equiv \sigma_1 \wedge ... \wedge \sigma_k$ for some $k \geq 1$ then the proof follows by induction. Thus, let $A \not\equiv \Omega$ and $\sigma \equiv \sigma_1 \to ... \to \sigma_k \to \phi$ for some $k \geq 1$.

- Let $A \equiv xA_1...A_m$ for some $m \in \mathbb{N}$.
  Clearly $\Omega \in \mathcal{A}(N)$ and $A \not\ll_c \Omega$, for all paths c; but Ω is not maximal along c in $\mathcal{A}(N)$, by hypotheses. Namely, there is $A' \in \mathcal{A}(N)$ such that $\Omega \ll^\circ_\epsilon A'$, i.e. $A' \not\equiv \Omega$, by definition of $\ll^\circ_c$. So, there is $n \geq \max\{k, m\}$ such that $A =_\eta \lambda y_1...y_n.xA_1...A_m y_1...y_n$ and $A' =_\eta \lambda y_1...y_n.xA'_1...A'_{m+n} \in \mathcal{A}(N)$ by hypothesis (a). Actually, all approximants of N but Ω have the shape of A'; furthermore, $N =_{\mathcal{A}\eta} \lambda y_1...y_n.xN_1...N_{m+n}$ by Property 11.1.20.(iii), where $A'_i \in \mathcal{A}(N_i)$ $(1 \leq i \leq m+n)$. There are two cases.
  - Let $k > m$. So $B \vdash_\infty \lambda y_1...y_n.xA_1...A_m y_1...y_n : \sigma_0 \to ... \to \sigma_k \to \phi$, by Property 11.1.8. By Property 11.1.36.(ii),
  $$B \vdash_\infty \lambda y_1...y_n.xA_1...A_m y_1...y_n : \sigma_0 \to ... \to \sigma_k \to \underbrace{\omega \to ... \to \omega}_{n-k} \to \phi.$$
  By Lemma 10.1.7.(vi)
  $$B[\sigma_1/y_1, ..., \sigma_k/y_k, \omega/y_{k+1}, ..., \omega/y_n] \vdash_\infty xA_1...A_m y_1...y_n : \phi.$$
  So $B[\sigma_1/y_1, ..., \sigma_k/y_k, \omega/y_{k+1}, ..., \omega/y_n] \vdash_\infty x : \sigma'_1 \to ... \to \sigma'_{n+m} \to \phi$, $B[\sigma_1/y_1, ..., \sigma_k/y_k, \omega/y_{k+1}, ..., \omega/y_n] \vdash_\infty A_i : \sigma'_i$ $(1 \leq i \leq m)$ and $B[\sigma_1/y_1, ..., \sigma_k/y_k, \omega/y_{k+1}, ..., \omega/y_n] \vdash_\infty y_i : \sigma'_{i+m}$ $(1 \leq i \leq n)$ by Lemma 10.1.7.(vii). Note that, both $\sigma_i \leq_\infty \sigma'_{i+m}$ $(1 \leq i \leq k)$ and $\omega \leq_\infty \sigma'_{i+m}$ $(k+1 \leq i \leq n)$ by Lemma 10.1.7.(ii). By Property 10.1.6
  $$\sigma'_1 \to ... \to \sigma'_{n+m} \to \phi \leq_\infty \sigma'_1 \to ... \to \sigma'_m \to \sigma_1 \to ... \to \sigma_k \to \underbrace{\omega \to ... \to \omega}_{n-k} \to \phi$$
  and by rule $(\leq_\infty)$
  $$B[\sigma_1/y_1, ..., \sigma_k/y_k, \omega/y_{k+1}, ..., \omega/y_n] \vdash_\infty x : \sigma'_1 \to ... \to \sigma'_m \to \sigma_1 \to ... \to \sigma_k \to \underbrace{\omega \to ... \to \omega}_{n-k} \to \phi.$$

If $A'_i \in \mathcal{A}(N)$ $(1 \leq i \leq m)$ then it would be clear that

$$A_i \not\lll_c A'_i \text{ implies both } \begin{cases} (a)\ A_i, A'_i \text{ are structurally similar in } c, \\ (b)\ A'_i \text{ is not maximal along } c \text{ in } \mathcal{A}(N_i). \end{cases}$$

Thus $B[\sigma_1/y_1, ..., \sigma_k/y_k, \omega/y_{k+1}, ..., \omega/y_n] \vdash_\infty N_i : \sigma'_i$ $(1 \leq i \leq m)$ by induction on A.

$B[\sigma_1/y_1, ..., \sigma_k/y_k, \omega/y_{k+1}, ..., \omega/y_n] \vdash_\infty y_i : \sigma_i$ $(1 \leq i \leq k)$, by rule (var); moreover, if $A'_i \in \mathcal{A}(N)$ $(m+1 \leq i \leq m+k)$ then it would be clear that

$$y_i \not\lll_c A'_i \text{ implies both } \begin{cases} (a)\ y_i, A'_i \text{ are structurally similar in } c, \\ (b)\ A'_i \text{ is not maximal along } c \text{ in } \mathcal{A}(N_i). \end{cases} \text{So}$$

$B[\sigma_1/y_1, ..., \sigma_k/y_k, \omega/y_{k+1}, ..., \omega/y_n] \vdash_\infty N_i : \sigma_i$ $(m+1 \leq i \leq m+k)$ by induction on σ in case $m = 0$, by induction on A otherwise. Moreover, $B[\sigma_1/y_1, ..., \sigma_k/y_k, \omega/y_{k+1}, ..., \omega/y_n] \vdash_\infty N_i : \omega$ $(m+k+1 \leq i \leq m+n)$ by rule (ω).

Hence $B[\sigma_1/y_1, ..., \sigma_k/y_k, \omega/y_{k+1}, ..., \omega/y_n] \vdash_\infty xN_1...N_{m+n} : \phi$ by $m+n$ applications of rule $(\to E)$, that implies $B \vdash_\infty \lambda y_1...y_n.xN_1...N_{m+n} : \sigma$ by n applications of rule $(\to I)$ and by an application of $(\leq_\infty)$. Since typings are preserved by $=_{\Lambda\eta}$, the proof follows.

– In case $k \leq m$ the proof is simpler.

- Let $A \equiv \lambda x.\bar{A}$ and $\sigma \equiv \mu \to \nu$ (i.e. $k \geq 1$). Clearly $\Omega \in \mathcal{A}(N)$ and $A \not\lll_c \Omega$, so by hypothesis (b) there is $A' \in \mathcal{A}(N)$ such that $\Omega \lll_c^\circ A'$ and $A' =_\eta \lambda x.A^*$. So by Property 11.1.20.(ii), $N =_{\Lambda\eta} \lambda x.N'$ and $A^* \in \mathcal{A}(N')$. Moreover, $B[\mu/x] \vdash_\infty \bar{A} : \nu$ by Lemma 10.1.7.(vi).
  Note that $\bar{A}$ and N' satisfy the hypothesis of this property (indeed, $\not\lll_c$ is defined modulo $=_{\Lambda\eta}$), thus $B[\mu/x] \vdash_\infty N' : \nu$ by induction. This implies $B \vdash_\infty \lambda x.N' : \mu \to \nu$, by rule $(\to I)$.
- Let $A \equiv \lambda x.\bar{A}$ and $k = 0$. Since $\phi \simeq_\infty \omega \to \phi$, the proof can be done as in the previous point. □

Example 11.1.53.
$\mathcal{A}(E_\infty) = \{\Omega, \lambda z_0 z_1.z_0\Omega, \lambda z_0 z_1.z_0(\lambda z_2.z_1(...(\lambda z_{n+1}.z_n\Omega)...)) \mid n \geq 0\}$.
It is easy to see that $I \not\lll A'$, for all $A' \in \mathcal{A}(E_\infty)$.
Let $A_n \equiv \lambda z_0 z_1.z_0(\lambda z_2.z_1(...(\lambda z_{n+1}.z_n\Omega)...))$, therefore $A_n \lll_c^\circ A_{n+1}$ where $c \equiv \underbrace{1.....1}_{n}$. It is easy to see that $B \vdash_\infty I : \sigma$ implies that there is $A \in \mathcal{A}(E_\infty)$ such that $B \vdash_\infty A : \sigma$, by the previous property.
This agrees with the Theorem 11.1.31.

The next lemma shows that, if $M \not\sqsubseteq_{\mathcal{H}} N$, there are always two approximants of them which are the "witness" of the difference between M and N, and such that the context semiseparating the two approximants can be transformed in a context descriminating the two terms.

Lemma 11.1.54. *$M \not\sqsubseteq_{\mathcal{H}} N$ implies that there are two approximants $A \in \mathcal{A}(M)$ and $A' \in \mathcal{A}(N)$, and a path c such that $A \not\lll_c A'$, and either A and A' are structurally different in c, or A' is maximal along c in $\mathcal{A}(N)$.*

Proof. $M \not\sqsubseteq_{\mathcal{H}} N$ means that there are a basis B and a type σ such that $B \vdash_\infty M : \sigma$ while $B \not\vdash_\infty N : \sigma$. By the approximation theorem, this means that there is $A \in \mathcal{A}(M)$ such that $B \vdash A : \sigma$, while there is not $A' \in \mathcal{A}(N)$, such that $B \vdash A' : \sigma$. This implies: $A \not\leq A'$, for all $A' \in \mathcal{A}(N)$.
The proof follows by Property 11.1.52. □

Now we are ready to prove the $\mathcal{H}$-discriminability theorem.

▼ **Proof of the $\mathcal{H}$-Discriminability Theorem** (Theorem 11.1.25 pag. 127).
Let $M \not\sqsubseteq_{\mathcal{H}} N$. Choose $A \in \mathcal{A}(M)$, $A' \in \mathcal{A}(N)$ and a path c satisfying Lemma 11.1.54. Hence $A, A' \Rrightarrow_D^c C[.]$ such that $C[A] \to_{\Lambda\Omega}^* I$ and $C[A'] \to_{\Lambda\Omega}^* \Omega$. Let $C'[.]$ be the context obtained by $C[.]$ by replacing every occurrence of Ω by a Λ-unsolvable term (say DD). We will prove that $C'[.]$ is a **H**-discriminating context between M and N, by induction on the derivation d of $A, A' \Rrightarrow^c C[.]$.
If the only used rule is an axiom, then if the axiom is either $(D5)$ or $(D6)$ or $(D7)$, the proof follows from the fact that M and N have both Λ-hnf, and by Property 11.1.20.(iii), their Λ-hnf's have the same shape respectively of A and A'. If axiom $(D1)$ was used, then $A' \equiv \Omega$ and A' is maximal along the empty path in $\mathcal{A}(N)$. Hence, N is Λ-unsolvable and $C[.]$ is the discriminating context.
If the last rule applied is $(D2)$, $(D3)$, $(D4)$, $(D8)$ or $(D9)$, then the proof follows by induction, always taking into account the fact that M and N both have Λ-hnf, and by Property 11.1.20.(iii), their Λ-hnf's have the same shape respectively of A and A'. ■

Example 11.1.55. Let $M \equiv \lambda x.xII$ and $N \equiv \lambda x.xE_\infty(DD)$.
Let B_ω be the basis such that $B_\omega(x) = \omega$, for all $x \in \mathrm{Var}$; it is easy to see that:

$$B_\omega \vdash_\infty M : (\omega \to (\phi \to \phi) \to \phi \to \phi) \to \phi \to \phi,$$
$$B_\omega \not\vdash_\infty N : (\omega \to (\phi \to \phi) \to \phi \to \phi) \to \phi \to \phi.$$

Hence $M \not\sqsubseteq_{\mathcal{H}} N$.
Two approximants respectively of $\lambda x.xII$ and $\lambda x.xE_\infty(DD)$ and a path satisfying the conditions of Lemma 11.1.54 are $\lambda x.x\Omega I$, $\lambda x.x\Omega\Omega$ and path 2. In fact, all approximants of $\lambda x.xE_\infty(DD)$ are of the shape $\lambda x.xA\Omega$, where $A \in \mathcal{A}(E_\infty)$, and so $\lambda x.x\Omega\Omega$ is maximal along path 2 in $\mathcal{A}(N)$.
$\lambda x.x\Omega I, \lambda x.x\Omega\Omega \Rrightarrow_D^2 C[.]$ where $C[.] \equiv \lambda y.(\lambda x.([.]x))U_2^2 y$; in fact,

$$\cfrac{\cfrac{\cfrac{\cfrac{}{y, \Omega \Rrightarrow_D^\epsilon \lambda y.[.]}\,(D1)}{\lambda y.y, \Omega \Rrightarrow_D^\epsilon \lambda y.[.]y}\,(D2)}{x\Omega I, x\Omega\Omega \Rrightarrow_D^2 \lambda y.(\lambda x.[.])U_2^2 y}\,(D8)}{\lambda x.x\Omega I, \lambda x.x\Omega\Omega \Rrightarrow_D^2 \lambda y.(\lambda x.([.]x))U_2^2 y}\,(D3)$$

It is easy to see that $C[\lambda x.x\Omega I] \to_{\Lambda\Omega} I$ and $C[\lambda x.xII] \Downarrow_{\mathbf{H}} I$, while $C[\lambda x.x\Omega\Omega] \to_{\Lambda\Omega} \Omega$ and $C[\lambda x.xE_\infty(DD)] \Uparrow_{\mathbf{H}}$.

Finally, the proof of the $\mathcal{H}$-characterization theorem can be done.

▼ **Proof of the $\mathcal{H}$-Characterization Theorem** (Theorem 11.1.30 pag. 128).

($M \sqsubseteq_{\mathcal{H}} N$ implies $\forall k \geq 0, M \sqsubseteq_k N$) We will prove that if there is k such that $M \not\sqsubseteq_k N$ then there are $A \in \mathcal{A}(M)$ and $A' \in \mathcal{A}(N)$ and a path c satisfying the conditions of Lemma 11.1.54. Then, by the proof of the $\mathcal{H}$-discriminability theorem, there is a context $C[.]$ discriminating between M and N, i.e. $C[M] \rightarrow^*_\Lambda I$ and $C[N]$ is a Λ-unsolvable. By the completeness of the model with respect to the **H**-operational semantics, this implies $M \not\sqsubseteq_{\mathcal{H}} N$. The existence of $A \in \mathcal{A}(M)$ and $A' \in \mathcal{A}(N)$ and a path c satisfying the conditions of Lemma 11.1.54 can be proved by induction on k. If $k = 0$, then there are two cases.

(i) N is Λ-unsolvable and M is Λ-solvable; take an $A \not\equiv \Omega$ and the only approximant of N, namely $A' \equiv \Omega$: so the empty path satisfies the given conditions.

(ii) Every Λ-hnf of M is of the shape $\lambda x_1...x_n.xM_1...M_m$, every Λ-hnf of N is of the shape $\lambda x_1...x_p.yN_1...N_q$, and either $x \not\equiv y$ or $| m - n | \neq | p - q |$. So every $A \in \mathcal{A}(M)$ and $A' \in \mathcal{A}(N)$ such that $A, A' \not\equiv \Omega$ and the empty path satisfy the given conditions.

If $k \geq 1$, then $M =_{\Lambda\eta} \lambda x_1...x_n.xM_1...M_m$, $N =_{\Lambda\eta} \lambda x_1...x_n.xN_1...N_m$ and $M_i \not\sqsubseteq_{k-1} N_i$, for some $1 \leq i \leq m$. Then by induction there are $A^* \in \mathcal{A}(M_i)$, $A^{**} \in \mathcal{A}(N_i)$ and a path c satisfying Lemma 11.1.54.
Hence, take approximants $A \in \mathcal{A}(M)$ and $A' \in \mathcal{A}(N)$ such that both

$$A =_\eta \lambda x_1...x_n.x \underbrace{\Omega...\Omega}_{i-1} A^* \underbrace{\Omega...\Omega}_{m-i} \text{ and } A' =_\eta \lambda x_1...x_n.x \underbrace{\Omega...\Omega}_{i-1} A^{**} \underbrace{\Omega...\Omega}_{m-i}:$$

so the path i, c satisfies the conditions.

($\forall k \geq 0, M \sqsubseteq_k N$ implies $M \sqsubseteq_{\mathcal{H}} N$) Assume $M \not\sqsubseteq_{\mathcal{H}} N$; by Lemma 11.1.54 there are $A \in \mathcal{A}(M)$, $A' \in \mathcal{A}(N)$ and there is a path c such that $A \not\lessdot_c A'$ and either A, A' are structurally different in c, or A' is maximal along c in $\mathcal{A}(N)$. Note that $A \not\equiv \Omega$, since $\Omega \ll A'$.
By induction on c, we prove that there is k such that $M \not\sqsubseteq_k N$.

- Let $c \equiv \epsilon$ and $A' \equiv \Omega$; so A' is maximal implies that N is Λ-unsolvable and so $M \not\sqsubseteq_0 N$.
- Let $A \equiv \lambda x_1 ... x_n.xA_1 ... A_m$ and $A' \equiv \lambda x_1 ... x_p.yA_1 ... A_q$, where either $|n - m| \neq |p - q|$ or $x \not\equiv y$; so $M \not\sqsubseteq_0 N$.
- Let $c \equiv i, c'$ and $A =_\eta \lambda x_1 ... x_n.xA_1 ... A_m$, $A' =_\eta \lambda x_1 ... x_n.xA_1 ... A_m$. Then $A_i \not\lessdot_{c'} A'_i$ and the proof follows by induction and by Property 11.1.20.(iii). ■

Remark 11.1.56. It is possible to transform the semiseparability algorithm in such a way that it always produces a head context. So, from the $\mathcal{H}$-characterization theorem and from the completeness of $\mathcal{H}$ with respect to the **H**-operational semantics it follows that **H** is head-discriminable.

11.2 The Model $\mathcal{N}$

In this section, we will introduce the filter model $\mathcal{N}$, which is fully abstract with respect to the **N**-operational semantics.

By keeping in mind Property 10.1.15 and by observing that, if N has Λ-nf and M has not Λ-nf then $M \prec_{\mathbf{N}} N$, in order to define a model correct with respect to the **N**-operational semantics, it is sufficient to ask for the property:

$$\forall M, N \text{ if } M \text{ has } \Lambda\text{-nf and } N \text{ has not } \Lambda\text{-nf then } N \sqsubset_{\mathcal{N}} M.$$

So we must define a legal type system ∇ based on a set of constants C such that there is a type σ characterizing the terms having Λ-nf. While the class of Λ-nf's is not closed under application, as the class of Λ-solvable terms, such a characterization is possible, thanks to the existence of a particular class of terms, which is a proper subclass of Λ-NF, and such that the two classes are mutually closed under application.

Definition 11.2.1. *A term M has a Λ-*persistent normal form *(Λ-pnf) if and only if it has Λ-normal form and moreover, for each sequence $\vec{N}$ of terms having Λ-nf, $M\vec{N}$ has a Λ-persistent normal form too.*

The class of Λ-persistent normal forms is not empty, since at least the variables belong to it.

Example 11.2.2. If $M \to_{\Lambda}^{*} \lambda x_1...x_n.zM_1...M_m \in \Lambda\text{-NF}$ and $z \notin \{x_1, ..., x_n\}$ then M has a Λ-pnf.

Now an alternative definition of Λ-normal forms can be given, thanks to the following property.

Property 11.2.3. A term M has a Λ-nf if and only if, for each sequence $\vec{N}$ of terms having Λ-pnf, $M\vec{N}$ has a Λ-nf too.

Proof. Let $\vec{N} \equiv N_1, ..., N_h$ $(h \geq 1)$.

($\Rightarrow$) Let $M \to_{\Lambda}^{*} \bar{M} \in \Lambda\text{-NF}$; we reason by induction on $\bar{M}$.
If $\bar{M} \equiv zM_1...M_m$ $(m \in \mathbb{N})$ then the proof is immediate; otherwise $\bar{M} \equiv \lambda x_1...x_n.zM_1...M_m$ $(n \geq 1)$. If $h \leq n$ then $M\vec{N} \to_{\Lambda}^{*} \lambda x_{h+1}\ldots x_n.\zeta M_1'\ldots M_m'$ where $\zeta \equiv z[N_1/x_1, \ldots, N_h/x_h]$ and $M_j' \equiv M_j[N_1/x_1, \ldots, N_h/x_h] =_{\Lambda} (\lambda x_1\ldots x_n.M_j)\vec{N}$ $(1 \leq j \leq m)$. Since N_i $(1 \leq i \leq h)$ has a Λ-pnf, by induction M_j' has a Λ-nf $(1 \leq j \leq m)$.
If $\zeta \equiv z$ then the proof is given. Otherwise $\zeta \equiv N_j$, for some j, and the proof follows from the definition of Λ-persistent normal form.
Let $h > n$, $\vec{P} \equiv N_1...N_n$, $\vec{Q} \equiv N_{n+1}...N_h$; so $M\vec{P}\vec{Q} \to_{\Lambda}^{*} \zeta M_1'\ldots M_m'\vec{Q}$, where $M_j' \equiv M_j[N_1/x_1, \ldots, N_n/x_n] =_{\Lambda} (\lambda x_1\ldots x_n.M_j)\vec{P}$ and $\zeta \equiv z[N_1/x_1, \ldots, N_n/x_n]$ $(1 \leq j \leq m)$. M_j' has a Λ-nf $(1 \leq j \leq m)$ by induction, while each term in $\vec{Q}$ has a Λ-nf too by hypothesis. If $\zeta \equiv z$ then the proof is trivial; otherwise it follows from the fact that ζ has Λ-pnf.

($\Leftarrow$) It is easy to check that Mx has Λ-nf if and only if M has Λ-nf, so the proof is done since x is a Λ-pnf. □

So the class of Λ-normal forms and the class of Λ-persistent normal forms are mutually closed under application.

We will take the set of constants $C_{\bowtie} = \{\phi, \psi, \omega\}$ and we will use the type constants ϕ and ψ, for characterizing respectively the class of Λ-normal forms and the class of Λ-persistent normal forms. Since a Λ-persistent normal form is a Λ-normal form too, it must be that $\psi \leq \phi$. Moreover, by definition of Λ-persistent normal form, it must be that $\psi \simeq \underbrace{\phi \to \ldots \to \phi}_{n} \to \psi$, and $\phi \simeq \underbrace{\psi \to \ldots \to \psi}_{n} \to \phi$, for all $n \geq 1$, by Property 11.2.3 ($n \in \mathbb{N}$). We want also $\omega \simeq \omega \to \omega$, since all terms must have a functional behavior.

Definition 11.2.4. *Let $C_{\bowtie} = \{\phi, \psi, \omega\}$ and $I(C_{\bowtie}) = T(C_{\bowtie})$.*
$\bowtie$ is the type system $< C_{\bowtie}, \leq_{\bowtie}, I(C_{\bowtie}) >$, where $\leq_{\bowtie}$ is the least intersection relation induced from the rules in Fig. 11.3.

$$\frac{}{\sigma \leq_{\bowtie} \omega}\,(a) \qquad \frac{}{\sigma \leq_{\bowtie} \sigma \wedge \sigma}\,(b) \qquad \frac{}{\sigma \wedge \tau \leq_{\bowtie} \sigma}\,(c) \qquad \frac{}{\sigma \wedge \tau \leq_{\bowtie} \tau}\,(c')$$

$$\frac{}{(\sigma \to \tau) \wedge (\sigma \to \pi) \leq_{\bowtie} \sigma \to (\tau \wedge \pi)}\,(d) \qquad \frac{}{\sigma \to \omega \leq_{\bowtie} \omega \to \omega}\,(g)$$

$$\frac{\sigma \leq_{\bowtie} \sigma', \tau \leq_{\bowtie} \tau'}{\sigma \wedge \tau \leq_{\bowtie} \sigma' \wedge \tau'}\,(e) \qquad \frac{\sigma' \leq_{\bowtie} \sigma, \tau \leq_{\bowtie} \tau'}{\sigma \to \tau \leq_{\bowtie} \sigma' \to \tau'}\,(f) \qquad \frac{}{\sigma \leq_{\bowtie} \sigma}\,(r) \qquad \frac{\sigma \leq \rho \quad \rho \leq \tau}{\sigma \leq_{\bowtie} \tau}\,(t)$$

$$\frac{}{\psi \leq_{\bowtie} \phi}\,(n0) \qquad \frac{}{\psi \leq_{\bowtie} \phi \to \psi}\,(n1) \qquad \frac{}{\phi \to \psi \leq_{\bowtie} \psi}\,(n2)$$

$$\frac{}{\phi \leq_{\bowtie} \psi \to \phi}\,(n3) \qquad \frac{}{\psi \to \phi \leq_{\bowtie} \phi}\,(n4) \qquad \frac{}{\omega \leq_{\bowtie} \omega \to \omega}\,(n5)$$

Fig. 11.3. $\bowtie$-intersection relation

Note that $\phi \simeq_{\bowtie} \psi \to \phi$, $\psi \simeq_{\bowtie} \phi \to \psi$ and $\omega \simeq_{\bowtie} \omega \to \omega$. Some key properties of the $\bowtie$-intersection relation are shown in the Sect. 11.2.1.

Theorem 11.2.5. *The type system $\bowtie$ is legal.*

Proof. The proof is in Sect. 11.2.1. □

So the model $\mathcal{N}$ can be defined.

Definition 11.2.6. *$\mathcal{N}$ is the $\lambda\Lambda$-model $< \mathcal{F}(\bowtie), \mathcal{F}(\bowtie), \circ_{\bowtie}, [\![.]\!]^{\mathcal{F}(\bowtie)} >$.*

We will prove that the type assignment system $\vdash_{\bowtie}$ characterizes the terms having normal form. Let B_{ψ} be the basis assigning ψ to each variable; the typing $B_{\psi} \vdash_{\bowtie} M : \phi$ can be proved if and only if M has Λ-normal form. So the correctness of $\mathcal{N}$ with respect to the **N**-operational semantics follows by Theorem 10.0.4.

Lemma 11.2.7. *If M has a Λ-normal form then there is a derivation proving $B_{\psi} \vdash_{\bowtie} M : \phi$, where $B_{\psi}(x) = \psi$ for every x.*

Proof. Since $\mathcal{N}$ is a $\lambda\Lambda$-model, by Lemma 10.1.13.(ii), the $\bowtie$-type assignment system is closed under Λ-reduction; so, let $M \equiv \lambda x_1...x_n.xM_1...M_m$ where $M_i \in \Lambda\text{-NF}$ $(1 \leq i \leq m)$, without loss of generality. The proof is given by induction on M. Thus $B_{\psi} \vdash_{\bowtie} M_i : \phi$ $(1 \leq i \leq m)$ by induction. But $B_{\psi} \vdash_{\bowtie} x : \psi$ by rule (var), hence $B_{\psi} \vdash_{\bowtie} x : \underbrace{\phi \to \ldots \to \phi}_{m} \to \psi$ by rule $(\leq_{\bowtie})$. We obtain $B_{\psi} \vdash_{\bowtie} xM_1...M_m : \psi$ by m application of the rule $(\to E)$, therefore $B_{\psi} \vdash_{\bowtie} \lambda x_1...x_n.xM_1...M_m : \underbrace{\psi \to \ldots \to \psi}_{n} \to \phi$, by rule $(\leq_{\bowtie})$ and n applications of rule $(\to I)$. Yet, by applying the rule $(\leq_{\bowtie})$ the proof is done. □

To prove the correctness of the model with respect to the **N**-operational semantics, we need an approximation theorem. We will prove that such a theorem holds for the same definition of approximants given for the model $\mathcal{H}$. First, we need to prove that the theory induced by the model $\mathcal{N}$ is a $\Lambda\eta$-theory.

Property 11.2.8. $\sim_{\mathcal{N}}$ is a $\Lambda\eta$-theory.

Proof. The theory induced by $\mathcal{N}$ is a Λ-theory by Corollary 10.0.3, since $\mathcal{N}$ is a $\lambda\Lambda$-model by Theorem 11.0.1. In order to prove that it is also a $\Lambda\eta$-theory, by Property 2.1.7, it is sufficient to prove that $I \sim_{\mathcal{N}} E$. We will prove both $I \sqsubseteq_{\mathcal{N}} E$ and $E \sqsubseteq_{\mathcal{N}} I$.

$(I \sqsubseteq_{\mathcal{N}} E)$ We prove that $B \vdash_{\bowtie} I : \sigma$ implies $B \vdash_{\bowtie} E : \sigma$.
By Property 11.2.27.(i), either $\sigma \not\simeq_{\bowtie} \omega$ or $\sigma \simeq_{\bowtie} \sigma_0 \wedge \ldots \wedge \sigma_n$ $(n \geq 0)$ such that $\forall i \leq n,\ \sigma_i \simeq_{\bowtie} \tau_1^i \to \ldots \to \tau_{m_i}^i \to \rho$, $m_i \in \mathbb{N}$ and $\rho \in \{\phi, \psi\}$. The case $\sigma \simeq_{\bowtie} \omega$ is trivial, by rule (ω); otherwise the proof can be done by induction on n.
Let $n = 0$, thus we can assume $\sigma \simeq_{\bowtie} \mu \to \nu \to \tau$, by Property 11.2.27.(ii). Note that $B \vdash_{\infty} I : \sigma$ implies $B \vdash_{\infty} I : \mu \to \nu \to \tau$, by rule $(\leq_{\infty})$; thus $B[\mu/x] \vdash_{\bowtie} x : \nu \to \tau$ by Lemma 10.1.7.(vi), and $\mu \leq_{\bowtie} \nu \to \tau$ by Lemma 10.1.7.(iii). We can build the following derivation:

$$\dfrac{\dfrac{\dfrac{\dfrac{\dfrac{\dfrac{}{B[\mu/x][\nu/y] \vdash_{\bowtie} x : \mu}(var) \quad \mu \leq_{\bowtie} \nu \to \tau}{B[\mu/x][\nu/y] \vdash_{\bowtie} x : \nu \to \tau}(\leq_{\bowtie}) \quad \dfrac{}{B[\mu/x][\nu/y] \vdash_{\bowtie} y : \nu}(var)}{B[\mu/x][\nu/y] \vdash_{\bowtie} xy : \tau}(\to E)}{B[\mu/x] \vdash_{\bowtie} \lambda y.xy : \nu \to \tau}(\to I)}{B \vdash_{\bowtie} \lambda xy.xy : \mu \to \nu \to \tau}(\to I) \quad \mu \to \nu \to \tau \leq_{\bowtie} \sigma}{B \vdash_{\bowtie} \lambda xy.xy : \sigma}(\leq_{\bowtie})$$

If $n \geq 1$ then the proof follows by inductive hypothesis.

($E \sqsubseteq_{\mathcal{N}} I$) We prove that $B \vdash_{\bowtie} E : \sigma$ implies $B \vdash_{\bowtie} I : \sigma$.
By Property 11.2.27.(i), either $\sigma \not\simeq_{\bowtie} \omega$ or $\sigma \simeq_{\bowtie} \sigma_0 \wedge ... \wedge \sigma_n$ $(n \geq 0)$ such that $\forall i \leq n,\ \sigma_i \simeq_{\bowtie} \tau_1^i \to ... \to \tau_{m_i}^i \to \rho$, $m_i \in \mathbb{N}$ and $\rho \in \{\phi, \psi\}$. The case $\sigma \simeq_{\bowtie} \omega$ is trivial, by rule (ω); otherwise the proof can be done by induction on n.
Let $n = 0$, thus we can assume $\sigma \simeq_{\bowtie} \mu \to \nu \to \tau$, by Property 11.2.27.(ii). Note that $B \vdash_{\bowtie} \lambda xy.xy : \sigma$ implies $B \vdash_{\bowtie} \lambda xy.xy : \mu \to \nu \to \tau$, by rule $(\leq_{\bowtie})$; thus, by Lemma 10.1.7.vi), this implies $B[\mu/x][\nu/y] \vdash_{\bowtie} xy : \tau$, which in turn implies $B[\mu/x][\nu/y] \vdash_{\bowtie} x : \theta \to \tau$ and $B[\mu/x][\nu/y] \vdash_{\bowtie} y : \theta$, for some θ, by Lemma 10.1.7.(vii). Hence, by Lemma 10.1.7.(iii), $\mu \leq_{\bowtie} \theta \to \tau$ and $\nu \leq_{\bowtie} \theta$ (so $\mu \leq_{\bowtie} \nu \to \tau$). Then we can build the following derivation:

$$\dfrac{\dfrac{\dfrac{\dfrac{}{B[\mu/x] \vdash_{\bowtie} x : \mu}(var) \quad \mu \leq_{\bowtie} \nu \to \tau}{B[\mu/x] \vdash_{\bowtie} x : \nu \to \tau}(\leq_{\bowtie})}{B \vdash_{\bowtie} \lambda x.x : \mu \to \nu \to \tau}(\to I) \quad \mu \to \nu \to \tau \leq_{\bowtie} \sigma}{B \vdash_{\bowtie} \lambda x.x : \sigma}(\leq_{\bowtie})$$

If $n \geq 1$ then the proof follows by inductive hypothesis. □

Let $\Lambda\Omega$ and the set $\mathcal{A}(M)$, for a term M, be defined as in Definitions 11.1.9 and 11.1.11. As before, we will use A, A' for ranging over the class of approximants. We can extend to $\Lambda\Omega$ the type assignment system $\vdash_{\bowtie}$, as we did for the system $\vdash_{\infty}$. It is immediate to verify that $\vdash_{\bowtie}$ is closed under $=_{\Lambda\Omega}$.

Theorem 11.2.9 ($\mathcal{N}$-Approximation).
$B \vdash_{\bowtie} M : \sigma$ if and only if $B \vdash_{\bowtie} A : \sigma$, for some $A \in \mathcal{A}(M)$.

Proof. The proof is in Sect. 11.2.3. □

As already observed, Λ-normal forms are approximants.

Property 11.2.10. M has Λ-normal form if and only if there is $A \in \mathcal{A}(M)$ having no occurrences of the constant Ω.

Proof. Trivial. □

Let B_ψ be the basis such that $B(x) = \psi$, for all $x \in$ Var, as in Lemma 11.2.7; moreover, let B_ω be the basis such that $B_\omega(x) = \omega$, for all $x \in$ Var.

Lemma 11.2.11. (i) *$B_\psi \vdash_{\bowtie} M : \phi$ if and only if M has Λ-normal form.*
(ii) *$B \vdash_{\bowtie} M : \sigma$ and $\sigma \not\simeq_{\bowtie} \omega$ if and only if M has a Λ-head normal form.*

Proof. (i) ($\Rightarrow$) By the approximation theorem, $B_\psi \vdash_{\bowtie} M : \phi$ implies there is $A \in \mathcal{A}(M)$ such that $B_\psi \vdash_{\bowtie} A : \phi$. We will prove that A does not contain occurrences of Ω, by induction on A. Hence M has Λ-normal form, by Property 11.2.10.

$A \equiv \Omega$ is not possible, since types derivable for Ω are equivalent to ω.
If $A \equiv \lambda x_1 ... x_n . z A_1 ... A_m$ and $B_\psi \vdash_{\bowtie} A : \phi$ then, by rule ($\leq_{\bowtie}$), $B_\psi \vdash_{\bowtie} A : \underbrace{\psi \to .... \to \psi}_{m} \to \phi$, and thus $B_\psi \vdash_{\bowtie} z A_1 ... A_m : \phi$ by Lemma 10.1.7.(vi).

By Lemma 10.1.7.(vii) this implies both (*) $B_\psi \vdash_{\bowtie} z : \sigma_1 \to ... \to \sigma_m \to \phi$, and (**) $B_\psi \vdash_{\bowtie} A_i : \sigma_i$, for some $\sigma_1, \ldots, \sigma_m$ $(1 \leq i \leq m)$. (*) implies $\psi \leq_{\bowtie} \sigma_1 \to ... \to \sigma_m \to \phi$ by Lemma 10.1.7.(iii) and so $\sigma_i \leq_{\bowtie} \phi$, since $\bowtie$ is legal and $\psi \simeq_{\bowtie} \underbrace{\phi \to ... \to \phi}_{m} \to \psi$.

Hence $B_\psi \vdash_{\bowtie} A_i : \phi$ $(1 \leq i \leq m)$, by rule ($\leq_{\bowtie}$). By inductive hypothesis A_i $(1 \leq i \leq m)$ has no occurrences of Ω and the proof is done.

($\Leftarrow$) By Lemma 11.2.7.

(ii) ($\Rightarrow$) If M is Λ-unsolvable then the only approximant of M is Ω, so the proof follows from the $\mathcal{N}$-approximation theorem.

($\Leftarrow$) Let $M \to_\Lambda^* \lambda x_1 \ldots x_n . z M_1 \ldots M_m$, for some $n, m \in \mathbb{N}$.
Let $\mu \equiv \underbrace{\omega \to \ldots \to \omega}_{m} \to \pi$, where $\pi \not\simeq \omega$. Clearly $B_\omega[\mu/z] \vdash_{\bowtie} M_i : \omega$ $(1 \leq i \leq m)$ by rule (ω); so $B_\omega[\mu/z] \vdash_{\bowtie} z M_1 \ldots M_m : \pi$ by m applications of rule ($\to E$). If $z \notin \mathrm{FV}(M)$ then, by applying n times the rule ($\to I$),

$$B_\omega[\mu/z] \vdash_{\bowtie} \lambda x_1 \ldots x_n . z M_1 \ldots M_m : \underbrace{\omega \to \ldots \to \omega}_{n} \to \pi.$$

So $B_\omega[\mu/z] \vdash_{\bowtie} M : \underbrace{\omega \to \ldots \to \omega}_{n} \to \pi$, since the type assignment system is closed under $=_\Lambda$. If $z \equiv x_k$ for some $k \leq n$ then, it is easy to see that

$$B_\omega \vdash_{\bowtie} M : \underbrace{\omega \to \ldots \to \omega}_{k-1} \to (\underbrace{\omega \to \ldots \to \omega}_{m} \to \pi) \to \underbrace{\omega \to \ldots \to \omega}_{n-k} \to \pi,$$

hence the proof follows by Theorem 11.2.23. □

The order relation $\sqsubseteq_{\mathcal{N}}$ between approximants can be axiomatized through the same syntactical relation $\ll$ defined in Definition 11.1.15 for the model $\mathcal{H}$. Namely $A \ll A'$ if and only if one of the two following cases arises:

(i) $A \equiv \Omega$;
(ii) $A =_\eta \lambda x_1 ... x_n.zA_1...A_m$ and $A' =_\eta \lambda x_1...x_n.zA'_1...A'_m$ and $A_i \ll A'_i$, for all i $(1 \leq i \leq m)$, for some $n, m \in \mathbb{N}$.

Lemma 11.2.12. *$A \sqsubseteq_\mathcal{N} A'$ if and only if $A \ll A'$.*

Proof. ($\Leftarrow$) We prove that, if $A \ll A'$, then $B \vdash_\bowtie A : \sigma$ implies $B \vdash_\bowtie A' : \sigma$. The proof is by induction on the definition of $\ll$, and it is similar to the proof of the ($\Rightarrow$) part of the proof of Lemma 11.1.18, where, instead of Properties 11.1.36.(ii) and 11.1.8, Properties 11.2.27.(ii) and 11.2.8 must be respectively used.
($\Rightarrow$) We prove, by contraposition, that $A \not\ll A'$ implies $A \not\sqsubseteq_\mathcal{N} A'$.
By the semiseparability theorem (Theorem 11.1.17), $A \not\ll A'$ implies there is a context $C[.]$ such that $C[A] \rightarrow^*_{\Lambda\Omega} I$ while $C[A'] \rightarrow^*_{\Lambda\Omega} \Omega$. Since the type assignment system is closed under $=_{\Lambda\Omega}$, this implies $C[A] \not\sqsubseteq_\mathcal{N} C[A']$, which in turn implies $A \not\sqsubseteq_\mathcal{N} A'$. □

Now we are able to prove the correctness of the model, with respect to the **N**-operational semantics.

Theorem 11.2.13 ($\mathcal{N}$-Correctness).
The model $\mathcal{N}$ is correct with respect to the **N***-operational semantics.*

Proof. By Property 10.1.15 it is sufficient to prove that, for all M, N, if M has Λ-nf and N has not Λ-nf then $M \not\sqsubseteq_\mathcal{N} N$.
Hence, either N is a Λ-unsolvable term or it has a Λ-head normal form. Then the proof follows respectively from Lemmas 11.2.11.(i) and 11.2.11.(ii). □

The model $\mathcal{N}$ also characterizes the class of Λ-persistent normal forms. Here we will prove just a partial characterization, namely that if a term can be assigned type ψ, from the basis B_ψ, then it has Λ-persistent normal form. The inverse implication has a rather technical proof, which is out of the scope of this book. The interested reader can read it in [30].

Let us notice that the fact that a term is a Λ-persistent normal form if and only if $B_\psi \vdash_\bowtie M : \psi$ implies that Λ-persistent normal forms are closed under substitution. An interesting consequence of this fact is that the set of Λ-persistent normal forms is a set of input values.

Theorem 11.2.14. *If $B_\psi \vdash_\bowtie M : \psi$ then M has a Λ-persistent normal form.*

Proof. $B_\psi \vdash_\bowtie M : \psi$ implies $B_\psi \vdash_\bowtie M : \underbrace{\phi \rightarrow \ldots \rightarrow \phi}_{n} \rightarrow \psi$ $(n \in \mathbb{N})$ by rule $(\leq_\bowtie)$. Let $\vec{N}$ be a sequence of normal forms, so by Lemma 11.2.11 $B_\psi \vdash_\bowtie N : \phi$, for every $N \in \vec{N}$. Hence $B_\psi \vdash_\bowtie M\vec{N} : \psi$ implies, by rule $(\leq_\bowtie)$, that $B_\psi \vdash_\bowtie M\vec{N} : \phi$, and so $M\vec{N}$ has a normal form. □

Using the same techniques as for model $\mathcal{H}$, we can use the correctness of the model $\mathcal{N}$ with respect to the **N**-operational semantics for proving some of its properties. The proofs of the following properties can be easily carried out in a similar way to the analogous properties for model $\mathcal{H}$.

Property 11.2.15. (i) $\approx_{\mathbf{N}}$ is a sensible theory.
(ii) $\approx_{\mathbf{N}}$ is fully extensional.
(iii) All fixed-point operators are equated by $\approx_{\mathbf{N}}$.

For proving that $\mathcal{N}$ is also complete with respect to the **N**-operational semantics, we can state a discriminability property between terms which are not in the $\sqsubseteq_{\mathcal{N}}$ relation, which is based on a minor variant of the semiseparability algorithm between approximants showed in Sect. 11.1.3.

Theorem 11.2.16 ($\mathcal{N}$-Discriminability).
If $M \not\sqsubseteq_{\mathcal{N}} N$ then there is a context $C[.]$ such that both $C[M]$ and $C[N]$ are closed and $C[M] \Downarrow_{\mathbf{N}}$ while $C[N] \Uparrow_{\mathbf{N}}$.

Proof. The proof is in Sect. 11.2.3. □

Theorem 11.2.17 ($\mathcal{N}$-Completeness).
The model $\mathcal{N}$ is complete with respect to the **N***-operational semantics.*

Proof. Let $M \not\sqsubseteq_{\mathcal{N}} N$. By the $\mathcal{N}$-discriminability theorem, there is a context $C[.]$ such that both $C[M]$ and $C[N]$ are closed and $C[M] \Downarrow_{\mathbf{N}}$ while $C[N] \Uparrow_{\mathbf{N}}$, which implies, by definition, that $M \not\preceq_{\mathbf{N}} N$. □

Finally, we can state the full abstraction result.

Corollary 11.2.18 ($\mathcal{N}$-Full abstraction).
The model $\mathcal{N}$ is fully abstract with respect to the **N***-operational semantics.*

Proof. By the $\mathcal{N}$-correctness and $\mathcal{N}$-completeness theorems. □

Thanks to the $\mathcal{N}$-discriminability theorem, a finite axiomatization of the preorder $\sqsubseteq_{\mathcal{H}}$ between terms can be given. Note that this characterization is simpler than the characterization of $\sqsubseteq_{\mathcal{H}}$.

Definition 11.2.19. *The relation $\sqsubseteq \subseteq \Lambda \times \Lambda$ is defined as follows:*
$M \sqsubseteq N$ if and only if one of the following conditions holds:

- *M is Λ-unsolvable;*
- *$M =_{\Lambda\eta} \lambda x_1...x_n.xM_1...M_m$, $N =_{\Lambda\eta} \lambda x_1...x_n.xN_1...N_m$, where $M_i \sqsubseteq N_i$, and M has Λ-nf implies N has Λ-nf $(1 \le i \le m)$.*

Theorem 11.2.20 ($\mathcal{N}$-Characterization).
$M \sqsubseteq_{\mathcal{N}} N$ if and only if $M \sqsubseteq N$.

Proof. The proof is in Sect. 11.2.3 □

Then the following corollary can be easily proved.

Corollary 11.2.21. *$M \sqsubseteq_{\mathcal{N}} N$ if and only if $\forall A \in \mathcal{A}(M)$, $\exists A' \in \mathcal{A}(N)$ such that $A \sqsubseteq_{\mathcal{N}} A'$.*

Note, for model $\mathcal{H}$, only the if implication of a similar corollary holds.

It seems that the **H** and **N** operational theories coincide. The next property shows that this is not true.

Property 11.2.22. $E_\infty \prec_{\mathbf{N}} I$.

Proof. Let $E_\infty \equiv Y(\lambda xyz.y(xz))$, where Y is a call-by-name fixed-point operator, so $\mathcal{A}(E_\infty) = \{\Omega, \lambda z_0 z_1.z_0\Omega, \lambda z_0 z_1.z_0(\lambda z_2.z_1(...(\lambda z_{n+1}.z_n\Omega)...)) \mid n \geq 0\}$ (see Example 11.1.53). For all $A' \in \mathcal{A}(E_\infty)$ it is easy to see that $I \not\ll A'$ and $A' \ll I$, hence $E_\infty \sqsubseteq_{\mathcal{N}} I$ but $I \not\sqsubseteq_{\mathcal{N}} E_\infty$. □

So the **N**-operational semantics is fully extensional, but it is not able to grasp the $=_\eta$ "up to infinite", as the model $\mathcal{H}$ does.

11.2.1 The $\leq_{\bowtie}$-Intersection Relation

First, let us prove that $\leq_{\bowtie}$ is well posed, i.e. different type constants in $C_{\bowtie}$ are not $\simeq_{\bowtie}$. The following theorem characterizes the syntactic shape of types that are $\simeq_{\bowtie} \omega$.

Theorem 11.2.23. *$\sigma \simeq_{\bowtie} \omega$ if and only if $\sigma \in \Omega^{\mathcal{N}}$, where*

$$\Omega^{\mathcal{N}} = \left\{ \sigma \in T(C_{\bowtie}) \,\middle|\, \begin{array}{l} \sigma \equiv \sigma_0 \wedge ..... \wedge \sigma_n \ (n \in \mathbb{N}), \textit{ where } \forall i \leq n \textit{ either} \\ \sigma_i \equiv \omega, \textit{ or } \sigma_i \equiv \pi_i \to \tau_i \textit{ and } \tau_i \simeq_{\bowtie} \omega, 1 \leq i \leq n \end{array} \right\}.$$

Proof. Note that $\omega \simeq_{\bowtie} \sigma$ if and only if $\omega \leq_{\bowtie} \sigma$, by rule (a).

$\Leftarrow$ The proof can be obtained from that of Theorem 11.1.32, taking into account that now rule $(h3)$ is named $(n5)$.

$\Rightarrow$ First we will prove that, if $\sigma \in \Omega^{\mathcal{N}}$ and $\sigma \leq_{\bowtie} \tau$ then, $\tau \in \Omega^{\mathcal{N}}$, by induction on the rules of $\leq_{\bowtie}$.

(a),(b), (c),(c'),(d),(e),(f),(g),(r),(t) See Theorem 11.1.32.
$(n0)$,$(n1)$,$(n2)$,$(n3)$,$(n4)$ Not possible.
$(n5)$ Corresponds to case $(h3)$ in Theorem 11.1.32.

The proof is done, since $\sigma \simeq_{\bowtie} \omega$ implies $\omega \leq_{\bowtie} \sigma$, but $\omega \in \Omega^{\mathcal{N}}$ implies $\sigma \in \Omega^{\mathcal{N}}$. □

Note that $\sigma_1 \to ... \to \sigma_n \to \omega \simeq_{\bowtie} \omega$, for all $\sigma_1, ..., \sigma_n$ $(n \in \mathbb{N})$.

Corollary 11.2.24. (i) $\phi \not\simeq_{\bowtie} \omega$,
(ii) $\psi \not\simeq_{\bowtie} \omega$.

Proof. For proving both points, it is sufficient to observe that both ϕ and ψ do not belong to the set $\Omega^{\mathcal{N}}$. □

In order to prove that $\phi \not\simeq_{\bowtie} \psi$, we need the following lemma.

Lemma 11.2.25. *$\sigma \in \Psi^{\mathcal{N}}$ and $\tau \leq_{\bowtie} \sigma$ imply $\tau \in \Psi^{\mathcal{N}}$, where*

$$\Psi^{\mathcal{N}} = \left\{ \sigma \in T(C_{\bowtie}) \,\middle|\, \begin{array}{l} \sigma \equiv \sigma_0 \wedge ..... \wedge \sigma_n \ (n \in \mathbb{N}),\ \textit{where} \\ \exists i \leq n \ \textit{such that}\ \sigma_i \equiv \pi_1 \to ... \to \pi_{h_i} \to \psi \ (h_i \in \mathbb{N}) \end{array} \right\}.$$

Proof. By induction on the definition of $\leq_{\bowtie}$. If the last applied rule is either (b), (c), (c′), (r), (n1) or (n2) the proof is obvious.
If the last applied rule is (f), then $\tau \equiv \mu \to \nu$ and $\sigma \equiv \rho \to \pi$, where $\rho \leq_{\bowtie} \mu$ and $\nu \leq_{\bowtie} \pi$; so, by induction, $\nu \in \Psi^{\mathcal{N}}$, which implies $\tau \in \Psi^{\mathcal{N}}$.
Cases (t) and (e) follow easily by induction. All other cases are not possible. □

Note that $\sigma \in \Psi^{\mathcal{N}}$ does not imply $\sigma \simeq_{\bowtie} \psi$.

Corollary 11.2.26. $\phi \not\simeq_{\bowtie} \psi$.

Proof. Since $\psi \leq_{\bowtie} \phi$, $\phi \simeq_{\bowtie} \psi$ if and only if $\phi \leq_{\bowtie} \psi$. But $\psi \in \Psi^{\mathcal{N}}$, so this would imply $\phi \in \Psi^{\mathcal{N}}$, by Lemma 11.2.25. But this is not possible, by the definition of $\Psi^{\mathcal{N}}$. □

The following property characterizes the shape of types in $T(C_{\bowtie})$, modulo $\simeq_{\bowtie}$. It will be extensively used in order to prove some properties about the type assignment system $\vdash_{\bowtie}$.

Property 11.2.27. (i) If $\sigma \not\simeq_{\bowtie} \omega$ then $\sigma \simeq_{\bowtie} \sigma_0 \wedge ... \wedge \sigma_n$ $(n \geq 0)$ such that $\forall i \leq n,\ \sigma_i \simeq_{\bowtie} \tau_1^i \to ... \to \tau_{m_i}^i \to \rho_i$ for some $m_i \in \mathbb{N}$, and $\rho_i \in \{\phi, \psi\}$.

(ii) If $\sigma \simeq_{\bowtie} \tau_1 \to ... \to \tau_n \to \rho$ $(n \geq 0)$, where $\rho \in \{\phi, \psi\}$, then for all $p \in \mathbb{N}$,

$$\sigma \simeq_{\bowtie} \tau_1 \to ... \to \tau_n \to \underbrace{\pi \to ... \to \pi}_{p} \to \rho,$$

where $\pi \in \{\phi, \psi\}$ but $\pi \not\equiv \rho$.

Proof. (i) By induction on σ. If $\sigma \equiv \omega, \phi, \psi$ then the proof is trivial.
Let $\sigma \equiv \mu \to \pi$; so, by induction $\pi \simeq_{\bowtie} \pi_0 \wedge ... \wedge \pi_h$ $(h \in \mathbb{N})$ such that $\forall i \leq h,\ \pi_i \simeq_{\bowtie} \nu_1^i \to ... \to \nu_{m_i}^i \to \rho$ and $\rho \in \{\phi, \psi\}$. By Lemma 10.1.4.(iii), $\sigma \simeq_{\bowtie} \sigma_0 \wedge ... \wedge \sigma_h$, where $\sigma_i \simeq_{\bowtie} \mu \to \nu_1^i \to ... \to \nu_{m_i}^i \to \rho$.
The case $\sigma \equiv \mu \wedge \pi$ follows by induction.

(ii) By rules $(n1)$, $(n2)$, $(n3)$, $(n4)$ and (f) it is easy to see that

$$\tau_1 \to ... \to \tau_n \to \rho \simeq_{\bowtie} \tau_1 \to ... \to \tau_n \to \pi \to \rho,$$

where $\pi \in \{\phi, \psi\}$ but $\pi \not\equiv \rho$, so the proof follows by induction on p. □

The following lemma implies the legality of $\bowtie$. Note that a type in $T(C_{\bowtie})$ has the following syntactical shape:

$$(\sigma_1 \to \tau_1) \wedge ... \wedge (\sigma_n \to \tau_n) \wedge c_1 \wedge ... \wedge c_m \quad \text{where} \quad m, n \geq 0, m+n \geq 1,$$

where $c_j \in \{\phi, \psi\}$ $(1 \leq j \leq m)$. Moreover, let us recall that every constant in $C_{\bowtie}$ is $\simeq_{\bowtie}$ to an arrow type, indeed $\omega \simeq \omega \to \omega$, $\phi \simeq_{\bowtie} \psi \to \phi$ and $\psi \simeq_{\bowtie} \phi \to \psi$.

Property 11.2.28. Let $n, m, p, q \geq 0$, $n+m \geq 1$, $p+q \geq 1$, and

$$(\sigma_1 \to \tau_1) \wedge ... \wedge (\sigma_n \to \tau_n) \wedge c_1 \wedge ... \wedge c_m \leq_{\bowtie} (\sigma'_1 \to \tau'_1) \wedge ... \wedge (\sigma'_p \to \tau'_p) \wedge d_1 \wedge ... \wedge d_q$$

where $c_i, d_j \in \{\omega, \phi, \psi\}$, $c_i \simeq_{\bowtie} \sigma_{n+i} \to \tau_{n+i}$ and $d_j \simeq_{\bowtie} \mu_{p+j} \to \nu_{p+j}$ $(\tau_{n+i}, \tau'_{p+j} \in \{\omega, \phi, \psi\}, 1 \leq i \leq m, 1 \leq j \leq q)$. If $\tau'_h \not\simeq_{\bowtie} \omega$ $(1 \leq h \leq p+q)$, then there is $\{i_1, ..., i_k\} \subseteq \{1, ..., n+m\}$ such that $\sigma_{i_1} \wedge ... \wedge \sigma_{i_k} \geq_{\bowtie} \sigma'_h$, $\tau_{i_1} \wedge ... \wedge \tau_{i_k} \leq_{\bowtie} \tau'_h$ and $\tau_j \not\simeq \omega$, for each $j \in \{i_1, ..., i_k\}$.

Proof. We reason by induction on the last rule of the derivation proving

$$(\sigma_1 \to \tau_1) \wedge ... \wedge (\sigma_n \to \tau_n) \wedge c_1 \wedge ... \wedge c_m \leq_{\bowtie} (\sigma'_1 \to \tau'_1) \wedge ... \wedge (\sigma'_p \to \tau'_p) \wedge d_1 \wedge ... \wedge d_q.$$

$(b), (r), (a)$ Obvious.
$(c), (c'), (d)$ Easy.
(e) By induction.
(f) $n = 1$, $m = 0$, $p = 1$ and $q = 0$. Then $\sigma_1 \to \tau_1 \leq_{\bowtie} \sigma'_1 \to \tau'_1$ if and only if $\sigma'_1 \leq_{\bowtie} \sigma_1$ and $\tau_1 \leq_{\bowtie} \tau'_1$.
$(n0), (n1), (n2), (n3), (n4)$ Immediate, since $\psi \simeq_{\bowtie} \phi \to \psi$ and $\phi \leq_{\bowtie} \psi \to \phi$.
$(n5), (g)$ Trivial.
(t) Let $\rho \in T(C_{\bowtie})$ be such that $(\sigma_1 \to \tau_1) \wedge ... \wedge (\sigma_n \to \tau_n) \wedge c_1 \wedge ... \wedge c_m \leq_{\bowtie} \rho$ and $\rho \leq_{\bowtie} (\sigma'_1 \to \tau'_1) \wedge ... \wedge (\sigma'_p \to \tau'_p) \wedge d_1 \wedge ... \wedge d_q$. ρ must be of the shape $(\mu_1 \to \nu_1) \wedge ... \wedge (\mu_t \to \nu_t) \wedge e_1 \wedge ... \wedge e_s$, where $e_i \in \{\omega, \phi, \psi\}$ and $e_i \simeq_{\bowtie} \mu_{t+i} \to \nu_{t+i}$ $(1 \leq i \leq s)$. By induction, for all h such that $\tau'_h \not\simeq_{\bowtie} \omega$, there is $\{i^h_1, ..., i^h_{k_h}\} \subseteq \{1, ..., t+s\}$ such that $\mu_{i^h_1} \wedge ... \wedge \mu_{i^h_{k_h}} \geq_{\bowtie} \sigma'_h$, $\nu_{i^h_1} \wedge ... \wedge \nu_{i^h_{k_h}} \leq_{\bowtie} \tau'_h$ and $\nu_{i^h_j} \not\simeq \omega$, for all $1 \leq j \leq k_h$. By induction, for every $\nu_{i^h_j}$, there is $\{r^{h,j}_1, ..., r^{h,j}_{w_{h,j}}\} \subseteq \{1, 2, ..., n+m\}$ such that $\sigma_{r^{h,j}_1} \wedge ... \wedge \sigma_{r^{h,j}_{w_{h,j}}} \geq_{\bowtie} \mu_{i^h_j}$ and $\tau_{r^{h,j}_1} \wedge ... \wedge \tau_{r^{h,j}_{w_{h,j}}} \leq_{\bowtie} \nu_{i^h_j}$.
So the proof follows from rule (e). □

▼ **Proof of Theorem 11.2.5** (pag. 145).

The legality of the type system $\bowtie$ is a particular case of Property 11.2.28. ■

11.2.2 Proof of $\mathcal{N}$-Approximation Theorem

Also in this case, the proof is very similar to the proof of the approximation theorem for the model $\mathcal{H}$. We will show here just the differences with respect that proof.

Definition 11.2.29. *Let the predicate* $App_{\mathcal{N}}$ *be so defined:* $App_{\mathcal{N}}(B, \sigma, M)$ *if and only if there is* $A \in \mathcal{A}(M)$ *such that* $B \vdash_{\bowtie} A : \sigma$.

In order to prove the ($\Leftarrow$) part of the theorem, we need to prove the following implication:

$$B \vdash_{\bowtie} M : \sigma \Rightarrow App_{\mathcal{N}}(B, \sigma, M).$$

We will build the proof in two steps. First, it will be proved that

$$B \vdash_{\bowtie} M : \sigma \Rightarrow Comp_{\mathcal{N}}(B, \sigma, M), \tag{11.3}$$

and then

$$Comp_{\mathcal{N}}(B, \sigma, M) \Rightarrow App_{\mathcal{N}}(B, \sigma, M), \tag{11.4}$$

where $Comp_{\mathcal{N}}(B, \sigma, M)$ (read: the term M is *computable* of type σ with respect to a basis B) is a property of the triple $< B, \sigma, M >$. We will prove Eq. (11.3) by induction on terms and Eq. (11.4) by induction on types.

A basis B is *finite* if and only if $B(y) \simeq_{\bowtie} \omega$ except in a finite number of variables. We will use $[\sigma_1/x_1, ..., \sigma_n/x_n]$ to denote a finite basis. By Lemma 10.1.7.(i), in this section we limit ourselves to consider only such a kind of basis.

Let B and B' be two basis. Recall that $B \cup B'$ denotes the basis such that, for every x, $B \cup B'(x) = B(x) \wedge B'(x)$ (remember that, for every type σ, $\sigma \wedge \omega \simeq \sigma$). The key point is the difference in the definition of the computability predicate.

Definition 11.2.30. *The predicate* $Comp_{\mathcal{N}}$ *is defined by induction on types as follows:*

- $Comp_{\mathcal{N}}(B, \omega, M)$ *is true;*
- $Comp_{\mathcal{N}}(B, \phi, M)$ *if and only if* $App_{\mathcal{N}}(B, \phi, M)$*;*
- $Comp_{\mathcal{N}}(B, \psi, M)$ *if and only if* $App_{\mathcal{N}}(B, \psi, M)$ *and* $\forall N \in \vec{N}$, $App_{\mathcal{N}}(B', \phi, N)$ *implies* $App_{\mathcal{N}}(B \cup B', \psi, M\vec{N})$*;*
- $Comp_{\mathcal{N}}(B, \sigma \rightarrow \tau, M)$ *if and only if* $\forall N$, $Comp_{\mathcal{N}}(B', \sigma, N)$ *implies* $Comp_{\mathcal{N}}(B \cup B', \tau, MN)$*;*
- $Comp_{\mathcal{N}}(B, \sigma \wedge \tau, M)$ *if and only if* $Comp_{\mathcal{N}}(B, \sigma, M)$ *and* $Comp_{\mathcal{N}}(B, \tau, M)$.

It can be proved that $Comp_{\mathcal{N}}$ is defined modulo $=_{\Lambda}$, as for the model $\mathcal{H}$.

Lemma 11.2.31. (i) *$App_{\mathcal{N}}(B, \sigma, x\vec{M})$ implies $Comp_{\mathcal{N}}(B, \sigma, x\vec{M})$.*
(ii) *$Comp_{\mathcal{N}}(B, \sigma, M)$ implies $App_{\mathcal{N}}(B, \sigma, M)$.*

Proof. As for model $\mathcal{H}$, the proof can be done by mutual induction on σ. □

$Comp_{\mathcal{N}}$ is closed under $\leq_{\bowtie}$.

Lemma 11.2.32. (i) *If $Comp_{\mathcal{N}}(B, \psi, M)$ then $Comp_{\mathcal{N}}(B, \phi, M)$.*
(ii) *$Comp_{\mathcal{N}}(B, \psi, M)$ if and only if $Comp_{\mathcal{N}}(B, \phi \to \psi, M)$.*
(iii) *$App_{\mathcal{N}}(B[\psi/z], \phi, M)$ and $Comp_{\mathcal{N}}(B', \psi, N)$ imply $App_{\mathcal{N}}(B \cup B', \phi, M[N/z])$.*
(iv) *$Comp_{\mathcal{N}}(B, \phi, M)$ if and only if $Comp_{\mathcal{N}}(B, \psi \to \phi, M)$.*
(v) *If $Comp_{\mathcal{N}}(B, \sigma, M)$ and $\sigma \leq_{\bowtie} \tau$ then $Comp_{\mathcal{N}}(B, \tau, M)$.*

Proof. (i) Trivial.
(ii) ($\Rightarrow$) We prove that $Comp_{\mathcal{N}}(B', \phi, N)$ implies $Comp_{\mathcal{N}}(B \cup B', \psi, MN)$ under the hypothesis that $Comp_{\mathcal{N}}(B, \psi, M)$.
$Comp_{\mathcal{N}}(B', \phi, N)$ implies $App_{\mathcal{N}}(B', \phi, N)$, by Lemma 11.2.31.(ii).
$Comp_{\mathcal{N}}(B, \psi, M)$ implies, by definition, that if $\forall Q \in \vec{Q}$, $App_{\mathcal{N}}(B', \phi, Q)$ then $App_{\mathcal{N}}(B \cup B', \psi, M\vec{Q})$. Hence $App_{\mathcal{N}}(B \cup B', \psi, MN)$ and $\forall P \in \vec{P}$, $App_{\mathcal{N}}(B', \phi, P)$ imply $App_{\mathcal{N}}(B \cup B', \psi, MN\vec{P})$.
Thus $Comp_{\mathcal{N}}(B \cup B', \psi, MN)$; so, by definition $Comp_{\mathcal{N}}(B, \phi \to \psi, M)$.

($\Leftarrow$) We prove that $App_{\mathcal{N}}(B, \psi, M)$ and if $\forall N \in \vec{N}$ $App_{\mathcal{N}}(B', \phi, N)$ then $App_{\mathcal{N}}(B \cup B', \psi, M\vec{N})$, under the hypothesis that $Comp_{\mathcal{N}}(B, \phi \to \psi, M)$.
$App_{\mathcal{N}}(B, \phi \to \psi, M)$, by Lemma 11.2.31.(ii); thus, by rule ($\leq_{\bowtie}$), we obtain both $App_{\mathcal{N}}(B, \psi, M)$ and $App_{\mathcal{N}}(B, \underbrace{\phi \to \ldots \to \phi}_{h} \to \psi, M)$, for all $h \in \mathbb{N}$. If $\vec{N} \equiv N_1 ... N_m$ is such that $App_{\mathcal{N}}(B', \phi, N_i)$ $(1 \leq i \leq m)$ then $Comp_{\mathcal{N}}(B', \psi, N)$ implies $App_{\mathcal{N}}(B \cup B', \psi, M\vec{N})$, by m applications of the rule $(\to E)$.
Hence, by definition $Comp_{\mathcal{N}}(B \cup B', \psi, M\vec{N})$.
(iii) $App_{\mathcal{N}}(B[\psi/z], \phi, M)$ means that there is $A \in \mathcal{A}(M)$ such that $B[\psi/z] \vdash_{\bowtie} A : \phi$. The proof is given by induction on A.
Clearly, $A \not\equiv \Omega$ since $B \not\vdash_{\bowtie} \Omega : \phi$. Let $A \equiv z$; the proof follows since $Comp_{\mathcal{N}}(B', \psi, N)$ implies $App_{\mathcal{N}}(B', \psi, N)$, therefore $App_{\mathcal{N}}(B', \phi, N)$ by rule ($\leq_{\bowtie}$), and $App_{\mathcal{N}}(B \cup B', \phi, M[N/z])$ by Lemma 10.1.7.(iii).
The case $A \in$ Var, but $A \not\equiv z$ is trivial.
Let $A \equiv zA_1 ... A_m$ $(m \geq 1)$, thus $M =_{\Lambda} zM_1 ... M_m$ and $A_i \in \mathcal{A}(M_i)$, for all $i \leq m$. Hence $B[\psi/z] \vdash_{\bowtie} z : \tau_1 \to ... \to \tau_m \to \phi$ and $B[\psi/z] \vdash_{\bowtie} A_i : \tau_i$ for some τ_i, by Lemma 10.1.7.(vi). Since $\underbrace{\phi \to ... \to \phi}_{m} \to \psi \leq_{\bowtie} \psi$ by rule ($\leq_{\bowtie}$) and $\psi \leq_{\bowtie} \tau_1 \to ... \to \tau_m \to \phi$ by Lemma 10.1.7.(ii), it follows that $\tau_i \leq_{\bowtie} \phi$ by Property 10.1.6. Thus $B[\psi/z] \vdash_{\bowtie} A_i : \phi$ and this implies $App_{\mathcal{N}}(B[\psi/z], \phi, M_i)$. Therefore $App_{\mathcal{N}}(B \cup B', \phi, M_i[N/z])$ by induction. So $Comp_{\mathcal{N}}(B', \psi, N)$ implies $App_{\mathcal{N}}(B \cup B', \phi, NM_1[N/z] ... M_m[N/z])$ by definition, i.e. $App_{\mathcal{N}}(B \cup B', \phi, M[N/z])$.

The case $A \equiv yA_1...A_m$ $(m \geq 1)$ where $y \not\equiv z$ is simpler.
Let $A \equiv \lambda y.A'$, for some $A' \in \mathcal{A}$, thus $M =_{\Lambda\eta} \lambda y.M'$ and $A' \in \mathcal{A}(M')$. Hence, $B[\psi/z] \vdash_{\bowtie} \lambda y.A' : \psi \to \phi$ by rule $(\leq_{\bowtie})$, and $B[\psi/z, \psi/y] \vdash_{\bowtie} A' : \phi$ by Lemma 10.1.7.(vi). Without loss of generality, let $B'(y) = \psi$; so the proof follows by induction on $App_{\mathcal{N}}(B[\psi/z, \psi/y], \phi, M')$.

(iv) $(\Rightarrow)$ We will prove that $Comp_{\mathcal{N}}(B', \psi, N)$ implies $Comp_{\mathcal{N}}(B \cup B', \phi, MN)$, under the hypothesis that $Comp_{\mathcal{N}}(B, \phi, M)$.
By definition $App_{\mathcal{N}}(B, \phi, M)$, so $App_{\mathcal{N}}(B, \psi \to \phi, M)$ by rule $(\leq_{\bowtie})$, which means there is $A \in \mathcal{A}(M)$ such that $B \vdash_{\bowtie} A : \psi \to \phi$.
If A is of the shape $\lambda x.A'$ then $M =_{\Lambda} \lambda x.M'$ and $A' \in \mathcal{A}(M')$; thus $B[\psi/x] \vdash_{\bowtie} A' : \phi$, by Lemma 10.1.7.(vi) and so $App_{\mathcal{N}}(B[\psi/x], \phi, M')$. Hence, $App_{\mathcal{N}}(B \cup B', \phi, M'[N/x])$ by the point (iii) of this Lemma. But $MN =_{\Lambda} M'[N/x]$, so $App_{\mathcal{N}}(B \cup B', \phi, MN)$ and by definition of $Comp_{\mathcal{N}}$ the proof follows. The case $A \equiv z\vec{A}$ is simpler.

$(\Leftarrow)$ If $Comp_{\mathcal{N}}(B, \psi \to \phi, M)$ then $App_{\mathcal{N}}(B, \psi \to \phi, M)$, by Lemma 11.2.31.(ii), hence $App_{\mathcal{N}}(B, \phi, M)$, by rule $(\leq_{\bowtie})$.

(v) By induction on the rules of $\leq_{\bowtie}$.
$(a), (b), (c), (c'), (e), (r), (t)$ Trivial.
(d) $Comp_{\mathcal{H}}(B, (\sigma_0 \to \tau_0) \wedge (\sigma_0 \to \pi_0), M)$ implies, by definition of $Comp_{\mathcal{N}}$, $Comp_{\mathcal{N}}(B, \sigma_0 \to \tau_0, M)$ and $Comp_{\mathcal{N}}(B, \sigma_0 \to \pi_0, M)$.
If $Comp_{\mathcal{N}}(B', \sigma_0, N)$ then $Comp_{\mathcal{N}}(B \cup B', \tau_0, MN)$ and $Comp_{\mathcal{N}}(B \cup B', \pi_0, MN)$; hence, $Comp_{\mathcal{N}}(B \cup B', \tau_0 \wedge \pi_0, MN)$ and $Comp_{\mathcal{N}}(B, \sigma_0 \to (\tau_0 \wedge \pi_0), M)$, by definition of $Comp_{\mathcal{N}}$.
(f) Let $\sigma_0' \leq_{\bowtie} \sigma_0$, $\tau_0 \leq_{\bowtie} \tau_0'$ and $Comp_{\mathcal{N}}(B, \sigma_0 \to \tau_0, M)$.
If $Comp_{\mathcal{N}}(B', \sigma_0', N)$ then $Comp_{\mathcal{N}}(B', \sigma_0, N)$ by induction, hence $Comp_{\mathcal{N}}(B \cup B', \tau_0, MN)$ by definition of $Comp_{\mathcal{N}}$. Again by induction $Comp_{\mathcal{N}}(B \cup B', \tau_0', MN)$, so the proof is done.
$(g), (n5)$ Easy, since $Comp_{\mathcal{N}}(B, \omega, MN)$ is always true.
$(n0), (n1), (n2), (n3), (n4)$ By using points (i), (ii) and (iii) of this lemma. □

Now we will prove the implication of Eq. (11.3).

Lemma 11.2.33. *Let* $\mathrm{FV}(M) \subseteq \{x_1, ..., x_n\}$ *and* $B = [\sigma_1/x_1, ..., \sigma_n/x_n]$. *If* $Comp_{\mathcal{N}}(B_i, \sigma_i, N_i)$ $(1 \leq i \leq n)$ *and* $B \vdash_{\bowtie} M : \tau$ *then*

$$Comp_{\mathcal{N}}(B_1 \cup ... \cup B_n, \tau, M[N_1/x_1, ..., N_n/x_n]).$$

Proof. Similar to the proof of Lemma 11.1.43, for model $\mathcal{H}$. □

▼ **Proof of $\mathcal{N}$-Approximation Theorem** (Theorem 11.2.9 pag. 147).
$(\Rightarrow)$ We prove that $B \vdash_{\bowtie} M : \sigma$ implies $B \vdash_{\bowtie} A : \sigma$, for some $A \in \mathcal{A}(M)$.
Clearly $Comp_{\mathcal{H}}([\tau/x], \tau, x)$ from Lemma 11.2.31.(i).
Let $\mathrm{FV}(M) \subseteq \{x_1, ..., x_n\}$, so we can assume $B = [\sigma_1/x_1, ..., \sigma_n/x_n]$ without loss of generality, by Lemma 10.1.7.(i). Therefore $B \vdash_{\bowtie} M : \sigma$ and

$Comp_{\mathcal{H}}([\sigma_i/x_i], \sigma_i, x_i)$ $(1 \leq i \leq n)$ imply $Comp_{\mathcal{H}}(B, \sigma, M)$ by Lemma 11.2.33. So, by Lemma 11.2.31.(ii) and by definition of $App_{\mathcal{N}}$, the proof is given.

($\Leftarrow$) By definition, there is M' such that $M =_\Lambda M'$, and A matches M' except at occurrences of Ω. A derivation of $B \vdash_{\bowtie} A : \sigma$ can be transformed into a derivation of $B \vdash_{\bowtie} M' : \sigma$ simply by replacing every subderivation

$$\overline{B \vdash_{\bowtie} \Omega : \omega}\ (\omega) \qquad \text{by} \qquad \overline{B \vdash_{\bowtie} N : \omega}\ (\omega),$$

where N are the subterms replaced by Ω in M'.
$B \vdash_{\bowtie} M' : \sigma$ implies $B \vdash_{\bowtie} M : \sigma$, since the type assignment system is closed under $=_\Lambda$ on terms, as a consequence of the fact that it induces a $\lambda\Lambda$-model, and the proof is given. ■

11.2.3 Proof of $\mathcal{N}$-Discriminability and $\mathcal{N}$-Characterization Theorems

In this section we use the notion of path c and of the relation $\ll_c$ between approximants, as defined in Sect. 11.1.3. Moreover, a minor modification of the semiseparability algorithm presented in Fig. 11.2 (pag.138) is introduced.

Let M and N be two terms such that $M \not\sqsubseteq_{\mathcal{N}} N$; it is always possible to find two approximants $A \in \mathcal{A}(M)$ and $A' \in \mathcal{A}(N)$ and a path c such that $A \not\ll_c A'$, and the context $C[.]$ such that $A, A' \Rrightarrow^c_D C[.]$ can be easily transformed in a context **N**-discriminating M and N.

In order to choose the correct approximants, we use the next definition.

Definition 11.2.34. *A path* c is deep on A *under the following conditions:*

- *if* $c \equiv \epsilon$ *then* $A \in \Lambda$*-NF, namely in* A *there are no occurrences of* Ω*;*
- *if* $c \equiv i, c'$ *then* $A =_\eta \lambda x_1..x_n.xA_1...A_m$ $(i \leq m)$ *and* c' *is deep on* A_i*.*

Example 11.2.35. Let $A \in \mathcal{A}(\lambda x.xE_\infty I)$; it is easy to see that ϵ is not deep on A, paths having 1 as prefix are not deep on A, while if $A \equiv \lambda x.xA_1A_2$ and $A_2 \not\equiv \Omega$, then every path having 2 as prefix is deep on A.

Note that if c is deep on A then c, c' is deep on A, for all paths c'.

Property 11.2.36. Let $B \vdash_{\bowtie} A : \sigma$ and let N be such that:

1. for each $A' \in \mathcal{A}(N)$, $A \not\ll_c A'$ implies A, A' are structurally similar along c and A' is not maximal along c in $\mathcal{A}(N)$;
2. c is deep on A implies that there is $A' \in \mathcal{A}(N)$ such that c is deep on A'.

Then $B \vdash_{\bowtie} N : \sigma$.

Proof. The proof is given by induction on the derivation proving $B \vdash_{\bowtie} A : \sigma$.

(var) Then $A \equiv x$, and the derivation is

$$\frac{}{B[\sigma/x] \vdash_{\bowtie} x : \sigma}\ (var)$$

Note that the path ϵ is deep on x, so there is $A^* \in \mathcal{A}(N)$ such that ϵ is deep on A^*. Hence $A^* \not\equiv \Omega$, i.e. A^* is of the shape $\lambda x_1...x_p.zA'_1...A'_q$. From the condition that $A \not\lll_c A^*$ implies A, A^* are structurally similar along c, it follows that it must be $x \equiv z$; otherwise it would be $A \not\lll_\epsilon A^*$, and A, A^* not structurally similar along ϵ, against the hypothesis. By the same reasoning, it must be $p = q$.
Clearly $A^* \in \Lambda$-NF implies that $x =_\eta A^*$ and $A'_i =_\eta x_i$; moreover, $N =_{\Lambda\eta} A^*$. Since typings are preserved by $=_{\Lambda\eta}$, the proof follows.

($\rightarrow I$) Then $A \equiv \lambda x.\bar{A}$, $\sigma \equiv \mu \rightarrow \nu$, and the derivation is

$$\frac{B[\mu/x] \vdash_{\bowtie} \bar{A} : \nu}{B \vdash_{\bowtie} \lambda x.\bar{A} : \mu \rightarrow \nu}\ (\rightarrow I)$$

Clearly $\Omega \in \mathcal{A}(N)$ and $A \not\lll_c \Omega$, so by hypothesis 1 there is $A' \in \mathcal{A}(N)$ such that $\Omega \lll^\circ_c A'$ and $A' =_\eta \lambda x.A^*$. Hence, by Property 11.1.20.(ii), $N =_{\Lambda\eta} \lambda x.N'$ and $A^* \in \mathcal{A}(N')$.
Note that $B[\mu/x] \vdash_{\bowtie} \bar{A} : \nu$ and N' satisfy the hypothesis of this property (indeed, $\not\lll_c$ is defined modulo $=_{\Lambda\eta}$), thus $B[\mu/x] \vdash_{\bowtie} N' : \nu$ by induction. This implies $B \vdash_{\bowtie} \lambda x.N' : \mu \rightarrow \nu$, by rule ($\rightarrow I$).

($\rightarrow E$) Then $A \equiv xA_1...A_m A_{m+1}$, since $A \in \mathcal{A}$, and the derivation is

$$\frac{B \vdash_{\bowtie} xA_1...A_m : \tau \rightarrow \sigma \qquad B \vdash_{\bowtie} A_{m+1} : \tau}{B \vdash_{\bowtie} xA_1...A_m A_{m+1} : \sigma}\ (\rightarrow I)$$

for some τ. Clearly $\Omega \in \mathcal{A}(N)$ and $A \not\lll_c \Omega$, so by hypothesis 1 there is $A' \in \mathcal{A}(N)$ such that $\Omega \lll^\circ_c A'$. Assume that for every $A' \in \mathcal{A}(N)$, if $A' \not\equiv \Omega$ then A' is of the shape $\lambda x_1...x_p.zA'_1...A'_q$. The given conditions assure us that $z \equiv x$, $q = m + 1 + p$ and $A'_{m+i+1} =_\eta x_i$ $(1 \le i \le p)$, i.e.

$$A' \equiv \lambda x_1...x_p.xA'_1...A'_{m+1}x_1...x_p =_\eta xA'_1...A'_{m+1}.$$

So $N =_{\Lambda\eta} xN_1...N_{m+1}$ and $A'_i \in \mathcal{A}(N_i)$ $(1 \le i \le m + 1)$. Note that, for A_i and every $A'_i \in \mathcal{A}(N_i)$, the conditions of the theorem are satisfied. So, by induction, $B \vdash_{\bowtie} N_{m+1} : \tau$ and $B \vdash_{\bowtie} xN_1...N_m : \tau \rightarrow \sigma$. So, by rule ($\rightarrow E$), $B \vdash_{\bowtie} xN_1...N_m N_{m+1} : \sigma$, and the proof follows by the fact that typings are preserved by $=_{\Lambda\eta}$.

$(\wedge I), (\wedge E), (\le_{\bowtie})$ The proof follows directly from the inductive hypothesis.

(ω) Trivial. □

We need a further definition.

Definition 11.2.37. *Let c a path and A an approximant:*

- *if $c \equiv \epsilon$ then c is defined on A;*
- *if $c \equiv k, c'$ then c is defined on A, whenever $A =_\eta \lambda x_1 ... x_p.zA_1 ... A_q$ $(q \geq k)$ and c' is defined on A.*

Example 11.2.38. 1 is defined on $\lambda z.x\Omega I$, 1 is defined on $\lambda z.xI\Omega$, but 1 is not defined on Ω.

Lemma 11.2.39. *Let $M \not\sqsubseteq_{\mathcal{N}} N$. Then there are $A \in \mathcal{A}(M)$, $A' \in \mathcal{A}(N)$ and a path c such that $A \not\lessdot_c A'$, c is defined on A, A' and one of the following conditions holds:*

1. *A and A' are structurally different in c;*
2. *A' is maximal along c in $\mathcal{A}(N)$;*
3. *c is deep on A but c is not deep on A'', for all $A'' \in \mathcal{A}(N)$.*

Proof. $M \not\sqsubseteq_{\mathcal{N}} N$ implies there are B, σ such that $B \vdash_{\bowtie} M : \sigma$ while $B \not\vdash_{\bowtie} N : \sigma$. By the $\mathcal{N}$-approximation theorem, this means there is $A \in \mathcal{A}(M)$ such that $B \vdash_{\bowtie} A : \sigma$. Then, by Property 11.2.36, there are three cases.

1. There is $A' \in \mathcal{A}(N)$ such that $A \not\lessdot_c A'$, and A and A' are structurally different along c. Then A and A' are the desired approximants.
2. There is $A' \in \mathcal{A}(N)$ such that $A \not\lessdot_c A'$, and A and A' are structurally similar along c, and A' is maximal along c in $\mathcal{A}(N)$. Then A and A' are the desired approximants.
3. Otherwise, c is deep in A and not deep in A'', for all $A'' \in \mathcal{A}(N)$. So each A'' such that $A \not\lessdot_c A''$ and c is defined on A'' can be chosen. □

The $\mathcal{N}$-semiseparability algorithm is presented in Fig. 11.4. The algorithm is defined as a formal system, proving statements of the shape: $A, A' \Rrightarrow_N^c C[.]$, where $A \not\lessdot_c A'$ and $C[.]$ is a context. The $\mathcal{N}$-semiseparablity algorithm is a minor modification of that we given in Fig. 11.2 (pag.138). Rule $(D1)$ has been divided in two rules, namely $(N0)$ and$(N1)$, according to the fact that the given path is deep or not in A.

For the sake of simplicity, in the $\mathcal{N}$-semiseparablity algorithm we assume that all bound and free variables have different names.

Theorem 11.2.40. (i) *If $A \not\lessdot_c A'$ then $A, A' \Rrightarrow_N^c C[.]$.*
(ii) *If $A, A' \Rrightarrow_N^c C[.]$ then $C[A]$ has Λ-normal form, while $C[A'] \to_{\Lambda\Omega}^* \Omega$.*

Proof. (i) Similar to the proof of Lemma 11.1.48.
(ii) The proof is carried out by induction on the derivation proving $A, A' \Rrightarrow_N^c C[.]$. In case the last applied rule is $(N0)$, then the proof follows from definition of deep. In case the last applied rule is $(N1)$, then obviously $C[A] \to_{\Lambda\Omega}^* I$, while $C[A'] \to_{\Lambda\Omega}^* \Omega$. In the remaining cases, the proof is similar to that of the Lemma 11.1.49. □

Let A, A' be approximants such that $A \not\ll_c A'$ and $r \geq \max\{\mathsf{args}(A), \mathsf{args}(A')\}$.

The rules of the system proving statements $A, A' \Rrightarrow_N^c C[.]$, are the following:

$$\frac{\epsilon \text{ is deep on } xA_1...A_m}{xA_1...A_m, \Omega \Rrightarrow_N^c [.]} \;(N0) \qquad \frac{\epsilon \text{ is not deep on } xA_1...A_m}{xA_1...A_m, \Omega \Rrightarrow_N^c (\lambda x.[.])O^m} \;(N1)$$

$$\frac{xA_1...A_m, \Omega \Rrightarrow_N^c C[.]}{\lambda x_1...x_n.xA_1...A_m, \Omega \Rrightarrow_N^c C[[.]x_1...x_n]} \;(D2)$$

$$\frac{p \leq q \qquad xA_1...A_m x_{p+1}...x_q, yA'_1...A'_n \Rrightarrow_N^c C[.]}{\lambda x_1...x_p.xA_1...A_m, \lambda x_1...x_q.yA'_1...A'_n \Rrightarrow_N^c C[[.]x_1...x_q]} \;(D3)$$

$$\frac{q < p \qquad xA_1...A_m, yA'_1...A'_n x_{q+1}...x_p \Rrightarrow_N^c C[.]}{\lambda x_1...x_p.xA_1...A_m, \lambda x_1...x_q.yA'_1...A'_n \Rrightarrow_N^c C[[.]x_1...x_p]} \;(D4)$$

$$\frac{n < m}{xA_1...A_m, xA'_1...A_n \Rrightarrow_N^c (\lambda x.[.])O^m \underbrace{I.....I}_{m-n-2} KI\Omega} \;(D5)$$

$$\frac{m < n}{xA_1...A_m, xA'_1...A'_n \Rrightarrow_N^c (\lambda x.[.])O^n \underbrace{I.....I}_{n-m-2} K\Omega I} \;(D6)$$

$$\frac{x \not\equiv y}{xA_1...A_m, yA'_1...A'_n \Rrightarrow_N^c (\lambda xy.[.])(\lambda x_1...x_m.I)(\lambda x_1...x_n.\Omega)} \;(D7)$$

$$\frac{\begin{array}{c} x \notin \mathrm{FV}(A_k) \cup \mathrm{FV}(A'_k) \\ A_k \not\ll A'_k \qquad A_k, A'_k \Rrightarrow_N^c C[.] \end{array}}{xA_1...A_m, xA'_1...A'_m \Rrightarrow_N^{k,c} C[(\lambda x.[.])U_m^k]} \;(D8)$$

$$\frac{\begin{array}{c} x \in \mathrm{FV}(A_k) \cup \mathrm{FV}(A'_k) \qquad A_k \not\ll A'_k \\ C_x^r[.] \equiv (\lambda x.[.])B^r \qquad \mathsf{nf}_{\Lambda\Omega}(C_x^r[A_k]), \mathsf{nf}_{\Lambda\Omega}(C_x^r[A'_k]) \Rrightarrow_N^c C[.] \end{array}}{xA_1...A_m, xA'_1...A'_m \Rrightarrow_N^{k,c} C[C_x^r[.] \underbrace{I.....I}_{r-m} U_r^k]} \;(D9)$$

Fig. 11.4. $\mathcal{N}$-Semiseparability algorithm

▼ **Proof of the $\mathcal{N}$-Discriminability Theorem** (Theorem 11.2.16 pag. 150).
Let $M \not\sqsubseteq_{\mathcal{N}} N$. Choose $A \in \mathcal{A}(M)$, $A' \in \mathcal{A}(N)$ and a path c satisfying the conditions of Lemma 11.2.39. Then $A, A' \Rrightarrow_N^c C[.]$. Let $C'[.]$ be the context obtained from $C[.]$ by replacing every occurrence of Ω by a Λ-unsolvable term, say DD. We will prove that $C'[.]$ is a context **N**-discriminating M and N. The proof will be done by induction on the derivation of $A, A' \Rrightarrow_N^c C[.]$. If the derivation coincides with an application of the axiom $(N0)$, then c is deep in A but not deep in A'. Thus M has Λ-nf while N does not have Λ-nf; so $[.]$ is a discriminating context.
If the derivation coincides with an application of the axiom $(N1)$, then $A' \equiv \Omega$ and A' is maximal along the empty path in $\mathcal{A}(N)$. Thus N is Λ-unsolvable and $(\lambda x.[.])O^m$ is the discriminating context.
If the used axiom is either $(D5)$, $(D6)$ or $(D7)$, the proof follows from the fact that M and N have both Λ-hnf, and by Property 11.1.20, their Λ-hnf's have the same shape respectively as A and A'. If the last rule applied is $(D2)$, $(D3)$, $(D4)$, $(D8)$ or $(D9)$, the proof follows by induction, always taking into account that M and N have both Λ-hnf, and that by Property 11.1.20, their Λ-hnf's have the same shape respectively as A and A'. ■

Example 11.2.41. (i) Let $M \equiv I$ and $N \equiv E_\infty$, where $E_\infty \equiv Y(\lambda xyz.y(xz))$. Clearly $I \in \mathcal{A}(M)$, and, for every $A' \in \mathcal{A}(N)$, for every path c, I and A' are structurally similar along c, and c is not maximal in A'. Moreover, every path c is deep on I, while c is not deep in A', for all $A' \in \mathcal{A}(N)$. So, a choice satisfying the conditions of Lemma 11.2.39 is $\lambda x.x \in \mathcal{A}(M)$, $\Omega \in \mathcal{A}(N)$ and $c \equiv \epsilon$. In fact,

$$\cfrac{\cfrac{}{x, \Omega \Rrightarrow_N^\epsilon \lambda x.[.]}\,(N0)}{\lambda x.x, \Omega \Rrightarrow_N^\epsilon \lambda x.[.]x}\,(D2)$$

Therefore $(\lambda x.Ix) \to_\Lambda (\lambda x.x)$ and $(\lambda x.Ix) \Downarrow_{\mathbf{N}}$, while $(\lambda x.E_\infty x) \Uparrow_{\mathbf{N}}$, since it has no Λ-nf.

(ii) Let $M \equiv \lambda xz.xzI$ and $N \equiv \lambda xz.xzE_\infty$. Then two approximants of M and N and a path c satisfying the conditions of Lemma 11.2.39 are respectively $A \equiv \lambda xz.xzI$, $A' \equiv \lambda xz.x\Omega\Omega$ and $c \equiv 2$.
Note that the A, A' and path $c' \equiv 1$ does not satisfy the requirements of Lemma 11.2.39. In fact, c' is not deep on all $A'' \in \mathcal{A}(N)$. Note that also $A \equiv \lambda xz.x\Omega I$, $A' \equiv \lambda xz.x\Omega\Omega$ and $c \equiv 2$ can be safely chosen.

▼ **Proof of the $\mathcal{N}$-Characterization Theorem** (Theorem 11.2.20 pag. 150).

($\Leftarrow$) We will prove that if $M \not\sqsubseteq N$ then there are $A \in \mathcal{A}(M)$ and $A' \in \mathcal{A}(N)$ and a path c satisfying the conditions of Lemma 11.2.39. By the proof of the $\mathcal{N}$-discriminability theorem, there is a context $C[.]$ **N**-discriminating M and N, i.e. $C[M] \Downarrow_{\mathbf{N}}$ while $C[N] \Uparrow_{\mathbf{N}}$. By the correctness of the model with respect to the **N**-operational semantics, this implies $M \not\sqsubseteq_{\mathcal{N}} N$. The existence of $A \in \mathcal{A}(M)$ and $A' \in \mathcal{A}(N)$ and a path c satisfying the conditions of Lemma 11.2.39 can be proved by induction on the definition of $\sqsubseteq$. There are four cases:

1. N is Λ-unsolvable and M is Λ-solvable; let $A \in \mathcal{A}(M)$ and $A \not\equiv \Omega$. Thus A, Ω and the empty path satisfy the given conditions.
2. $M =_{\Lambda\eta} \lambda x_1 ... x_n.xM_1...M_m$, $N =_{\Lambda\eta} \lambda x_1...x_p.yM_1...M_q$, and M has Λ-nf while N has not Λ-nf; so $M \in \mathcal{A}(M)$, every $A' \in \mathcal{A}(N)$ such that $A' \not\equiv \Omega$ and the empty path satisfy the given conditions.
3. Each Λ-hnf of M is of the shape $\lambda x_1...x_n.xM_1...M_m$, each Λ-hnf of N is of the shape $\lambda x_1...x_p.yN_1...N_q$, and either $x \not\equiv y$ or $\mid m-n \mid \neq \mid p-q \mid$. So every $A \in \mathcal{A}(M)$ and $A' \in \mathcal{A}(N)$ such that $A, A' \not\equiv \Omega$ and the empty path satisfy the given conditions.
4. Let $M =_{\Lambda\eta} \lambda x_1...x_n.xM_1...M_m$, $N =_{\Lambda\eta} \lambda x_1...x_n.xN_1...N_m$ and $M_k \not\sqsubseteq N_k$, for some k ($1 \leq k \leq m$). Then by induction there are $A^* \in \mathcal{A}(M_k)$, $A^{**} \in \mathcal{A}(N_k)$ and a path c satisfying Lemma 11.2.39. Hence $A \in \mathcal{A}(M)$ and $A' \in \mathcal{A}(M')$ having the following shapes,

$$A =_\eta \lambda x_1...x_n.x \underbrace{\Omega...\Omega}_{k-1} A^* \underbrace{\Omega...\Omega}_{m-k},$$
$$A' =_\eta \lambda x_1...x_n.x \underbrace{\Omega...\Omega}_{k-1} A^{**} \underbrace{\Omega...\Omega}_{m-k},$$

and the path k, c satisfy the conditions.

($\Rightarrow$) Assume $M \not\sqsubseteq_{\mathcal{N}} N$; by Lemma 11.2.39 there are $A \in \mathcal{A}(M)$, $A' \in \mathcal{A}(N)$, and there is a path c such that $A \not\ll_c A'$ and either A, A' are structurally different in c, or A' is maximal along c in $\mathcal{A}(N)$ or there is not A'' maximal along c, c' in $\mathcal{A}(N)$, for all c', and c is deep in A and not deep in A'. Note that $A \not\equiv \Omega$, since $\Omega \ll A'$.
By induction on c, we will prove that $M \not\sqsubseteq N$.
Let $c \equiv \epsilon$. Let $A' \equiv \Omega$. If A' is maximal then N is Λ-unsolvable, and so $M \not\sqsubseteq N$. If there is not A'' maximal along c, c' in $\mathcal{A}(N)$, for all c', and c is deep in A and not deep in A', then M has Λ-nf while N has not Λ-nf, and so $M \not\sqsubseteq N$. Let $A \equiv \lambda x_1 \ldots x_n.xA_1 \ldots A_m$ and $A' \equiv \lambda x_1 \ldots x_p.yA_1 \ldots A_q$, where either $|n-m| \neq |p-q|$ or $x \not\equiv y$. Then $M \not\sqsubseteq N$ by definition. Let $c \equiv i, c'$. Then $A =_\eta \lambda x_1 \ldots x_n.xA_1 \ldots A_m$ and $A' =_\eta \lambda x_1 \ldots x_n.xA_1 \ldots A_m$ where $A_i \not\ll_c A'_i$, and the proof follows by induction. ∎

11.3 The Model $\mathcal{L}$

In this section we will introduce a filter model correct with respect to the **L**-operational semantics. Property 10.1.15 says that a filter model such that

> for all M, N, if $M \Downarrow_{\mathbf{L}}$ (i.e. it has Λ-lazy head normal form) and $N \Uparrow_{\mathbf{L}}$ (i.e. it has not Λ-lazy head normal form) then $N \sqsubset_{\mathcal{L}} M$.

is correct with respect to the **L**-operational semantics.

So we define a legal type theory, say $\angle$, based on a set of constants $C_{\angle}$ such that, for every M with Λ-lhnf and N without Λ-lhnf, there is a basis B and at least one type σ such that $B \vdash_{\angle} M : \sigma$ while not $B \vdash_{\angle} N : \sigma$. Since terms without Λ-lhnf are all and only the Λ-unsolvable terms of order 0, following the same approach as for the previous two models, it seems natural to characterize them by assigning them only type ω. Let us recall that **L** induces a not sensible Λ-theory; in fact, $DD \Uparrow_{\mathbf{L}}$ while $\lambda x.DD \Downarrow_{\mathbf{L}}$. Since, by rules (ω) and $(\rightarrow I)$, $\lambda x.DD$ can always be assigned type $\omega \rightarrow \omega$, a natural choice is to characterize the convergent terms by this type. This allows us to have as set of type constants just the singleton $\{\omega\}$. Clearly the inequality $\omega \leq \omega \rightarrow \omega$, which holds in the two previous models, is no longer correct in this setting.

Definition 11.3.1. *Let $C_{\angle} = \{\omega\}$ and $I(C_{\angle}) = T(C_{\angle})$.*
$\angle$ is the type system $< C_{\angle}, \leq_{\angle}, I(C_{\angle}) >$, where $\leq_{\angle}$ is the least intersection relation induced by the rules of Fig. 11.5.

$$\frac{}{\sigma \leq_{\angle} \omega}\ (a) \qquad \frac{}{\sigma \leq_{\angle} \sigma \wedge \sigma}\ (b) \qquad \frac{}{\sigma \wedge \tau \leq_{\angle} \sigma}\ (c) \qquad \frac{}{\sigma \wedge \tau \leq_{\angle} \tau}\ (c')$$

$$\frac{}{(\sigma \rightarrow \tau) \wedge (\sigma \rightarrow \pi) \leq_{\angle} \sigma \rightarrow (\tau \wedge \pi)}\ (d) \qquad \frac{\sigma \leq_{\angle} \sigma', \tau \leq_{\angle} \tau'}{\sigma \wedge \tau \leq_{\angle} \sigma' \wedge \tau'}\ (e)$$

$$\frac{\sigma' \leq_{\angle} \sigma, \tau \leq_{\angle} \tau'}{\sigma \rightarrow \tau \leq_{\angle} \sigma' \rightarrow \tau'}\ (f) \qquad \frac{}{\sigma \rightarrow \omega \leq_{\angle} \omega \rightarrow \omega}\ (g) \qquad \frac{}{\sigma \leq_{\angle} \sigma}\ (r) \qquad \frac{\sigma \leq \rho \quad \rho \leq \tau}{\sigma \leq_{\angle} \tau}\ (t)$$

Fig. 11.5. $\angle$-intersection relation

Note that $\leq_{\angle}$ is the minimum intersection relation, and we already noticed that $\angle$ is a legal type system.

Definition 11.3.2. *$\mathcal{L}$ is the $\lambda\Lambda$-model $< \mathcal{F}(\angle), \mathcal{F}(\angle), \circ_{\angle}, [\![.]\!]^{\mathcal{F}(\angle)} >$.*

Now we will state some properties of the $\angle$-intersection relation that will be useful in this chapter. Namely, it will be proved that ω is $\not\simeq_{\angle}$ to any arrow type, and the general shape of a type, modulo $\simeq_{\angle}$ will be shown. To do this, we need to characterize the set of types that are $\simeq \omega$.

Theorem 11.3.3. *$\sigma \simeq_{\angle} \omega$ if and only if $\sigma \equiv \underbrace{\omega \wedge ..... \wedge \omega}_{n}$ $(n \geq 1)$.*

Proof. Note that $\omega \simeq_{\angle} \sigma$ if and only if $\omega \leq_{\angle} \sigma$, by rule (a).

($\Leftarrow$) The proof follows by rules (b), (e) and (t).

($\Rightarrow$) It is easy to prove, by induction on the rules of $\leq_{\angle}$, that if $\sigma \equiv \underbrace{\omega \wedge ..... \wedge \omega}_{k}$, for some $k \geq 1$, and $\sigma \leq_{\angle} \tau$ then $\tau \equiv \underbrace{\omega \wedge ..... \wedge \omega}_{h}$, for some $h \geq 1$. Then the proof is done, since $\sigma \simeq_{\angle} \omega$ implies $\omega \leq_{\angle} \sigma$, so $\sigma \equiv \underbrace{\omega \wedge ..... \wedge \omega}_{n}$, for some $n \geq 1$. □

Corollary 11.3.4. *$\sigma \equiv \mu \to \nu$ implies $\sigma \not\simeq_{\angle} \omega$.*

The following property states a characterization of the shape of types in the theory $\angle$.

Property 11.3.5. (i) If $\sigma \not\simeq_{\angle} \omega$ then $\sigma \simeq_{\angle} \sigma_0 \wedge ... \wedge \sigma_n$ such that $\forall i \leq n$, $\sigma_i \simeq_{\angle} \tau_1^i \to ... \to \tau_{m_i}^i \to \omega \to \omega$, for some $n, m_i \in \mathbb{N}$.
(ii) There is $p \in \mathbb{N}$ such that $\underbrace{\omega \to ... \to \omega}_{k} \to \omega \to \omega \leq_{\angle} \sigma$, for all $k \geq p$.
(iii) If $\sigma \not\simeq_{\angle} \omega$ then $\sigma \leq_{\angle} \omega \to \omega$.

Proof. (i) By induction on σ. The case $\sigma \equiv \omega$ is against the hypothesis. If $\sigma \equiv \pi \wedge \tau$ then the proof follows by induction. Let $\sigma \equiv \tau \to \pi$. If $\pi \simeq_{\angle} \omega$ then by rules (g), (f) and (a) it is easy to check that $\omega \to \omega \simeq_{\angle} \tau \to \pi$. Otherwise, by induction $\pi \simeq_{\angle} \pi_0 \wedge ... \wedge \pi_k$ for some $k \in \mathbb{N}$ and $\pi_i \simeq_{\angle} \pi_1^i \to ... \to \pi_{k_i}^i \to \omega \to \omega$ for some $k_i \in \mathbb{N}$. Hence, by Lemma 10.1.4.(iii), $\sigma \simeq_{\angle} (\tau \to \pi_0) \wedge ... \wedge (\tau \to \pi_k)$, and the proof is done.
(ii) If $\sigma \simeq_{\angle} \omega$ then $p = 0$, and the proof follows from rule (a). Otherwise, by the previous point, $\sigma \simeq_{\angle} \sigma_0 \wedge ... \wedge \sigma_n$ $(n \in \mathbb{N})$ such that $\forall i \leq n$, $\sigma_i \simeq_{\angle} \tau_1^i \to ... \to \tau_{m_i}^i \to \omega \to \omega$ for some $m_i \in \mathbb{N}$. Notice that

$$\underbrace{\omega \to ... \to \omega}_{m_i} \to \omega \to \omega \leq_{\angle} \tau_1^i \to ... \to \tau_{m_i}^i \to \omega \to \omega$$
$$\underbrace{\omega \to ... \to \omega}_{m_i} \to \omega \to \omega \to \omega \leq_{\angle} \underbrace{\omega \to ... \to \omega}_{m_i} \to \omega \to \omega$$

so, posing $p = \max\{m_1, ..., m_n\}$, the proof is easy.
(iii) By point (i), $\sigma \simeq_{\angle} \sigma_0 \wedge ... \wedge \sigma_n$ $(n \in \mathbb{N})$ such that $\forall i \leq n$, $\sigma_i \simeq_{\angle} \tau_1^i \to ... \to \tau_{m_i}^i \to \omega \to \omega$ for some $m_i \in \mathbb{N}$. It is easy to see that $\sigma_i \leq_{\angle} \tau_1^i \to \omega \leq_{\angle} \omega \to \omega$ $(1 \leq i \leq n)$ by rules (a), (f) and (g), so the proof follows by rule (b). □

In the related type assignment system, every term having Λ-lhnf can be assigned at least the type $\omega \to \omega$.

Lemma 11.3.6. *If M has Λ-lhnf then $B \vdash_{\angle} M : \omega \to \omega$, for some basis B.*

Proof. Let $M \equiv \lambda x.P$ and let B_ω be a basis such that $B_\omega(x) = \omega$, for all $x \in \text{Var}$. So $B \vdash_\angle P : \omega$ by rule (ω), and $B \vdash_\angle \lambda x.P : \omega \to \omega$ by rule $(\to I)$.
Let $M \equiv xM_1...M_m$. Let B be such that $B(x) = \underbrace{\omega \to ... \to \omega}_{m} \to \omega \to \omega$.

Clearly $B \vdash_\angle M_i : \omega$ $(1 \leq i \leq m)$ by rule (ω), therefore by rule $(\to E)$, $B \vdash_\angle xM_1...M_m : \omega \to \omega$.
The proof is done, since $\mathcal{L}$ is a $\lambda\Lambda$-model, hence it is closed under $=_\Lambda$. □

In order to show that this model is correct with respect to the **L**-operational semantics, we must show that if M has no Λ-lhnf, i.e. it is a Λ-unsolvable term of order 0, and $B \vdash_\angle M : \sigma$ then $\sigma \simeq_\angle \omega$. To prove this, we need an approximation theorem.

The notion of approximant needed for studying the lazy evaluation is different from that used in the previous sections.

Definition 11.3.7. *Let $\Lambda\Omega$ be defined as in Definition 11.1.9 pag. 123.*

(i) *The* $\angle\Omega$-reduction $(\to_{\angle\Omega})$*is defined as the contextual closure of the following rule:*

$$\Omega M \to \Omega.$$

(ii) *The* $\Lambda\angle\Omega$-reduction $(\to_{\Lambda\angle\Omega})$ *is the contextual closure of the following rules:*

$$(\lambda x.M)N \to M[N/x], \text{ for all } N \in \Lambda\Omega, \qquad \Omega M \to \Omega.$$

$\to^*_{\angle\Omega}$ *denotes the symmetric and transitive closure of* $\to_{\angle\Omega}$.
The η-reduction $(\to_\eta)$ can be directly applied to the language $\Lambda\Omega$ (see Definition 1.3.7). $M \in \Lambda\Omega$ is in $\Lambda\angle\Omega$-normal form ($\Lambda\angle\Omega$-nf) if and only if it does not contain $\Lambda\angle\Omega$-redexes.

Note that the reduction rule $\lambda x.\Omega \to_\Omega \Omega$, which has been used in both models $\mathcal{H}$ and $\mathcal{N}$, is no longer correct for the lazy semantics.

Definition 11.3.8. *The set of* $\angle\Omega$-approximants *of a term M is defined as follows:*

$$\mathcal{A}^\angle(M) = \left\{ A \;\middle|\; \begin{array}{c} \exists M' \text{ such that } M =_\Lambda M' \text{ and } A \text{ is a } \Lambda\angle\Omega\text{-normal form} \\ \text{obtained from } M' \text{ by replacing some subterms with } \Omega. \end{array} \right\}$$

In this chapter, we will simply call approximant a $\Lambda\angle\Omega$-normal form.

Example 11.3.9. Some sets of approximants are shown.

$$\mathcal{A}^\angle(I) = \{\Omega, \lambda x.\Omega, \lambda x.x\};$$
$$\mathcal{A}^\angle(D) = \{\Omega, \lambda x.\Omega, \lambda x.xx, \lambda x.x\Omega\};$$
$$\mathcal{A}^\angle(DD) = \{\Omega\};$$
$$\mathcal{A}^\angle\big(K(\lambda x.x(II))(DD)\big) = \{\Omega, \lambda x.\Omega, \lambda x.x\Omega, \lambda x.x(\lambda x.\Omega), \lambda x.xI\}.$$

Theorem 11.3.10 ($\mathcal{L}$-Approximation).
$B \vdash_{\angle} M : \sigma$ if and only if $B \vdash_{\angle} A : \sigma$, for some $A \in \mathcal{A}^{\angle}(M)$.

Proof. The proof is in Sect. 11.3.1. □

The following property shows that the theory of the model $\mathcal{L}$ is not sensible.

Property 11.3.11.
(i) If M is Λ-unsolvable of order n then $\mathcal{A}^{\angle}(M) = \{\lambda x_1...x_p.\Omega \mid 0 \le p \le n\}$.
(ii) If M is Λ-unsolvable of infinite order then $\mathcal{A}^{\angle}(M) = \{\lambda x_1...x_p.\Omega \mid p \in \mathbb{N}\}$.
(iii) Let M and N be Λ-unsolvable respectively of order p and q ($p, q \in \mathbb{N}$). If $p < q$ then $M \sqsubset_{\mathcal{L}} N$.

Proof. Easy. □

Corollary 11.3.12. *$M \Downarrow_{\mathbf{L}}$ if and only if $B \vdash_{\angle} M : \omega \to \omega$, for some B.*

Proof. By Lemma 11.3.6, by Property 11.3.11 and by the approximation theorem. □

Thus we can prove the correctness of the model.

Theorem 11.3.13 ($\mathcal{L}$-Correctness).
The model $\mathcal{L}$ is correct with respect to the **L***-operational semantics.*

Proof. From Property 10.1.15 and Corollary 11.3.12 . □

The correctness implies some properties of the **L**-operational semantics.

Property 11.3.14. (i) $M \to_{\eta} N$ implies $N \preceq_{\mathbf{L}} M$.
(ii) All call-by-name fixed-point operators are equated in **L**.
(iii) Let Z be a call-by-name fixed-point operator. $M \preceq_{\mathbf{L}} ZK$, for all $M \in \Lambda$.
(iv) All Λ-unsolvable of the same order are equated in **L**.

Proof. (i) It is sufficient to prove that $I \preceq_{\mathbf{L}} E$. Let $B \vdash_{\angle} I : \sigma$; by Lemma 11.3.5.(i), either $\sigma \simeq_{\angle} \omega$ or $\sigma \simeq_{\angle} (\mu_1 \to \nu_1) \wedge ... \wedge (\mu_k \to \nu_k)$, for some $k \in \mathbb{N}$. The case $\sigma \simeq_{\angle} \omega$ is trivial, so let $\sigma \simeq_{\angle} (\mu_1 \to \nu_1) \wedge ... \wedge (\mu_k \to \nu_k)$. But $B \vdash_{\angle} I : \mu_i \to \nu_i$ by rule $(\le_{\angle})$, therefore by Lemma 10.1.7.(vi) $B[\mu_i/x] \vdash_{\angle} x : \nu_i$, and by Lemma 10.1.7.(ii) $\mu_i \le_{\angle} \nu_i$ $(i \le k)$. We prove that, if $\mu_i \le_{\angle} \nu_i$ then $B \vdash_{\angle} E : \mu_i \to \nu_i$; thus the proof follows by rule $(\wedge I)$. The proof is given by induction on μ_i.
If $\mu_i \simeq_{\angle} \omega$ then $\nu_i \simeq_{\angle} \omega$, so it is easy to see that $B \vdash_{\angle} E : \omega \to \omega$, by rules (ω) and $(\to I)$. If $\mu_i \equiv \tau_i \to \rho_i$ then $B[\mu_i/x, \tau_i/y] \vdash_{\angle} xy : \rho_i$, so $B \vdash_{\angle} \lambda xy.xy : \mu_i \to \mu_i$, by applying rule $(\to I)$ twice. Finally, by rule $(\le_{\angle})$, $B \vdash_{\angle} \lambda xy.xy : \mu_i \to \nu_i$. If μ_i is an intersection type then the proof follows by induction.
Note that this inclusion is strict. In fact, $B \vdash_{\angle} \lambda xy.xy : \omega \to \omega \to \omega$, for all B, while this typing is not derivable for I.

(ii) Similar to the proof of Theorem 11.1.24.

(iii) By definition of call-by-name fixed-point, $ZM =_\Lambda M(ZM)$, for all terms M, so $ZK =_\Lambda K(ZK) =_\Lambda \lambda x.ZK$. So $\mathcal{A}^{\angle}(ZK) = \{\lambda x_1...x_p.\Omega \mid 0 \leq p\}$, and $\underbrace{\omega \to ... \to \omega}_{p} \to \omega \to \omega \in [\![ZK]\!]_\rho^{\mathcal{F}(\angle)}$ for all environments ρ and natural numbers p.
The proof follows by Property 11.3.5.(ii) and by Correctness.

(iv) By Property 11.3.11.(iii) and by correctness. □

It is interesting to compare the proof of point (i) of the previous lemma with the proof of the Lemma 6.3.10, proving the same statement. The denotational proof is much easier, this fact is a witness of the powerful of the type assignment system for proving properties of filter models.

The model $\mathcal{L}$ is not fully abstract with respect to the **L**-operational semantics. In fact, an incompleteness result holds.

Theorem 11.3.15 ($\mathcal{L}$-Incompleteness).
The model $\mathcal{L}$ is incomplete with respect to the **L***-operational semantics.*

Proof. The proof is in Sect. 11.3.2. □

There is no a filter $\lambda\Lambda$-model fully abstract with respect to the **L**-operational semantics, as will be proved in the next theorem.

Theorem 11.3.16. *There is no a filter $\lambda\Lambda$-model that is fully abstract with respect to the* **L***-operational semantics.*

Proof. The proof is in Sect.11.3.2. □

Until now, a syntactical axiomatization of the relation $\preceq_{\mathbf{L}}$ has not been found. Note that the $\sim_{\mathcal{L}}$ relation has an unusual behaviour under application, as can be seen from the following example, first proved in [2].

Example 11.3.17. $x \not\sim_{\mathcal{L}} \lambda y.xy$ but $xx \sim_{\mathcal{L}} x(\lambda y.xy)$.
$x \not\sim_{\mathcal{L}} \lambda y.xy$ follows from the fact that, if B is a basis such that $B(x) = \omega$, then $B \not\vdash_{\angle} x : \omega \to \omega$, while $B \vdash_{\angle} \lambda y.xy : \omega \to \omega$.
To prove $xx \sim_{\mathcal{L}} x(\lambda y.xy)$, by Property 11.3.14.(i), it is sufficient to prove $x(\lambda y.xy) \sqsubseteq_{\mathcal{L}} xx$, namely that $B \vdash_{\angle} x(\lambda y.xy) : \sigma$ implies $B \vdash_{\angle} xx : \sigma$.
Without loss of generality let $\sigma \not\simeq_{\nabla} \omega$; thus $B \vdash_{\angle} x(\lambda y.xy) : \sigma$ implies, by Lemma 10.1.7.(vii), that $B \vdash_{\angle} x : \rho \to \sigma$ and $B \vdash_{\angle} \lambda y.xy : \rho$, for some ρ.
Moreover, $B(x) \leq_{\angle} \rho \to \sigma$ by Lemma 10.1.7.(ii).
If $\rho \simeq_{\angle} \omega$ then $\rho \to \sigma \leq_{\angle} \omega \to \sigma$ thus

$$\dfrac{\dfrac{\dfrac{}{B \vdash_{\angle} x : B(x)}\,(var)}{B \vdash_{\angle} x : \omega \to \sigma}\,(\leq_{\angle}) \qquad \dfrac{}{B \vdash_{\angle} x : \omega}\,(\omega)}{B \vdash_{\angle} xx : \sigma}\,(\to E)$$

Otherwise, $\rho \simeq_{\angle} \sigma_0 \wedge \ldots \wedge \sigma_n$ $(n \geq 1)$ such that $\forall i \leq n,\ \sigma_i \simeq_{\angle} \mu_i \to \nu_i$ by Property 11.3.5.(i) Hence $B[\mu_i/y] \vdash_{\angle} xy : \nu_i$ by Lemma 10.1.7.(vi), and by Lemma 10.1.7.(vii), $B[\mu_i/y] \vdash_{\angle} y : \tau_i$ and $B[\mu_i/y] \vdash_{\angle} x : \tau_i \to \nu_i$, for some τ_i. So by Lemma 10.1.7.(ii), $\mu_i \leq_{\angle} \tau_i$ and $B(x) \leq_{\angle} \tau_i \to \nu_i$, hence by rule (f) of the $\angle$-intersection relation, $B(x) \leq_{\angle} \mu_i \to \nu_i$ $(i \leq n)$; it is easy to see that $B(x) \leq_{\angle} \sigma_0 \wedge \ldots \wedge \sigma_n$. So it is possible to build the following derivation:

$$\dfrac{\dfrac{\dfrac{}{B \vdash_{\angle} x : B(x)}\,(var)}{B \vdash_{\angle} x : \sigma_0 \wedge \ldots \wedge \sigma_n \to \sigma}\,(\leq_{\angle}) \qquad \dfrac{\dfrac{}{B \vdash_{\angle} x : B(x)}\,(var)}{B \vdash_{\angle} x : \sigma_0 \wedge \ldots \wedge \sigma_n}\,(\leq_{\angle})}{B \vdash_{\angle} xx : \sigma}\,(\to E)$$

11.3.1 Proof of $\mathcal{L}$-Approximation Theorem

We define a computability predicate as for the previous models. The notion of approximants is different from the previous one given in Definition 11.1.39, since it takes into account the different behaviour of the $\vdash_{\angle}$ type assignment system with respect to $\vdash_{\infty}$.

A basis B is *finite* if and only if $B(y) \simeq_{\angle} \omega$ except in a finite number of variables. We will use $[\sigma_1/x_1, \ldots, \sigma_n/x_n]$ to denote a finite basis. By Lemma 10.1.7.(i), in this section we limit ourselves to consider only such a kind of basis.

Let B and B' be two basis. $B \cup B'$ denotes the basis such that, for every x, $B \cup B'(x) = B(x) \wedge B'(x)$ (remember that, for every type σ, $\sigma \wedge \omega \simeq \sigma$).

Definition 11.3.18. (i) *$App_{\mathcal{L}}(B, \sigma, M)$ if and only if there is $A \in \mathcal{A}^{\angle}(M)$ such that $B \vdash_{\angle} A : \sigma$.*
(ii) *The predicate $Comp_{\mathcal{L}}$ is defined by induction on types as follows:*
- *$Comp_{\mathcal{L}}(B, \omega, M)$ is true;*
- *$Comp_{\mathcal{L}}(B, \sigma \to \tau, M)$ where $\tau \simeq_{\angle} \omega$, if and only if $App_{\mathcal{L}}(B, \omega \to \omega, M)$;*
- *$Comp_{\mathcal{L}}(B, \sigma \to \tau, M)$ where $\tau \not\simeq_{\angle} \omega$, if and only if* $\forall N$, *$Comp_{\mathcal{L}}(B', \sigma, N)$ implies $Comp_{\mathcal{L}}(B \cup B', \tau, MN)$;*
- *$Comp_{\mathcal{L}}(B, \sigma \wedge \tau, M)$ if and only if $Comp_{\mathcal{L}}(B, \sigma, M)$ and $Comp_{\mathcal{L}}(B, \tau, M)$.*

To prove the $(\Rightarrow)$ part of the $\mathcal{L}$-approximation theorem, we will prove, in the usual way, that $B \vdash_{\angle} M : \sigma$ implies $Comp_{\mathcal{L}}(B, M, \sigma)$, which in turn implies $App(B, \sigma, M)$.

It is easy to check, by induction on σ, that $Comp_{\mathcal{L}}(B, \sigma, M)$ and $M =_{\Lambda} M'$ imply $Comp_{\mathcal{L}}(B, \sigma, M')$; by induction on σ, the proof is easier than that of Lemma 11.1.40 pag. 133.

Note that typings are not preserved by the η-reduction, as we observed in the proof of Property 11.3.14.(i). This property was used for proving the approximation theorem in both models $\mathcal{H}$ and $\mathcal{N}$. Here a weak version of this property holds, just for approximants of a particular shape, but it is sufficient for the rest of the proof.

Property 11.3.19. Let $A \equiv \lambda z.xA_1...A_m z$, where $z \notin \mathrm{FV}(xA_1...A_m)$. If $B \vdash_{\angle} A : \sigma \to \tau$, with $\tau \not\simeq_{\angle} \omega$ then $B \vdash_{\angle} xA_1...A_m : \sigma \to \tau$.

Proof. $B \vdash_{\angle} A : \sigma \to \tau$ and $\tau \not\simeq_{\angle} \omega$ imply $B[\sigma/z] \vdash_{\angle} xA_1...A_m z : \tau$, by Lemma 10.1.7.(vi), so $B[\sigma/z] \vdash_{\angle} xA_1...A_m : \epsilon \to \tau$ and $B[\sigma/z] \vdash_{\angle} z : \epsilon$ for some ϵ, by Lemma 10.1.7.(vii). Thus $\sigma \leq_{\angle} \epsilon$, so $\epsilon \to \tau \leq_{\angle} \sigma \to \tau$; hence by rule $(\leq_{\angle})$ $B[\sigma/z] \vdash_{\angle} xA_1...A_m : \sigma \to \tau$.
$B \vdash_{\angle} xA_1...A_m : \sigma \to \tau$, by Lemma 10.1.7.(i) since $z \notin \mathrm{FV}(xA_1...A_m)$. □

Lemma 11.3.20. (i) *$App_{\mathcal{L}}(B, \sigma, x\vec{M})$ implies $Comp_{\mathcal{L}}(B, \sigma, x\vec{M})$.*
(ii) *$Comp_{\mathcal{L}}(B, \sigma, M)$ implies $App_{\mathcal{L}}(B, \sigma, M)$.*

Proof. The proof is done by mutual induction on σ. The only non obvious case is when $\sigma \equiv \tau \to \rho$ and $\rho \not\simeq_{\angle} \omega$.

(i) We will prove that $Comp_{\mathcal{L}}(B', \tau, N)$ implies $Comp_{\mathcal{L}}(B \cup B', \rho, x\vec{M}N)$, thus $Comp_{\mathcal{L}}(B, \tau \to \rho, x\vec{M})$ follows by definition.
$Comp_{\mathcal{L}}(B', \tau, N)$ implies $App_{\mathcal{L}}(B', \tau, N)$, by induction on (ii). By hypothesis $App_{\mathcal{L}}(B, \tau \to \rho, x\vec{M})$, so $App_{\mathcal{L}}(B \cup B', \rho, x\vec{M}N)$ by rule $(\to E)$, since $\vec{A} \in \mathcal{A}^{\angle}(\vec{M})$ and $A' \in \mathcal{A}^{\angle}(N)$ imply $x\vec{A}A' \in \mathcal{A}^{\angle}(x\vec{M}N)$; hence, $Comp_{\mathcal{L}}(B \cup B', \rho, x\vec{M}N)$ by induction.

(ii) Let $z \notin \mathrm{FV}(M)$ and $B(z) \simeq_{\angle} \omega$. Note that both $z \in \mathcal{A}^{\angle}$ and $[\tau/z] \vdash_{\angle} z : \tau$, hence $App_{\mathcal{L}}([\tau/z], \tau, z)$. Thus by induction on (i), $Comp_{\mathcal{L}}([\tau/z], \tau, z)$. $Comp_{\mathcal{L}}(B, \tau \to \rho, M)$ and $Comp_{\mathcal{L}}([\tau/z], \tau, z)$ imply $Comp_{\mathcal{L}}(B[\tau/z], \rho, Mz)$ and this implies $App_{\mathcal{L}}(B[\tau/z], \rho, Mz)$, by induction; which means there is $A \in \mathcal{A}^{\angle}(Mz)$ such that $B[\tau/z] \vdash_{\angle} A : \rho$. The case $A \equiv \Omega$ is not possible, since by the hypothesis $\rho \not\simeq_{\angle} \omega$. By rule $(\to I)$, $B \vdash_{\angle} \lambda z.A : \tau \to \rho$. By definition of the $\angle$-approximants of a term, $A \in \mathcal{A}^{\angle}(Mz)$ implies $\lambda z.A \in \mathcal{A}^{\angle}(\lambda z.Mz)$. Now there are two cases.
   1. If M is of order 0 then A is of the shape $xA_1...A_m z$, where $z \notin \mathrm{FV}(xA_1...A_m)$ and $xA_1...A_m \in \mathcal{A}^{\angle}(M)$. So $B \vdash_{\angle} xA_1...A_m : \tau \to \rho$, by Property 11.3.19 and the proof is given.
   2. Otherwise $M =_{\Lambda} \lambda y.M'$, so $\lambda z.Mz =_{\Lambda} \lambda z.M'[z/y] =_{\alpha} \lambda y.M'$, which implies $\lambda z.A \in \mathcal{A}^{\surd}(M)$, and the proof is given. □

Lemma 11.3.21. *If $Comp_{\mathcal{L}}(B, \sigma, M)$ and $\sigma \leq_{\angle} \tau$ then $Comp_{\mathcal{L}}(B, \tau, M)$.*

Proof. By induction on the rules of $\leq_{\angle}$. The more complex case is the rule (f), so let $\pi' \leq_{\angle} \pi$, $\tau \leq_{\angle} \tau'$ and $Comp_{\mathcal{L}}(B, \pi \to \tau, M)$.
If $\tau \simeq_{\angle} \omega$ then $\tau' \simeq_{\angle} \omega$, so the proof is immediate. If $\tau, \pi \not\simeq_{\angle} \omega$ then the proof follows by induction. If $\tau \not\simeq_{\angle} \omega$ but $\tau' \simeq_{\angle} \omega$ then, by Lemma 11.3.20.(ii), $App_{\mathcal{L}}(B, \pi \to \tau, M)$. By definition of $App_{\mathcal{L}}$, there is an $A \in \mathcal{A}^{\angle}(M)$ such that $B \vdash_{\angle} A : \pi \to \tau$, thus $B \vdash_{\angle} A : \omega \to \omega$, and the proof follows by the definition of $Comp_{\mathcal{L}}$. □

The following lemma will be used in the proof of the $\mathcal{L}$-approximation theorem.

Lemma 11.3.22. *Let* $\mathrm{FV}(M) \subseteq \{x_1, ..., x_n\}$ *and* $B = [\sigma_1/x_1, ..., \sigma_n/x_n]$. *If* $Comp_{\mathcal{L}}(B_i, \sigma_i, N_i)$ $(1 \leq i \leq n)$ *and* $B \vdash_{\angle} M : \tau$ *then*

$$Comp_{\mathcal{L}}(B_1 \cup ... \cup B_n, \tau, M[N_1/x_1, ..., N_n/x_n]).$$

Proof. By induction on the derivation $B \vdash_{\angle} M : \tau$. □

▼ **Proof of $\mathcal{L}$-Approximation Theorem** (Theorem 11.3.10 pag. 166).

($\Rightarrow$) Clearly $Comp_{\mathcal{L}}([\tau/x], \tau, x)$, by Lemma 11.3.20.(i).
Let $\mathrm{FV}(M) \subseteq \{x_1, ..., x_n\}$, so we can assume $B = [\sigma_1/x_1, ..., \sigma_n/x_n]$ without loss of generality, by Lemma 10.1.7.(i). Therefore $B \vdash_{\angle} M : \sigma$ and $Comp_{\mathcal{L}}([\sigma_i/x_i], \sigma_i, x_i)$ $(1 \leq_{\angle} i \leq_{\angle} n)$ imply $Comp(B, \sigma, M)$ by Lemma 11.3.22, which in turn implies $App_{\mathcal{L}}(B, \sigma, M)$, by Lemma 11.3.20.(ii).

($\Leftarrow$) By definition, there is M' such that $M =_{\Lambda} M'$ and A matches M' except at occurrences of Ω. A derivation of $B \vdash_{\angle} A : \sigma$ can be transformed into a derivation of $B \vdash_{\angle} M' : \sigma$, simply by replacing every subderivation

$$\dfrac{}{B \vdash_{\angle} \Omega : \omega}\,(\omega) \qquad \text{by} \qquad \dfrac{}{B \vdash_{\angle} N : \omega}\,(\omega),$$

where N is the subterm replaced by Ω in M'. $B \vdash_{\angle} M' : \sigma$ implies $B \vdash_{\angle} M : \sigma$, since the type assignment system is closed under $=_{\Lambda}$ on terms as a consequence of the fact that it induces a $\lambda\Lambda$-model, so the proof is given. ■

11.3.2 Proof of Theorems 11.3.15 and 11.3.16

Let $L_0 \equiv \lambda x.x(x(\lambda x.DD)(DD))(\lambda x.DD)$,
$L_1 \equiv \lambda x.x(\lambda y.x(\lambda x.DD)(DD)y)(\lambda x.DD)$.

We already showed, in Sect. 6.3, that $L_0 \approx_{\mathbf{L}} L_1$. Now we will prove $L_0 \not\sim_{\mathcal{L}} L_1$, so $\mathcal{L}$ is not complete with respect to the $\mathbf{L}$-operational semantics.

▼ **Proof of $\mathcal{L}$-Incompleteness Theorem** (Theorem 11.3.15 pag. 167).

Let us prove that $L_1 \not\sim_{\mathcal{L}} L_0$; namely we will show that, for every basis B, $B \vdash_{\angle} L_1 : ((\omega \to \omega) \to (\omega \to \omega) \to \omega \to \omega) \to \omega \to \omega$, while L_0 has no such a typing. Let $\sigma \equiv (\omega \to \omega) \to (\omega \to \omega) \to \omega \to \omega$.

$$\dfrac{\dfrac{\dfrac{}{B[\sigma/x] \vdash_{\angle} x : \sigma}\,(var) \quad \dfrac{\dfrac{}{B[\sigma/x][\omega/y] \vdash_{\angle} x(\lambda x.DD)(DD)y : \omega}\,(\omega)}{B[\sigma/x] \vdash_{\angle} \lambda y.x(\lambda x.DD)(DD)y : \omega \to \omega}\,(\to I)}{B[\sigma/x] \vdash_{\angle} x(\lambda y.x(\lambda x.DD)(DD)y) : (\omega \to \omega) \to \omega \to \omega}\,(\to E) \quad \dfrac{\dfrac{}{(B[\sigma/x])[\omega/x] \vdash_{\angle} DD : \omega}\,(\omega)}{B[\sigma/x] \vdash_{\angle} \lambda x.DD : \omega \to \omega}\,(\to I)}{\dfrac{B[\sigma/x] \vdash_{\angle} x(\lambda y.x(\lambda x.DD)(DD)y)(\lambda x.DD) : \omega \to \omega}{B \vdash_{\angle} L_1 : ((\omega \to \omega) \to (\omega \to \omega) \to \omega \to \omega) \to \omega \to \omega}\,(\to I)}\,(\to E)$$

On the other hand, assume $B \vdash_{\angle} L_0 : \sigma \to \omega \to \omega$; so, by Lemma 10.1.7.(vi) $B[\sigma/x] \vdash_{\angle} x(x(\lambda x.DD)(DD))(\lambda x.DD) : \omega \to \omega$.
By Lemma 10.1.7.(vii), there is μ such that $B[\sigma/x] \vdash_{\angle} (\lambda x.DD) : \mu$ and $B[\sigma/x] \vdash_{\angle} x(x(\lambda x.DD)(DD)) : \mu \to \omega \to \omega$.
By Lemma 10.1.7.(vii), there is ν such that $B[\sigma/x] \vdash_{\angle} x(\lambda x.DD)(DD) : \nu$ and $B[\sigma/x] \vdash_{\angle} x : \nu \to \mu \to \omega \to \omega$.
By Lemma 10.1.7.(ii), $\sigma \leq_{\angle} \nu \to \mu \to \omega \to \omega$, hence $\nu, \mu \leq_{\angle} \omega \to \omega$ by Property 10.1.6; but $B[\sigma/x] \vdash_{\angle} \lambda x.DD : \mu$ implies, by the $\mathcal{L}$-approximation theorem and Property 11.3.11, that $\mu \geq_{\angle} \omega \to \omega$, so $\mu \simeq_{\angle} \omega \to \omega$ and moreover $\nu \not\simeq_{\angle} \omega$.
By Lemma 10.1.7.(vii), there is π such that $B[\sigma/x] \vdash_{\angle} x(\lambda x.DD) : \pi \to \nu$ and

$$B[\sigma/x] \vdash_{\angle} DD : \pi. \tag{11.5}$$

Again by Lemma 10.1.7.(vii), there is τ such that $B[\sigma/x] \vdash_{\angle} x : \tau \to \pi \to \nu$ and $B[\sigma/x] \vdash_{\angle} \lambda x.DD : \tau$. By Lemma 10.1.7.(ii), $\sigma \leq_{\angle} \tau \to \pi \to \nu$, so by Property 10.1.6 it follows that $\pi \leq_{\angle} \omega \to \omega$, which together with Eq. (11.5) and Corollary 11.3.12 is absurd. ■

In order to prove the nonexistence of a fully abstract filter model for the **L**-operational semantics, we will prove that in every such model, if it did exist, the two terms L_0 and L_1 would be denotationally different, so reaching an absurdum. Let us notice that, while we said that a natural choice for a call-by-name filter $\lambda\Lambda$-model is $T(C) = I(C)$, a priori we cannot exclude that a different choice could be made. So, in order to prove the nonexistence of a fully abstract filter model, we must prove it in the case of every (correct) choice of the set of input types.

First, we need some properties.

Property 11.3.23. Let U be any closed Λ-unsolvable term of order 0, and let the type system $\nabla \equiv < C, \leq_\nabla, I(C) >$ induce a $\lambda\Lambda$-model $\mathcal{M}$ that is fully abstract with respect to the **L**-operational semantics.

(i) $U \sqsubseteq_{\mathcal{M}} \lambda x.U \sqsubseteq_{\mathcal{M}} \lambda y.x(\lambda x.DD)(DD)y$.
(ii) There is $\theta \in I(C)$, such that $B \vdash_\nabla \lambda x.U : \theta$ but $B \not\vdash_\nabla U : \theta$, for all bases B; furthermore, $B \vdash_\nabla \lambda y.x(\lambda x.DD)(DD)y : \theta$, for all bases B.
(iii) If θ is the type considered in point (ii) then $\theta \to \theta, \theta \to \theta \to \theta \in I(C)$.

Proof. (i) Clearly $U \prec_{\mathbf{L}} \lambda x.U \prec_{\mathbf{L}} \lambda y.x(\lambda x.DD)(DD)y$. So the proof follows from the definition of a fully abstract model.
(ii) By point (i), by the fact that all Λ-unsolvables of the same order are equated in **L** (by Property 11.3.14.(iv)) and by definition of filter model, there is $\theta \in T(C)$ such that $B \vdash_\nabla \lambda x.U : \theta$ but $B \not\vdash_\nabla U : \theta$, for every basis B (since U is closed). So, always by point (i) and by Property 10.1.13.(ii), for every B, $B \vdash_\nabla \lambda y.x(\lambda x.DD)(DD)y : \theta$. Note that U is

an input value for the $\lambda\Lambda$-calculus, so by definition of filter $\lambda\Lambda$-model, there is a type $\alpha \in I(C)$ such that $\alpha \in [\![U]\!]^{\mathcal{M}}_{\rho}$, for all environments ρ. If we assume $\theta \notin I(C)$ then $\alpha \leq_{\nabla} \theta$, by conditions on input types, thus $B \vdash_{\nabla} U : \theta$, against hypotheses. Hence it must be that $\theta \in I(C)$.

(iii) Since U is an input value for the $\lambda\Lambda$-calculus, there exists a type $\alpha \in I(C)$ such that $\alpha \in [\![U]\!]^{\mathcal{M}}_{\rho}$, for all environments ρ. If $\theta \to \theta \notin I(C)$ then $\alpha \leq_{\nabla} \theta \to \theta$, thus $B \vdash_{\nabla} U : \theta \to \theta$. But $B \vdash_{\nabla} \lambda x.U : \theta$ and $\theta \in I(C)$ imply that $B \vdash_{\nabla} U(\lambda x.U) : \theta$, by rule $(\to E)$; this is an absurdum, since $U(\lambda x.U)$ is itself a closed Λ-unsolvable term of degree 0, by Property 1.2.18.(ii). In a similar way it is easy to check that $\theta \to \theta \to \theta \in I(C)$. □

▼ ***Proof of the Theorem 11.3.16*** (pag.167).

Let $\nabla =< C, \leq_{\nabla}, I(C) >$ be a legal type system, inducing a filter $\lambda\Lambda$-model that is fully abstract with respect to the **L**-operational semantics. Let θ be the input type considered in Property 11.3.23 and note that $\theta \not\simeq_{\nabla} \omega$.
Let σ denote $\theta \to \theta \to \theta$; since $\sigma \in I(C)$, $B[\sigma/x] \vdash_{\nabla} \lambda y.x(\lambda x.DD)(DD)y : \theta$ and $B \vdash_{\nabla} DD : \theta$, by Property 11.3.23.(ii); so $B \vdash_{\nabla} L_1 : \sigma \to \theta$, by rules $(\to E)$ and $(\to I)$. Now, let us assume $B \vdash_{\nabla} L_0 : \sigma \to \theta$.
$B[\sigma/x] \vdash_{\nabla} x(x(\lambda x.DD)(DD))(\lambda x.DD) : \theta$, by Lemma 10.1.7.(vi); thus there is an input type μ such that $B[\sigma/x] \vdash_{\nabla} x(x(\lambda x.DD)(DD)) : \mu \to \theta$ and $B[\sigma/x] \vdash_{\nabla} \lambda x.DD : \mu$, by Lemma 10.1.7.(vii).
If $\mu \to \theta \simeq_{\nabla} \omega$ then $\mu \to \omega \leq_{\nabla} \mu \to \theta$, since $\mu \to \omega \leq_{\nabla} \omega$ by rule (a) of intersection relations. By Property 10.1.6, $\omega \leq_{\nabla} \theta$ and thus $\theta \simeq_{\nabla} \omega$, which is not possible, since $B \not\vdash_{\nabla} U : \theta$; hence $\mu \to \theta \not\simeq_{\nabla} \omega$.
So $B[\sigma/x] \vdash_{\nabla} x(\lambda x.DD)(DD) : \nu$ and $B[\sigma/x] \vdash_{\nabla} x : \nu \to \mu \to \theta$ for some $\nu \in I(C)$, by Lemma 10.1.7.(vii). By Lemma 10.1.7.(ii), $\sigma \leq_{\nabla} \nu \to \mu \to \theta$, hence $\nu, \mu \leq_{\nabla} \theta$ by Property 10.1.6. Again by Lemma 10.1.7.(vii), there is $\pi \in I(C)$ such that $B[\sigma/x] \vdash_{\nabla} x(\lambda x.DD) : \pi \to \nu$ and

$$B[\sigma/x] \vdash_{\nabla} DD : \pi. \tag{11.6}$$

If $\pi \to \nu \simeq_{\nabla} \omega$ then $\pi \to \omega \leq_{\nabla} \omega \leq_{\nabla} \pi \to \nu$ thus, by Property 10.1.6, $\omega \leq_{\nabla} \nu$, which would be in contradiction with $\theta \not\simeq_{\nabla} \omega$, hence we assume $\pi \to \nu \not\simeq_{\nabla} \omega$. Yet by Lemma 10.1.7.(vii), there is $\tau \in I(C)$ such that $B[\sigma/x] \vdash_{\nabla} x : \tau \to \pi \to \nu$ and $B[\sigma/x] \vdash_{\nabla} (\lambda x.DD) : \tau$. Hence by Lemma 10.1.7.(ii), $\sigma \leq_{\nabla} \tau \to \pi \to \nu$, so by Lemma 10.1.6, it follows that $\theta \leq_{\nabla} \nu$ and $\pi, \tau \leq_{\nabla} \theta$. But $\pi \leq_{\nabla} \theta$ and Eq. (11.6) imply an absurd, since from rule $(\leq_{\nabla})$, it follows $B[\sigma/x] \vdash_{\nabla} DD : \theta$, against Property 11.3.23.(ii). ■

11.4 A Fully Abstract Model for the L-Operational Semantics

It was proved, in Theorem 11.3.16, that there is not a filter model that is fully abstract with respect to the **L**-operational semantics. But we will show now that it is possible to build, starting from the model $\mathcal{L}$, the desired fully abstract model. In a very general way, the idea is to start from the model $\mathcal{L}$, to build a space of filters that is a subspace of $\mathcal{F}(\angle)$, consisting just of filters that are interpretations of closed terms, and then to identify those filters that are interpretations of terms **L**-operationally equivalent. The so-obtained model is no longer a filter model, and the interpretation of a term is no longer the collection of the types that can be assigned to it. The fully abstract model amounts just to the closed term model of the Λ-theory **L**, equipped with a partial order relation, which is obtained from a preorder on terms, defined in a stratified way by using the type assignment $\vdash_{\angle}$.

Definition 11.4.1. (i) $\trianglelefteq_\sigma$ *is a relation on* Λ^0 *defined as follows:*

- $M \trianglelefteq_\omega N$ *is true;*
- $M \trianglelefteq_{\sigma\to\tau} N$*, where* $\tau \simeq_{\angle} \omega$*, if and only if*
  $B \vdash_{\angle} M : \omega \to \omega$ *implies* $B \vdash_{\angle} N : \omega \to \omega$*, for all bases* B*;*
- $M \trianglelefteq_{\sigma\to\tau} N$*, where* $\tau \not\simeq_{\angle} \omega$*, if and only if*
  $\forall P \in \Lambda^0$*,* $B \vdash_{\angle} P : \sigma$ *implies* $MP \trianglelefteq_\tau NP$*;*
- $M \trianglelefteq_{\sigma\wedge\tau} N$ *if and only if both* $M \trianglelefteq_\sigma N$ *and* $M \trianglelefteq_\tau N$*.*

(ii) $M \trianglelefteq N$ *if and only if* $M \trianglelefteq_\sigma N$*, for all* σ*.*

The next property will be useful in order to better understand the previous definition.

Property 11.4.2. There is $P \in \Lambda^0$ such that $B \vdash_{\angle} P : \sigma$, for all B and σ.

Proof. We will prove, by induction on σ, that there is P of the shape: $\lambda x_1 \ldots x_n.DD$, for $n \geq 0$, to which σ can be assigned.
If $\sigma \simeq_{\angle} \omega$, then $B \vdash_{\angle} DD : \sigma$, by rules (ω) and $(\leq_{\angle})$.
If $\sigma \simeq_{\angle} \omega \to \omega$, then, by rule (ω), $B[\tau/x] \vdash_{\angle} DD : \omega$, and, by rule $(\to I)$, $B \vdash_{\angle} \lambda x.DD : \tau \to \omega$. The proof follows by rule $(\leq_{\angle})$.
If $\sigma \equiv \mu \to \nu$, where $\nu \not\simeq_{\angle} \omega$, then by induction there is $P \in \Lambda^0$ such that $B \vdash_{\angle} P : \nu$. Hence $B[\mu/x] \vdash_{\angle} P : \nu$, and the proof follows by rule $(\to I)$.
Let $\sigma \equiv \mu \wedge \nu$. By induction, there are $\lambda x_1 \ldots x_p.DD$ and $\lambda x_1 \ldots x_q.DD$ such that $B \vdash_{\angle} \lambda x_1 \ldots x_p.DD : \mu$ and $B \vdash_{\angle} \lambda x_1 \ldots x_q.DD : \nu$. Let $n = \max\{p, q\}$: $\lambda x_1 \ldots x_n.DD$ is the desired term by Lemma 11.3.5.(ii). □

If a type can be assigned to a closed term, then it is said to be *inhabited.* Note that although in the model $\mathcal{L}$ all types are inhabited, this does not imply that all filters are inhabited, i.e. it does not imply that every filter is the interpretation of a closed term. Indeed, the filter

$$\uparrow\{(\omega \to \omega) \to (\omega \to \omega) \to \omega \to \omega\}$$

is not the interpretation of any term, since the reader can check that every term having type $(\omega \to \omega) \to (\omega \to \omega) \to \omega \to \omega$ also has the type $\omega \to \omega \to \omega \to \omega$, which is not in the filter. If this filter were inhabited, then it would be $L_0 \not\approx_{\mathbf{L}} L_1$ (see Sect. 11.3.2).

Property 11.4.3. Let $M, N \in \Lambda^0$.

(i) If $M \sqsubseteq_{\mathcal{L}} N$ then $M \trianglelefteq N$.
(ii) $\trianglelefteq$ is reflexive.
(iii) $\trianglelefteq$ is transitive.

Proof. (i) We will prove that $M \not\trianglelefteq N$ implies $M \not\sqsubseteq_{\mathcal{L}} N$. By definition, $M \not\trianglelefteq N$ means there is σ such that $M \not\trianglelefteq_\sigma N$. The proof is given by induction on σ.
Clearly $\sigma \not\simeq_\angle \omega$, since by definition $M \trianglelefteq_\omega N$ is true. If $\sigma \equiv \mu \to \nu$ where $\nu \simeq_\angle \omega$, then $B \vdash_\angle M : \omega \to \omega$ and $B \not\vdash_\angle N : \omega \to \omega$ by definition 11.4.1, so the proof is immediate.
If $\sigma \equiv \mu \to \nu$ where $\nu \not\simeq_\angle \omega$, then there is $P \in \Lambda^0$ such that $MP \not\trianglelefteq_\nu NP$, by definition of $\trianglelefteq$. Hence, $MP \not\sqsubseteq_{\mathcal{L}} NP$ by induction, so $M \not\sqsubseteq_{\mathcal{L}} N$ by Lemma 10.1.13.(i).
If $\sigma \equiv \mu \wedge \nu$ then the proof follows by induction.
(ii) We will prove that $M \trianglelefteq_\sigma M$, for all σ, by induction on σ. The case ω is obvious. Let $\sigma \equiv \mu \to \nu$; the case $\nu \simeq_\angle \omega$ is obvious. Let $\nu \not\simeq_\angle \omega$, thus $P \in \Lambda^0$ and $B \vdash_\angle P : \mu$ imply $MP \trianglelefteq_\nu MP$, by induction; the proof follows by definition of $\trianglelefteq$. The case $\sigma \equiv \mu \wedge \nu$ follows by induction.
(iii) We will prove that $\trianglelefteq_\sigma$ is transitive, for all σ, by induction on σ.
The cases ω and $\mu \to \nu$ where $\nu \simeq_\angle \omega$ are obvious.
Let $\sigma \equiv \mu \to \nu$, $\nu \not\simeq_\angle \omega$ and $M \trianglelefteq_{\mu\to\nu} N \trianglelefteq_{\mu\to\nu} P$. If $Q \in \Lambda^0$ and $B \vdash_\angle Q : \mu$ then $MQ \trianglelefteq_\nu NQ \trianglelefteq_\nu PQ$, and by induction $MQ \trianglelefteq_\nu PQ$. Thus, the proof follows by the definition of $\trianglelefteq_{\mu\to\nu}$.
The case $\sigma \equiv \mu \wedge \nu$ follows by induction. □

In next two lemmas it is proved that $\trianglelefteq$ corresponds exactly to the operational inclusion $\preceq_{\mathbf{L}}$.

Lemma 11.4.4. *Let $M, N \in \Lambda^0$.*
$M \trianglelefteq N$ if and only if $M\vec{P} \trianglelefteq_{\omega\to\omega} N\vec{P}$ for each sequence of closed terms $\vec{P}$.

Proof. ($\Leftarrow$) We will prove that $M \not\trianglelefteq N$ implies that there is a closed sequence of terms $\vec{P}$ such that $M\vec{P} \not\trianglelefteq_{\omega\to\omega} N\vec{P}$. By hypothesis there is a type σ such that $M \not\trianglelefteq_\sigma N$, so the proof is done by induction on σ.
If $\sigma \simeq_\angle \omega$ then $\sigma \equiv \underbrace{\omega \wedge ..... \wedge \omega}_{n}$ $(n \geq 1)$, by Theorem 11.3.3; but, since $M \trianglelefteq_\omega N$ by definition, this is not possible.
Let $\sigma \not\simeq_\angle \omega$. If $\sigma \equiv \mu \to \nu$ where $\nu \simeq_\angle \omega$, then the proof is vacuous. If $\sigma \equiv \mu \to \nu$ where $\nu \not\simeq_\angle \omega$, then there is $P \in \Lambda^0$ such that $MP \not\trianglelefteq_\nu NP$, so the proof follows by induction. If $\sigma \equiv \mu \wedge \nu$ then the proof follows by induction.
($\Rightarrow$) We will prove that if there is a sequence of closed terms $\vec{P}$ and a type $\tau \not\simeq_\angle \omega$ such that $M\vec{P} \not\trianglelefteq_\tau N\vec{P}$ then $M \not\trianglelefteq N$, by induction on $\|\vec{P}\|$.
If $\|\vec{P}\| = 0$ then the proof is trivial, so let $\|\vec{P}\| \geq 1$ and $\vec{P} \equiv \vec{Q}Q'$. But

$B \vdash_{\angle} Q' : \omega$ by rule (ω) implies $M\vec{Q} \not\trianglelefteq_{\omega \to \tau} N\vec{Q}$ by definition of $\trianglelefteq$; so the proof follows by induction. □

Note that, although $\lambda x.DD \not\trianglelefteq_{\omega \to \omega} DD$, $\lambda x.DD \trianglelefteq_{\omega \to \omega \to \omega} DD$; in fact for each $P \in \Lambda^0$, $B \vdash_{\angle} P : \omega$, and so, by definition of $\trianglelefteq$, $(\lambda x.DD)P \trianglelefteq_{\omega \to \omega} (DD)P$ is true. Hence $M \trianglelefteq_{\sigma} N$ and $\sigma \leq_{\angle} \tau$ does not imply $M \trianglelefteq_{\tau} N$.

Lemma 11.4.5. *Let $M, N \in \Lambda^0$.*
$M \preceq_{\mathbf{L}} N$ if and only if $M\vec{P} \trianglelefteq_{\omega \to \omega} N\vec{P}$, for each sequence of closed terms $\vec{P}$.

Proof. Remember that, for every term Q, $Q \Downarrow_{\mathbf{L}}$ if and only if $B \vdash_{\angle} Q : \omega \to \omega$, by Corollary 11.3.12.

($\Rightarrow$) Let $\vec{P}$ be a sequence of closed terms and let B be a basis.
If $M \preceq_{\mathbf{L}} N$ then $M\vec{P} \Downarrow_{\mathbf{L}}$ implies $N\vec{P} \Downarrow_{\mathbf{L}}$; therefore, by Corollary 11.3.12, $B \vdash_{\angle} M\vec{P} : \omega \to \omega$ implies $B \vdash_{\angle} N\vec{P} : \omega \to \omega$. Hence, by definition of $\trianglelefteq_{\omega \to \omega}$ the proof is done.

($\Leftarrow$) Let $M\vec{P} \trianglelefteq_{\omega \to \omega} N\vec{P}$, for each sequence of closed terms $\vec{P}$. We will prove that, if $C[M], C[N] \in \Lambda^0$ and $C[M] \Downarrow_{\mathbf{L}}$, then $C[N] \Downarrow_{\mathbf{L}}$, for all contexts $C[.]$. The proof is done by induction on the derivation proving $C[M] \Downarrow_{\mathbf{L}}$. If the last applied rule is either (var) or $(lazy)$, then the proof is immediate. If the last applied rule is $(head)$, then there are two cases, according to the possible shape of $C[.]$.

- $C[.] \equiv [.]C_1[.]...C_m[.]$ $(m \in \mathbb{N})$.

  If $m = 0$ then $M \Downarrow_{\mathbf{L}}$ implies $B \vdash_{\angle} M : \omega \to \omega$, so $B \vdash_{\angle} N : \omega \to \omega$ by definition of $\trianglelefteq_{\omega \to \omega}$, and the proof follows by Corollary 11.3.12.

  Now, let $m \geq 1$ and $D[.] \equiv MC_1[.]...C_m[.]$. Clearly $D[M] \equiv C[M]$ and $D[.] \equiv (\lambda z.M_0)\vec{M}C_1[.]...C_m[.]$ $(m \in \mathbb{N})$, since $M \in \Lambda^0$.
  If $\|\vec{M}\| = 0$ then let $D^*[.] \equiv M_0[C_1[.]/z]C_2[.]...C_m[.]$, otherwise let $D^*[.] \equiv M_0[M_1/z]\vec{R}C_1[.]...C_m[.]$ where $\vec{M} \equiv M_1\vec{R}$. In all cases $D^*[M] \Downarrow_{\mathbf{L}}$ and by induction $D^*[N] \Downarrow_{\mathbf{L}}$, so $D[N] \Downarrow_{\mathbf{L}}$ by rule $(head)$. But $MC_1[N]...C_m[N] \Downarrow_{\mathbf{L}}$ implies $B \vdash_{\angle} MC_1[N]...C_m[N] : \omega \to \omega$, so by hypothesis $B \vdash_{\angle} NC_1[N]...C_m[N] : \omega \to \omega$. Hence, $NC_1[N]...C_m[N] \Downarrow_{\mathbf{L}}$ by Corollary 11.3.12.
- $C[.] \equiv (\lambda y.C_0[.])C_1[.]...C_m[.]$ $(m \in \mathbb{N})$.
  The case $m = 0$ is not possible, otherwise the proof follows by induction on the derivation proving $C_0[M][C_1[M]/y]C_2[M]...C_m[M] \Downarrow_{\mathbf{L}}$. □

Theorem 11.4.6. *$M \trianglelefteq N$ if and only if $M \preceq_{\mathbf{L}} N$, for all $M, N \in \Lambda^0$.*

Proof. By Lemmas 11.4.4 and 11.4.5. □

The next definition overloads the meaning of $\trianglelefteq$ on a subset of filters.

Definition 11.4.7. *Let $f, g \in \mathcal{F}^0(\angle)$ and let ρ be an environment.*
$f \trianglelefteq g$ if and only if $M, N \in \Lambda^0$, $[\![M]\!]_{\rho}^{\mathcal{F}(\angle)} = f$ and $[\![N]\!]_{\rho}^{\mathcal{F}(\angle)} = g$ imply $M \trianglelefteq N$. Moreover, $f \triangleq g$ if and only if $f \trianglelefteq g$ and $g \trianglelefteq f$.

Note that if M is closed then $[\![M]\!]_\rho^{\mathcal{L}} = [\![M]\!]_{\rho'}^{\mathcal{L}}$, for all ρ, ρ'; moreover, if M, N are closed then $[\![M]\!]_\rho^{\mathcal{L}} = [\![N]\!]_{\rho'}^{\mathcal{L}}$ implies $M \trianglelefteq N$ and $N \trianglelefteq M$, by Property 11.4.3.(i). Now we can define the new $\lambda\Lambda$-model.

Definition 11.4.8. *Let $f, g \in \mathcal{F}^0(\angle)$.*

(i) *$[f]$ is the equivalence class of f with respect to the equivalence relation $\overset{\triangle}{=}$, while $\mathcal{F}^0_{\underline{\triangle}}$ is the set of of equivalence classes induced from $\overset{\triangle}{=}$ on $\mathcal{F}^0(\angle)$.*

(ii) *$\circ_{\underline{\triangle}} : \mathcal{F}^0_{\underline{\triangle}} \times \mathcal{F}^0_{\underline{\triangle}} \to \mathcal{F}^0_{\underline{\triangle}}$ is defined as $[f] \circ_{\underline{\triangle}} [g] = [f \circ_\angle g]$, for all $[f], [g] \in \mathcal{F}^0_{\underline{\triangle}}$.*

(iii) *The interpretation function $[\![.]\!]^{\mathcal{LL}} : \Lambda \times (\text{Var} \to \mathcal{F}^0_{\underline{\triangle}}) \to \mathcal{F}^0_{\underline{\triangle}}$ is defined as:*
$$[\![M]\!]_\zeta^{\mathcal{LL}} = [[\![M]\!]_\rho^{\mathcal{F}(\angle)}], \text{ where } \rho \text{ is such that } \rho(x) \in \zeta(x) \text{ for all } x \in \text{Var}.$$

(iv) *Let $\mathcal{LL}$ be the quadruple: $< \mathcal{F}^0_{\underline{\triangle}}, \mathcal{F}^0_{\underline{\triangle}}, \circ_{\underline{\triangle}}, [\![.]\!]^{\mathcal{LL}} >$.*

Note that the interpretation is defined for open terms too.

Property 11.4.9. Let $M, N, P, Q \in \Lambda^0$. If $M \trianglelefteq N$ and $P \trianglelefteq Q$ then $MP \trianglelefteq NQ$.

Proof. Clearly $M \preceq_\mathbf{L} N$ and $P \preceq_\mathbf{L} Q$ imply $MP \preceq_\mathbf{L} NQ$, therefore the proof follows by Theorem 11.4.6. $\square$

It is easy to see that $\circ_{\underline{\triangle}}$ is well defined, by using the previous property.

Lemma 11.4.10. *$\mathcal{LL}$ is a $\lambda\Lambda$-model.*

Proof. We check that $\mathcal{LL}$ satisfies the conditions of Definition 10.0.1. If $\zeta \in (\text{Var} \to \mathcal{F}^0_{\underline{\triangle}})$ then let ρ be such that $\rho(x) \in \zeta(x)$ for all $x \in \text{Var}$.

1. $[\![x]\!]_\zeta^{\mathcal{LL}} = [[\![x]\!]_\rho^{\mathcal{F}(\angle)}] = [\rho(x)] = \zeta(x)$.

2. $[\![MN]\!]_\zeta^{\mathcal{LL}} = [[\![MN]\!]_\rho^{\mathcal{F}(\angle)}] = [[\![M]\!]_\rho^{\mathcal{F}(\angle)} \circ_\angle [\![N]\!]_\rho^{\mathcal{F}(\angle)}] = [[\![M]\!]_\rho^{\mathcal{F}(\angle)}] \circ_{\underline{\triangle}} [[\![N]\!]_\rho^{\mathcal{F}(\angle)}] = [\![M]\!]_\zeta^{\mathcal{LL}} \circ_{\underline{\triangle}} [\![N]\!]_\zeta^{\mathcal{LL}}$.

3. $[\![\lambda x.M]\!]_\zeta^{\mathcal{LL}} \circ_{\underline{\triangle}} d = [[\![\lambda x.M]\!]_\zeta^{\mathcal{F}(\angle)}] \circ_{\underline{\triangle}} d = [[\![\lambda x.M]\!]_\zeta^{\mathcal{F}(\angle)} \circ_\angle f] = [[\![M]\!]_{\zeta[f/x]}^{\mathcal{F}(\angle)}] = [\![M]\!]_{\zeta[d/x]}^{\mathcal{LL}}$, for all $d \in \mathcal{F}^0_{\underline{\triangle}}$ and $f \in d$.

4. Let $[\![M]\!]_{\zeta[d/x]}^{\mathcal{LL}} = [\![N]\!]_{\zeta'[d'/x']}^{\mathcal{LL}}$, where $d, d' \in \mathcal{F}^0_{\underline{\triangle}}$, $f \in d$ and $f' \in d'$. $[\![M]\!]_{\rho[f/x]}^{\mathcal{F}(\angle)} = [\![N]\!]_{\rho[f'/x']}^{\mathcal{F}(\angle)}$ since $\mathcal{L}$ is a model, so $[[\![M]\!]_{\rho[f/x]}^{\mathcal{F}(\angle)}] = [[\![N]\!]_{\rho[f'/x']}^{\mathcal{F}(\angle)}]$, therefore $[[\![\lambda x.M]\!]_\rho] = [[\![\lambda x'.N]\!]_{\rho'}]$ thus $[\![\lambda x.M]\!]_\zeta^{\mathcal{LL}} = [\![\lambda x'.N]\!]_{\zeta'}^{\mathcal{LL}}$.

5. Trivial. $\square$

Since $\trianglelefteq$ is a preorder on $\mathcal{F}^0(\angle)$ it induces a partial order on $\mathcal{F}^0_{\underline{\triangle}}$.

Definition 11.4.11. *Let $M \sqsubseteq_{\mathcal{LL}} N$ denote $[\![M]\!]_\zeta^{\mathcal{LL}} \trianglelefteq [\![N]\!]_\zeta^{\mathcal{LL}}$, for each $\zeta \in (\text{Var} \to \mathcal{F}^0_{\underline{\triangle}})$. Moreover, let $M \sim_{\mathcal{LL}} N$ denote $M \sqsubseteq_{\mathcal{LL}} N$ and $N \sqsubseteq_{\mathcal{LL}} M$.*

Consequently, the model $\mathcal{LL}$ induces a partial order on the interpretation of terms (not only closed terms).

Property 11.4.12. Let $M, N \in \Lambda^0$. $M \sqsubseteq_{\mathcal{LL}} N$ if and only if $M \trianglelefteq N$.

Proof. Let $\zeta \in (\mathrm{Var} \to \mathcal{F}_{\underline{\triangle}}^0)$ and let ρ be such that $\rho(x) \in \zeta(x)$ for all $x \in \mathrm{Var}$. $M \sqsubseteq_{\mathcal{LL}} N$ if and only if $[\![M]\!]_\zeta^{\mathcal{LL}} \trianglelefteq [\![N]\!]_\zeta^{\mathcal{LL}}$ if and only if $[[\![M]\!]_\rho^{\mathcal{F}(\triangleleft)}] \trianglelefteq [[\![N]\!]_\rho^{\mathcal{F}(\triangleleft)}]$ if and only if $[\![M]\!]_\rho^{\mathcal{F}(\triangleleft)} \trianglelefteq [\![N]\!]_\rho^{\mathcal{F}(\triangleleft)}$ if and only if $M \trianglelefteq N$. □

The correctness is easy.

Theorem 11.4.13 ($\mathcal{LL}$-Correctness).
*The model $\mathcal{LL}$ is correct with respect to the **L**-operational semantics.*

Proof. We will prove that $M \sqsubseteq_{\mathcal{LL}} N$ implies $M \preceq_{\mathbf{L}} N$, by definition of correctness. $M \sqsubseteq_{\mathcal{LL}} N$ implies $C[M] \sqsubseteq_{\mathcal{LL}} C[N]$, for each closing context $C[.]$, by Property 10.0.2.(v). Thus $C[M] \trianglelefteq C[N]$, by Lemma 11.4.12; hence $C[M] \trianglelefteq_{\omega \to \omega} C[N]$, thus $B \vdash_{\angle} C[M] : \omega \to \omega$ implies $B \vdash_{\angle} C[N] : \omega \to \omega$, for all bases B. So, by Corollary 11.3.12, $C[M] \Downarrow_{\mathbf{L}}$ implies $C[N] \Downarrow_{\mathbf{L}}$, and so $M \preceq_{\mathbf{L}} N$. □

The following theorem implies the full abstraction of $\mathcal{LL}$ with respect to the **L**-operational semantics.

Theorem 11.4.14 ($\mathcal{LL}$-Completeness).
*The model $\mathcal{LL}$ is complete with respect to the **L**-operational semantics.*

Proof. We will prove $\not\sqsubseteq_{\mathcal{LL}}$ implies $\not\preceq_{\mathbf{L}}$.
$M \not\sqsubseteq_{\mathcal{LL}} N$ means $[\![M]\!]_\zeta^{\mathcal{LL}} \not\trianglelefteq [\![N]\!]_\zeta^{\mathcal{LL}}$, for some $\zeta \in (\mathrm{Var} \to \mathcal{F}_{\underline{\triangle}}^0)$.
Since the codomain of ζ is $\mathcal{F}_{\underline{\triangle}}^0$, if $\mathrm{FV}(M) \cup \mathrm{FV}(N) = \{x_1, ..., x_m\}$ then there are $P_i \in \Lambda^0$ such that $\zeta(x_i) = [[\![P_i]\!]_\rho^{\mathcal{F}(\triangleleft)}]$. Thus, let $\mathbf{s}$ be such that $\mathbf{s}(x_i) = P_i$ $(1 \le i \le m)$, hence $\mathbf{s}(M), \mathbf{s}(N) \in \Lambda^0$.
$[\![\mathbf{s}(M)]\!]_{\zeta'}^{\mathcal{LL}} \not\trianglelefteq [\![\mathbf{s}(N)]\!]_{\zeta'}^{\mathcal{LL}}$, for all $\zeta' \in (\mathrm{Var} \to \mathcal{F}_{\underline{\triangle}}^0)$ by Property 10.0.2.(iv), since $\mathbf{s}(M)$ and $\mathbf{s}(N)$ are closed. Therefore $\mathbf{s}(M) \not\sqsubseteq_{\mathcal{LL}} \mathbf{s}(N)$ and, by Lemma 11.4.12, $\mathbf{s}(M) \not\trianglelefteq \mathbf{s}(N)$. So there is a sequence of closed terms $\vec{Q}$ such that $\mathbf{s}(M)\vec{Q} \not\trianglelefteq_{\omega \to \omega} \mathbf{s}(N)\vec{Q}$, by Lemma 11.4.4.
Let $C[.] \equiv (\lambda x_1 ... x_m.[.])\mathbf{s}(x_1)...\mathbf{s}(x_m)\vec{Q}$; clearly $C[M], C[N] \in \Lambda^0$ and, moreover it is such that $C[M] \Downarrow_{\mathbf{L}}$ and $C[N] \Uparrow_{\mathbf{L}}$, so $M \not\preceq_{\mathcal{LL}} N$. □

Hence, $\mathcal{LL}$ is fully abstract with respect to the **L**-operational semantics.

Corollary 11.4.15. *If $M \not\preceq_{\mathbf{L}} N$ then there is a head context separating M and N.*

Proof. Immediate, by the proof of Theorem 11.4.14. □

The technique used here for building the fully abstract model of the **L**-operational semantics is similar to that used in [71] and [44], for different calculi. The use of intersection types and filter models allows for the application of such techniques to a wider class of models. A further fully abstract model for the **L**-operational semantics, based on a variant of the game semantics, was built in [42].

11.5 Crossing Models

It can be interesting to ask for the behaviour of the three filter $\lambda\Lambda$-models we defined, with respect to all the considered call-by-name operational semantics.

11.5.1 The Model $\mathcal{H}$

In Sect. 11.1, we already proved that $\mathcal{H}$ is fully abstract with respect to the **H**-operational semantics.

It can be readily noticed that $\mathcal{H}$ is not correct with respect to the **N**-operational semantics. In fact, by Property 11.2.22 and by Theorem 11.1.31, $I \sqsubseteq_{\mathcal{H}} E_\infty$, but $I \not\preceq_{\mathbf{N}} E_\infty$.

On the other hand, $\mathcal{H}$ is complete with respect to the **N**-operational semantics, namely $M \preceq_{\mathbf{N}} N$ implies $M \sqsubseteq_{\mathcal{H}} N$. Let us prove that $M \not\sqsubseteq_{\mathcal{H}} N$ implies $M \not\preceq_{\mathbf{N}} N$. Let $M \not\sqsubseteq_{\mathcal{H}} N$; so by the correctness of $\mathcal{H}$ with respect to **H**-operational semantics, this implies that there is a closing context $C[.]$ such that $C[M]$ has Λ-hnf, while $C[N]$ has not Λ-hnf. So there are $P_1 ... P_n$ such that $C[M]P_1...P_n =_\Lambda I$, while $C[N]P_1...P_n$ has no Λ-hnf. Since I is a Λ-nf and a term without Λ-hnf in particular does not have Λ-nf, $C'[.] \equiv C[.]P_1...P_n$ is a context such that $C'[M] \Downarrow_{\mathbf{N}}$ while $C'[N] \Uparrow_{\mathbf{N}}$, so $M \not\preceq_{\mathbf{N}} N$.

$\mathcal{H}$ is not correct with respect to the **L**-operational semantics. In fact, $\lambda y.xy \sqsubseteq_{\mathcal{H}} x$ while $\lambda y.xy \not\preceq_{\mathbf{L}} x$, as shown in Example 11.3.17. Another counterexample is the pair of terms DD and $\lambda x.DD$; indeed, by Corollary 11.1.21 and Property 11.3.11, $\lambda x.DD \sqsubseteq_{\mathcal{H}} DD$ but $\lambda x.DD \not\preceq_{\mathbf{L}} DD$.

$\mathcal{H}$ is not complete with respect to the **L**-operational semantics. In fact, for every call-by-name fixed-point operator Z, ZK is a Λ-unsolvable term of infinite order, so $ZK \sqsubseteq_{\mathcal{H}} I$ while $ZK \not\preceq_{\mathbf{L}} I$ (see Property 11.3.14.(iii)).

But if we take into account the equivalence relation, we have that $\approx_{\mathbf{L}}$ implies $\sim_{\mathcal{H}}$. In fact, the following lemma holds, which, together with the completeness of $\mathcal{H}$ with respect to the **H**-operational semantics, proves this implication.

Lemma 11.5.1. *If $M \approx_{\mathbf{L}} N$ then $M \approx_{\mathbf{H}} N$.*

Proof. We prove that $M \not\approx_{\mathbf{H}} N$ implies $M \not\approx_{\mathbf{L}} N$. By hypothesis there is a context $C[.]$ such that $C[M], C[N] \in \Lambda^0$, $C[M] \Downarrow_{\mathbf{H}}$ and $C[N] \Uparrow_{\mathbf{H}}$ (or vice versa). Thus there is a sequence of terms $P_1...P_n$ such that $C[M]P_1...P_n =_\Lambda I$ and $C[N]P_1...P_n$ is Λ-unsolvable, since the unsolvability is closed under head contexts. If $C[N]P_1...P_n$ is Λ-unsolvable of the finite order p then let $C^*[.] \equiv C[.]P_1...P_n \underbrace{I.....I}_{p}$. Clearly $C^*[M] \Downarrow_{\mathbf{L}}$ while $C^*[N] \Uparrow_{\mathbf{L}}$, so $M \not\approx_{\mathbf{L}} N$.

Otherwise, $C[N]P_1...P_n$ must be a Λ-unsolvable of infinite order, therefore let $C^*[.] \equiv C[.]P_1...P_n(DD)$. Clearly $C^*[N] \Downarrow_{\mathbf{L}}$ while $C^*[M] \Uparrow_{\mathbf{L}}$, so $M \not\approx_{\mathbf{L}} N$. □

11.5.2 The Model $\mathcal{N}$

In Sect. 11.2, we already proved $\mathcal{N}$ is fully abstract with respect to the **N**-operational semantics.

Lemma 10.0.4 and Property 11.2.11.(ii) imply that $\mathcal{N}$ is correct with respect to the **H**-operational semantics. On the other hand, it is easy to see that it is not complete; in fact, by Property 11.2.22 and by Theorem 11.1.31, $I \preceq_{\mathbf{H}} E_\infty$ while $I \not\sqsubseteq_{\mathcal{N}} E_\infty$.

Finally, $\mathcal{N}$ is not correct with respect to the **L**-operational semantics; in fact, by Property 11.2.15.(ii) and by Example 11.3.17, $\lambda y.xy \sqsubseteq_{\mathcal{N}} x$, but $\lambda y.xy \not\preceq_{\mathbf{L}} x$. On the other hand, $\mathcal{N}$ is not complete with respect to the **L**-operational semantics; in fact, $I \preceq_{\mathbf{L}} ZK$ while $I \not\sqsubseteq_{\mathcal{N}} ZK$, for every call-by-name fixed-point operator Z (see Property 11.3.14.(iii)). Note that ZK is a Λ-unsolvable term of infinite order, hence it is easy to see $ZK \sqsubset_{\mathcal{N}} I$.

11.5.3 The Model $\mathcal{L}$

In Sect. 11.3, we already proved that $\mathcal{L}$ is correct but not complete with respect to the **L**-operational semantics.

$\mathcal{L}$ is not correct with respect to both **H** and **N**-operational semantics; in fact, $I \sqsubseteq_{\mathcal{L}} ZK$ (see Property 11.3.14.(iii)) while $I \not\preceq_{\mathbf{H}} ZK$ and $I \not\preceq_{\mathbf{N}} ZK$, for every call-by-name fixed point operator Z, by Corollary 11.1.21 and by Property 11.2.15.(i).

$\mathcal{L}$ is not complete with respect to both **H** and **N**-operational semantics; in fact, $\lambda x.DD \preceq_{\mathbf{H}} DD$ and $\lambda x.DD \preceq_{\mathbf{N}} DD$ while $\lambda x.DD \not\sqsubseteq_{\mathcal{L}} DD$.

12. Call-by-Value Denotational Semantics

For modeling the $\lambda\Gamma$-calculus, we must reflect in the model the fact that the set Γ of input values is a proper subset of the whole set Λ. In the setting of filter $\lambda\Gamma$-models, this implies that every type system ∇ inducing a filter $\lambda\Gamma$-model must be such that $\mathcal{I}(\nabla) \subset \mathcal{F}(\nabla)$.

Since ω is the universal type it cannot characterize any property of terms; note that from $B \vdash_\nabla \lambda x.M : \omega \to \sigma$ we cannot conclude $B \vdash_\nabla (\lambda x.M)N : \sigma$ for every N in a correct $\lambda\Gamma$-model. Indeed, $(\lambda x.M)(DD) \Uparrow_{\mathbf{V}}$ and so every type σ such that $B \vdash_\nabla (\lambda x.M)(DD) : \sigma$ must be such that $\sigma \notin I(C)$. Thus the type $\omega \to \sigma$ cannot have a meaningful applicative power. This is simply formalized by assuming $\omega \notin I(C)$, since the condition on the rule $(\to E)$ of the intersection type assignment system, namely

$$\frac{\sigma \in I(C) \quad B \vdash_\nabla M : \sigma \to \tau \quad B \vdash_\nabla N : \sigma}{B \vdash_\nabla MN : \tau}\,(\to E)\,.$$

The simplest choice in order to build a filter $\lambda\Gamma$-model is to choose a legal type system $\nabla =< C, \leq_\nabla, I(C) >$ such that $I(C) = \{\sigma \in T(C) \mid \sigma \not\leq_\nabla \omega\}$.

12.1 The Model $\mathcal{V}$

In this section, we will define a filter $\lambda\Gamma$-model that is correct with respect to the $\mathbf{V}$-operational semantics. In the $\lambda\Gamma$-calculus, there are terms, as $(\lambda z.D)(xI)D$, that are Γ-unsolvable of order 0 but Γ-normal form. So $(\lambda z.D)(xI)D \Downarrow_{\mathbf{V}}$ and $DD \Uparrow_{\mathbf{V}}$, while they are $\approx_{\mathbf{V}}$. This implies that Property 10.1.15 cannot be a guideline for building a correct model. But we can use the fact, proved in Property 7.1.10, that if M is potentially Γ-valuable and N is not potentially Γ-valuable then $N \prec_{\mathbf{V}} M$.

Property 12.1.1. Let $\mathcal{F}$ be a filter model such that

P potentially Γ-valuable and Q not potentially Γ-valuable imply $Q \sqsubset_{\mathcal{F}} P$, for all $P, Q \in \Lambda$.

Then $\mathcal{F}$ is correct with respect to $\preceq_{\mathbf{V}}$.

Proof. If $Q \sqsubset_{\mathcal{F}} P$ then $P \not\sqsubseteq_{\mathcal{F}} Q$, so the hypothesis implies the following statement:

"$P \sqsubseteq_{\mathcal{F}} Q$ and P is potentially Γ-valuable imply Q is potentially Γ-valuable."

Let $M \sqsubseteq_{\mathcal{F}} N$, so by Property 10.1.13 $C[M] \sqsubseteq_{\mathcal{F}} C[N]$, for each closing context $C[.]$; hence, by hypothesis, if $C[M]$ is potentially Γ-valuable then $C[N]$ is potentially Γ-valuable too. Since $C[M], C[N] \in \Lambda^0$, the potential Γ-valuability coincides with the Γ-valuability, and $C[M] \Downarrow_{\mathbf{V}}$ implies $C[N] \Downarrow_{\mathbf{V}}$; hence $M \preceq_{\mathbf{V}} N$. □

So, we want a legal type system $\surd$, based on a set of constants $C_{\surd}$, such that there is at least one type σ and a basis B such that, if M is potentially Γ-valuable then $B \vdash_{\surd} M : \sigma$, otherwise $B \not\vdash_{\surd} M : \sigma$. A term is potentially Γ-valuable if and only if it has $\Xi\ell$-normal form, so it is natural to characterize terms without $\Xi\ell$-normal form by assigning them only type ω. Moreover, the **V**-operational semantics performs a lazy evaluation (so it cannot be $\omega \simeq_{\surd} \omega \to \omega$), and this allows us to use the type $\omega \to \omega$ to characterize $\Xi\ell$-normal form.

Definition 12.1.2. *$\surd$ is the type system $<C_{\surd}, \leq_{\surd}, I(C_{\surd})>$ where $C_{\surd} = \{\omega\}$,*

$$I(C_{\surd}) = \{\sigma_0 \wedge ... \wedge \sigma_n \mid \exists k \leq n \ \ \exists \sigma, \tau \in T(C) \ \ \sigma_k \equiv \sigma \to \tau\},$$

and $\leq_{\surd}$ is the intersection relation induced by the rules in Fig. 12.1.

$$\frac{}{\sigma \leq_{\surd} \omega}\ (a) \qquad \frac{}{\sigma \leq_{\surd} \sigma \wedge \sigma}\ (b) \qquad \frac{}{\sigma \wedge \tau \leq_{\surd} \sigma}\ (c) \qquad \frac{}{\sigma \wedge \tau \leq_{\surd} \tau}\ (c')$$

$$\frac{}{(\sigma \to \tau) \wedge (\sigma \to \pi) \leq_{\surd} \sigma \to (\tau \wedge \pi)}\ (d) \qquad \frac{\sigma \leq_{\surd} \sigma',\ \tau \leq_{\surd} \tau'}{\sigma \wedge \tau \leq_{\surd} \sigma' \wedge \tau'}\ (e)$$

$$\frac{\sigma' \leq_{\surd} \sigma,\ \tau \leq_{\surd} \tau'}{\sigma \to \tau \leq_{\surd} \sigma' \to \tau'}\ (f) \qquad \frac{}{\sigma \to \omega \leq_{\surd} \omega \to \omega}\ (g) \qquad \frac{}{\sigma \leq_{\surd} \sigma}\ (r) \qquad \frac{\sigma \leq_{\surd} \rho,\ \rho \leq_{\surd} \tau}{\sigma \leq_{\surd} \tau}\ (t)$$

$$\frac{}{(\omega \to \omega) \to \tau \leq_{\surd} \omega \to \tau}\ (v)$$

Fig. 12.1. $\surd$-Intersection relation

Some properties of the $\leq_{\surd}$-intersection relation are proved in Sect. 12.1.1. In particular notice that $(\omega \to \omega) \to \tau \simeq_{\surd} \omega \to \tau$, for all $\tau \in T(C_{\surd})$.

By the definition of $I(C_{\surd})$ and the rule $(\to E)$, a type of the shape $\omega \to \sigma$ does not have applicative power. So a typing as $B[\omega \to \sigma/x] \vdash_{\surd} xM : \sigma$

cannot be proved, for all M, while it is a correct typing in every other type assignment system we have seen before.

Let $\mathcal{I}(\surd) = \{f \in \mathcal{F}(\surd) \mid f \neq\uparrow \{\omega\}\ \}$, so the type system $\surd$ induces a $\lambda\Gamma$-model, by Theorems 10.1.11 and 12.1.3.

Theorem 12.1.3. (i) *The type system* $\surd$ *is legal.*
(ii) *If* $M \in \Gamma$ *then* $[\![M]\!]_\rho \in \mathcal{I}(C_\surd)$, *for all environments* ρ.

Proof. (i) The proof is in Sect. 12.1.1.
(ii) $\omega \to \omega \in \mathcal{I}(\surd)$; thus it is sufficient to check that $B \vdash_\surd M : \omega \to \omega$, for all $M \in \Gamma$ and all bases B. Let $M \equiv \lambda x.P$ and $\sigma \in I(C_\surd)$, so

$$\dfrac{\dfrac{\dfrac{}{B[\sigma/x] \vdash_\surd P : \omega}(\omega)}{B \vdash_\surd \lambda x.P : \sigma \to \omega}(\to I) \qquad \sigma \to \omega \leq_\surd \omega \to \omega}{B \vdash_\surd \lambda x.P : \omega \to \omega}(\leq_\surd)$$

If $\sigma \in I(C_\surd)$ then $\sigma \not\leq_\surd \omega$, hence $\sigma \leq_\surd \omega \to \omega$ by Property 12.1.28.(ii). Let $M \equiv x$, so

$$\dfrac{\dfrac{}{B \vdash_\surd x : B(x)}(var) \qquad B(x) \leq_\surd \omega \to \omega}{B \vdash_\surd x : \omega \to \omega}(\leq_\surd)$$

□

The next definition is well given by Corollary 12.1.27.

Definition 12.1.4. $\mathcal{V}$ *is the* $\lambda\Gamma$*-model* $< \mathcal{F}(\surd), \mathcal{I}(\surd), \circ_\surd, [\![.]\!]^{\mathcal{F}(\surd)} >$.

Since $\mathcal{V}$ is a $\lambda\Gamma$-model the intepretation of terms is closed under $=_\Gamma$; hence, by Corollary 10.1.13, typings are closed under $=_\Gamma$. We will prove that they are closed under the $\Xi\ell$-reduction too. First of all, let us prove that a $\Xi\ell$-nf can always be assigned the type $\omega \to \omega$.

Lemma 12.1.5. *If* M *is a* $\Xi\ell$*-nf then* $B \vdash_\surd M : \omega \to \omega$, *for some basis* B.

Proof. By induction on M. If $M \in \Gamma$ then the proof is similar to that of Theorem 12.1.3.(ii). Let $M \equiv xM_1...M_m$ $(m \geq 1)$ where M_i is a $\Xi\ell$-nf, so by induction $B_i \vdash_\surd M_i : \omega \to \omega$ $(1 \leq i \leq m)$.
Let B be a basis such that $B(y) = B_1(y) \wedge ... \wedge B_m(y)$ for $y \not\equiv x$, while

$$B(x) = B_1(x) \wedge ... \wedge B_m(x) \wedge \underbrace{(\omega \to \omega) \to ... \to (\omega \to \omega)}_{m} \to \omega \to \omega.$$

$B \vdash_\surd M_i : \omega \to \omega$, by Lemma 10.1.7.(iii); so $B \vdash_\surd M : \omega \to \omega$, by rules (var), $(\leq_\surd)$ and m applications of rule $(\to E)$. □

Theorem 12.1.6. (i) *$B \vdash_{\surd} M : \sigma$ and $M \to_{\Xi\ell} N$ imply $B \vdash_{\surd} N : \sigma$;*
(ii) *$B \vdash_{\surd} M : \sigma$ and $N \to_{\Xi\ell} M$ imply there is B' such that, for all x, $B'(x) \leq_{\surd} B(x)$, and $B' \vdash_{\surd} N : \sigma$.*

Proof. The proof is in Sect. 12.1.2. □

In order to understand the difference between the two points of the previous theorem, consider the two terms I and $(\lambda zt.t)(xy)$. They are $=_{\Xi\ell}$ but they are not $\sim_{\mathcal{V}}$. Let B_ω be the basis such that, for all x, $B_\omega(x) = \omega$; thus

$$B_\omega \vdash_{\surd} I : ((\omega \to \omega) \to \omega \to \omega) \to (\omega \to \omega) \to \omega \to \omega$$

nevertheless, in order to assign the same type to the term $(\lambda zt.t)(xy)$ it is necessary to change the basis, choosing, for example, a basis B such that $B(x) = (\omega \to \omega) \to \omega \to \omega$ and $B(y) = \omega \to \omega$. So the following corollary holds for closed terms only.

Corollary 12.1.7. *If $M, N \in \Lambda^0$ and $M =_{\Xi\ell} N$ then $M \sim_{\mathcal{V}} N$.*

Proof. From Theorem 12.1.6, since the typings of a closed term are independent of the basis. □

So the following property holds.

Property 12.1.8. If M is potentially Γ-valuable then $B \vdash_{\surd} M : \omega \to \omega$, for some basis B.

Proof. M potentially valuable means M has $\Xi\ell$-nf. The proof follows from Lemma 12.1.5 and Theorem 12.1.6.(ii). □

In order to prove that a term that is not potentially Γ-valuable cannot be assigned the type $\omega \to \omega$ from any basis, we need an approximation theorem. As usual, we extend the language by adding a constant Ω to the formation rules of terms; hence we define some new reduction rules on the so-obtained language.

Definition 12.1.9. *Let $\Lambda\Omega$ be defined as in Definition 11.1.9.*

(i) *$\to_{\surd\Omega}$ is defined as the contextual closure of the following rules:*

$$\Omega M \to \Omega, \qquad M\Omega \to \Omega.$$

(ii) *The set of* approximated input values *is $\Gamma_\Omega = \mathrm{Var} \cup \{\lambda x.M \mid M \in \Lambda\Omega\}$. The $\Gamma\Omega$-*reduction *($\to_{\Gamma\Omega}$) is the contextual closure of the following rules:*

$$\Omega M \to \Omega, \qquad M\Omega \to \Omega,$$

$$(\lambda x.M)N \to M[N/x] \quad \text{if } N \in \Gamma_\Omega.$$

$\to^*_{\Gamma\Omega}$ is the symmetric and transitive closure of $\to_{\Gamma\Omega}$. The η-reduction can be directly applied to the language $\Lambda\Omega$ (see Definition 1.3.7 pag. 23). $M \in \Lambda\Omega$ is in $\Gamma\Omega$-normal form ($\Gamma\Omega$-nf) if and only if it does not contain $\Gamma\Omega$-redexes.

(iii) *Let* $\Xi_\Omega = \Gamma_\Omega \cup \{xM_1...M_m \mid M_i \in \Xi_\Omega \ \ (i \leq m)\}$. *The* $\Xi\ell\Omega$-reduction ($\rightarrow_{\Xi\ell\Omega}$) *is the applicative closure (see Definition 3.1.3 pag. 36) of the following rules:*

$$\Omega M \rightarrow \Omega, \qquad M\Omega \rightarrow \Omega,$$
$$(\lambda x.M)N \rightarrow M[N/x] \quad \textit{if } N \in \Xi_\Omega$$

$\rightarrow^*_{\Xi\ell\Omega}$ *is the symmetric and transitive closure of* $\rightarrow_{\Xi\ell\Omega}$. $M \in \Lambda\Omega$ *is in* $\Xi\ell\Omega$*-normal form (*$\Xi\ell\Omega$*-nf) if and only if it does not contain* $\Xi\ell\Omega$*-redexes. Note that* M *is a* $\Xi\ell\Omega$*-normal form if and only if* $M \in \Xi_\Omega \cup \{\Omega\}$.

The first reduction rule defined in point (i) of Definition 12.1.9, namely $\Omega M \rightarrow \Omega$, reflects the lazy behaviour of the Γ-calculus, while $M\Omega \rightarrow \Omega$ reflects its call-by-value behaviour. Note that the $\Xi\ell\Omega$-reduction is lazy, since it is closed under application, but not under abstraction. As usual, terms of $\Lambda\Omega$ will be considered modulo $=_\alpha$.

Example 12.1.10. The term $\lambda x.x(II)\Omega$ is a $\Xi\ell\Omega$-normal form, but it is not a $\Gamma\Omega$-normal form. Indeed, its subterm II is not a $\Xi\ell\Omega$-redex (since it occurs under the scope of a λ-abstraction), but it is a $\Gamma\Omega$-redex.
$\lambda x.x(II)\Omega \rightarrow_{\Gamma\Omega} \lambda x.\Omega$, which is a $\Gamma\Omega$-normal form.

The type assignment system of Definition 10.1.1 can be applied to $\Lambda\Omega$ without modifications. It is easy to see that to the term Ω only the type ω can be assigned, by using rule (ω). The intuitive interpretation of the constant Ω is that it represents a term with an unknown behaviour. The interpretation function is naturally extended to $\Lambda\Omega$, i.e. the interpretation of a term of $\Lambda\Omega$ is the set of types that can be assigned to it.

Definition 12.1.11. *The set of* $\surd\Omega$-approximants *of a term* M *is defined as follows:*

$$\mathcal{A}^{\surd}(M) = \left\{ A \,\middle|\, \begin{array}{l} \exists M' \textit{ such that } M =_\Gamma M' \textit{ and } A \textit{ is a } \Gamma\Omega\textit{-nf} \\ \textit{obtained from } M' \textit{ by replacing some subterms by } \Omega. \end{array} \right\}$$

The set of upper approximants *of a term* M *is defined as follows:*

$$\mathcal{U}^{\surd}(M) = \{U \mid \exists A \in \mathcal{A}^{\surd}(M) \textit{ such that } A \rightarrow^*_{\Xi\ell\Omega} U \in \Xi\ell\Omega\textit{-NF.}\}$$

Example 12.1.12. Some sets of approximants are shown.

- $\mathcal{A}^{\surd}((\lambda y.I)(DD)) = \mathcal{U}^{\surd}((\lambda y.I)(DD)) = \{\Omega\}$.
- $\mathcal{A}^{\surd}((\lambda zt.t)(xy)) = \{\Omega, (\lambda z.\Omega)(xy), (\lambda zt.\Omega)(xy), (\lambda zt.t)(xy)\}$,
  $\mathcal{U}^{\surd}((\lambda zt.t)(xy)) = \{\Omega, \lambda t.\Omega, \lambda t.t\}$ and note that $(\lambda zt.t)\Omega \notin \mathcal{U}^{\surd}((\lambda zt.t)(xy))$.
- $\mathcal{A}^{\surd}((\lambda zx.xD)(yI)D) \supseteq \{\Omega, (\lambda z.\Omega)(yI)(\lambda x.\Omega), (\lambda zx.xD)(yI)D\}$,
  $\mathcal{U}((\lambda zx.xD)(yI)D) = \{\Omega\}$.
- $\mathcal{A}^{\surd}(\lambda z.II) = \mathcal{U}^{\surd}(\lambda z.II) = \{\Omega, \lambda z.\Omega, \lambda zx.\Omega, \lambda zx.x\}$.

- $\mathcal{A}^{\surd}((\lambda xyz.yI)(uv)I) \supseteq \{(\lambda xyz.yI)(uv)I\}$.
  $\mathcal{U}((\lambda xyz.yI)(uv)I) \supseteq \{\lambda z.II\}$.

Approximants can be inductively defined.

Definition 12.1.13. *The set $\mathcal{A}^{\surd}$ of approximants can be inductively defined as follows:*

- $\Omega \in \mathcal{A}^{\surd}$;
- *If* $A_j \in \mathcal{A}^{\surd}$ *and* $A_j \not\equiv \Omega$ *then* $\lambda x_1...x_n.xA_1...A_m \in \mathcal{A}^{\surd}$ $(1 \leq j \leq m$ *and* $n, m \in \mathbb{N})$.
- *If* $A, A_j \in \mathcal{A}^{\surd}$, $A_1 \notin \Gamma_\Omega$ *and* $A_j \not\equiv \Omega$ *then* $\lambda x_1...x_n.(\lambda x.A)A_1...A_m \in \mathcal{A}^{\surd}$ $(1 \leq j \leq m,\ m \geq 1$ *and* $n \in \mathbb{N})$.

Approximants will be ranged over by A, A', possibly indexed. Upper approximants will be ranged over by U, U', possibly indexed.

Property 12.1.14. $A \in \mathcal{A}^{\surd}$ if and only if there is M such that $A \in \mathcal{A}^{\surd}(M)$.

Proof. Easy. □

An approximation theorem holds, relating the interpretation of a term to the intepretations both of its approximants and of its upper approximants.

Theorem 12.1.15 ($\mathcal{V}$-Approximation).

(i) $B \vdash_{\surd} M : \sigma$ *if and only if* $B \vdash_{\surd} A : \sigma$, *for some* $A \in \mathcal{A}^{\surd}(M)$.
(ii) $B \vdash_{\surd} M : \sigma$ *if and only if* $B' \vdash_{\surd} U : \sigma$ *where* $U \in \mathcal{U}^{\surd}(M)$, *for some basis* B, B' *such that* $B(x) \leq_{\surd} B'(x)$, *for all* x.

Proof. Both proofs are in Sect. 12.1.3. □

The next property implies $\mathcal{A}^{\surd}(M) = \{\Omega\}$ if and only if M is not Γ-valuable, for each closed term M. It also implies $\mathcal{U}^{\surd}(M) = \{\Omega\}$ if and only if M is not potentially Γ-valuable, for each term M (not necessarily closed).

Property 12.1.16. (i) M is Γ-valuable if and only if there is $A \in \mathcal{A}^{\surd}(M)$ such that $A \in \Gamma_\Omega$.
(ii) M is potentially Γ-valuable if and only if there is $U \in \mathcal{U}^{\surd}(M)$ such that $U \not\equiv \Omega$.

Proof. (i) ($\Leftarrow$) If $x \in \mathcal{A}^{\surd}(M)$ then $M =_\Gamma x$, by definition of $\mathcal{A}^{\surd}(M)$. If $\lambda x.A \in \mathcal{A}^{\surd}(M)$ then there is $M' \in \Lambda$ such that $M =_\Gamma \lambda x.M'$ and $A \in \mathcal{A}^{\surd}(M')$, by definition of $\mathcal{A}^{\surd}(M)$. ($\Rightarrow$) Trivial, by definition of Γ-valuable terms and by definition of $\mathcal{A}^{\surd}(M)$.

(ii) ($\Leftarrow$) If there is $U \in \mathcal{U}^{\surd}(M)$ such that $U \not\equiv \Omega$ then $\exists A \in \mathcal{A}^{\surd}(M)$ such that $A \to^*_{\Xi\ell\Omega} U \in \Xi\ell\Omega$-NF. As done in the proof of Lemma 3.1.35, it is possible to show that there is $r \in \mathbb{N}$ and $\bar{A}^r \in \Gamma_\Omega$ such that both $A[O^r/x_1, ..., O^r/x_n] \to^*_\Gamma \bar{A}^r$ and $U[O^r/x_1, ..., O^r/x_n] \to^*_\Gamma \bar{A}^r$, where $O^r \equiv \lambda x_1 \ldots x_{r+1}.x_{r+1}$. Since $\bar{A}^r \in \mathcal{A}^{\surd}(M[O^r/x_1, ..., O^r/x_n])$, by the point (i) of this Lemma the proof is done.
($\Rightarrow$) By Property 12.1.8 and by point (ii) of the $\mathcal{V}$-approximation theorem. □

Note that point (i) of the $\mathcal{V}$-approximation theorem is not sufficient in order to give a complete characterization of the not potentially Γ-valuable terms, through the syntactical shape of the approximants. In fact, $(\lambda zx.xD)(yI)D$ is not potentially Γ-valuable, but it is an approximant, as shown in Example 12.1.12.

Moreover, point (ii) of the $\mathcal{V}$-approximation theorem cannot be used to characterize the equivalence in the model, if we want to take into account also open terms (see Corollary 12.1.7).

Lemma 12.1.17. *Let M be not potentially Γ-valuable.*
If $B \vdash_{\surd} M : \sigma$ then $\sigma \simeq_{\surd} \omega$.

Proof. M not potentially Γ-valuable implies $\mathcal{U}^{\surd}(M) = \{\Omega\}$, by Property 12.1.16.(ii). Hence, the result follows from point (ii) of the $\mathcal{V}$-approximation theorem. □

Now we are able to state the correctness of the model.

Theorem 12.1.18 ($\mathcal{V}$-Correctness).
The model $\mathcal{V}$ is correct with respect to the **V***-operational semantics.*

Proof. By Property 12.1.1, it is sufficient to check that M is potentially Γ-valuable and N is not potentially Γ- valuable imply $N \not\sqsubseteq_{\mathcal{V}} M$. The proof follows by Property 12.1.8 and Lemma 12.1.17. □

The model gives also a (partial) characterization of the Γ-solvable terms.

Theorem 12.1.19. (i) *Let M be either a Γ-solvable term or a Γ-unsolvable term of infinite Γ-order. Then for all $p \geq 0$ there is a basis B and types $\sigma_1, ..., \sigma_p$ such that $B \vdash_{\surd} M : \sigma_1 \to \sigma_2 \to ... \to \sigma_p \to \omega$.*
(ii) *Let M be a Γ-unsolvable term of Γ-order p. If $B \vdash_{\surd} M : \sigma$ then for some $k \geq 1$, $\sigma \equiv \sigma_1 \wedge ... \wedge \sigma_k$, where $\sigma_i \equiv \tau_1 \to \tau_2 \to ... \to \tau_n \to \omega$, and $n \leq p$.*

Proof. (i) By Property 3.1.15 (pag. 39), M is Γ-solvable if and only if there are terms $M_1, .., M_n$ ($n \in \mathbb{N}$) such that $M \to^*_{\Xi\ell} \lambda x_1.M_1$, $M_i \to^*_{\Xi\ell} \lambda x_{i+1}.M_{i+1}$ ($1 \leq i \leq n$) and $M_n \equiv xP_1...P_m$ where $P_i \in \Xi$ for some $m \in \mathbb{N}$. Let M be Γ-solvable. The proof will be given by induction on n. If $n = 0$, then choose a basis B such that

$$B(x) = \underbrace{(\omega \to \omega) \to ... \to (\omega \to \omega)}_{m} \to \sigma_1 \to \sigma_2 \to ... \to \sigma_p \to \omega.$$

Then $B \vdash_{\surd} xP_1...P_m : \sigma_1 \to \sigma_2 \to ... \to \sigma_p \to \omega$, by rule $(\to E)$ and Property 12.1.8. The inductive step is easy, using the basic case, Theorem 12.1.6 and rule $(\to I)$.
Let M be Γ-unsolvable of Γ-order ∞. For every $i \geq 1$, $M \to^*_{\Xi\ell} \lambda x_1.M_1$, $M_i \to^*_{\Xi\ell} \lambda x_{i+1}.M_{i+1}$. So choose $i = p$, and obviously there is a typing $B \vdash_{\surd} \lambda x_p.M_p : \omega \to \omega$, by Property 12.1.8. Then the proof is similar to the previous case.

(ii) By induction on p. If $p = 0$, then M is not potentially Γ-valuable, and the result follows from Lemma 12.1.17. The case $p \geq 0$ follows easily from the definition of Γ-solvability and of Γ-order. □

As the previous theorem shows, the $\mathcal{V}$-model does not permit us to discriminate between Γ-solvable terms and Γ-unsolvable terms of infinite order. An intersection type assignment system giving a complete characterization of the Γ-solvable terms is shown in [73].

Some properties of the **V**-operational semantics can be proved by the approximation theorem.

Property 12.1.20. (i) The theory $\approx_{\mathbf{V}}$ is operationally extensional.

(ii) All call-by-value fixed-point operators are equated in $\mathcal{V}$.

(iii) All call-by-value recursion operators are equated in $\mathcal{V}$.

(iv) Let Z be a call-by-value recursion operator. Then every term M is such that $M \preceq_{\mathbf{V}} ZK$.

Proof. (i) It is sufficient to prove that $I \sim_{\mathcal{V}} E$, then the result follows by Property 3.1.18 and Lemma 8.1.9. The proof of $I \sim_{\mathcal{V}} E$ can be easily constructed in a way similar to the proof of Lemma 11.1.8, taking into account the differences between the two type assignment systems, namely Property 12.1.28 must be used in place of Property 11.1.36. The proof follows by correctness.

(ii) Let Z be a fixed-point operator, i.e. $ZM =_{\Gamma} M(ZM)$, for all Γ-valuable terms M. Then $\mathcal{A}^{\surd}(Z) = \{\Omega, \lambda x.\Omega\}$.

(iii) Let Z be a call-by-value recursion operator, i.e. $ZM =_{\Gamma} M(\lambda z.ZMz)$, for all Γ-valuable terms M.
Then $\mathcal{A}^{\surd}(Z) = \{\Omega, \lambda z.\Omega, \lambda x.x(\lambda z_0.(...(\lambda z_n.\Omega)...)z_0) \mid n \geq 1\}$.

(iv) $ZK =_{\Gamma} K(\lambda z.ZKz) =_{\Gamma} \lambda yz.ZKz$. So $\mathcal{A}^{\surd}(ZK) = \{\lambda x_1...x_n.\Omega \mid n \in \mathbb{N}\}$. As a result, it is easy to check that $B \vdash_{\surd} ZK : \sigma$, for all $\sigma \in T(C_{\surd})$ and for all bases B. □

As we anticipated at the end of Sect. 3.1, the notion of fixed-point in the call-by-value setting is in some sense meaningless. In fact, it follows from the proof of Property 12.1.20.(ii), that a fixed point operator Z is such that, for every Γ-valuable term M, ZM is Γ-unsolvable of order 0.

The model $\mathcal{V}$ induces a not semisensible Γ-theory (this notion was defined in Definition 1.3.4). In order to prove this result, we need to show that there is an infinite class of approximants which behaves, with respect to typing, as free variables.

Lemma 12.1.21. *Let $A_0 \equiv \lambda z.\Omega$ and $A_{n+1} \equiv \lambda z.(\lambda x.A_n)(xz)$.*

(i) *For all $\sigma \not\simeq_{\surd} \omega$ there is n such that $B[\sigma/x] \vdash_{\surd} A_n : \sigma$.*
(ii) *$B[\sigma/x] \vdash_{\surd} A_n : \tau$ implies $\sigma \leq_{\surd} \tau$.*

Proof. (i) $\sigma \simeq_{\surd} \sigma_0 \wedge ... \wedge \sigma_n$ ($n \in \mathbb{N}$), where $\sigma_i \simeq_{\surd} \tau_1^i \to ... \to \tau_{m_i}^i \to \omega \to \omega$ ($m_i \in \mathbb{N}, i \leq n$) by Property 12.1.28.(iii). Moreover, without loss of generality, we can assume that $\tau_r^i \not\simeq_{\surd} \omega$ for all r, by Property 12.1.28.(i). We will show that if $p_i \geq m_i$ then $B[\sigma_i/x] \vdash_{\surd} A_{p_i} : \sigma_i$ ($i \leq n$), by induction on m_i. Since $r \geq \max\{p_1, ..., p_n\}$ implies $B[\sigma_i/x] \vdash_{\surd} A_r : \sigma_i$ ($i \leq n$), the proof follows by rule $(\wedge I)$ and Lemma 10.1.7.(iii).
If $m_i = 0$ then the proof is trivial, by Lemma 12.1.5. Otherwise
$B[\tau_2 \to ... \to \tau_{m_i} \to \omega \to \omega/x] \vdash_{\surd} A_p : \tau_2 \to ... \to \tau_{m_i} \to \omega \to \omega$
where $p \geq m_i - 1$, by induction. By Lemma 10.1.7.(i), it follows that
$B[\tau_2 \to ... \to \tau_{m_i} \to \omega \to \omega/x, \tau_1/z] \vdash_{\surd} A_p : \tau_2 \to ... \to \tau_{m_i} \to \omega \to \omega$.
So $B[\tau_1/z] \vdash_{\surd} \lambda x.A_p : (\tau_2 \to ... \to \tau_{m_i} \to \omega \to \omega) \to \tau_2 \to ... \to \tau_{m_i} \to \omega \to \omega$ by rule $(\to I)$. Thus, always by Lemma 10.1.7.(i),

$$B[\tau_1/z, \tau_1 \to ... \to \tau_{m_i} \to \omega \to \omega/x] \vdash_{\surd} \lambda x.A_p : (\tau_2 \to ... \to \tau_{m_i} \to \omega \to \omega) \to \tau_2 \to ... \to \tau_{m_i} \to \omega \to \omega$$

and by rule $(\to E)$
$B[\tau_1/z, \tau_1 \to ... \to \tau_{m_i} \to \omega \to \omega/x] \vdash_{\surd} (\lambda x.A_p)(xz) : \tau_2 \to ... \to \tau_{m_i} \to \omega \to \omega$.
Finally, $B[\tau_1 \to ... \to \tau_{m_i} \to \omega \to \omega/x] \vdash_{\surd} \lambda z.(\lambda x.A_p)(xz) : \sigma$.

(ii) By induction on n. If $n = 0$ then the proof is obvious, since $\sigma \in I(C_{\surd})$. Let $n \geq 1$. If $\tau \simeq_{\surd} \omega \to \omega$ $\tau \simeq_{\surd} \omega \to \omega$ then the proof is obvious. Otherwise, $B[\sigma/x] \vdash_{\surd} A_n : \tau$ implies, by Lemma 10.1.7.(iv), $\tau \geq_{\surd} \tau_1 \wedge ... \wedge \tau_r$ ($r \geq 1$) where $\tau_i \equiv \mu_i \to \nu_i$, $B[\sigma/x, \mu_i/z] \vdash_{\surd} (\lambda x.A_{n-1})(xz) : \nu_i$ and $\mu_i \not\simeq_{\surd} \omega$ ($1 \leq i \leq r$). Since $\tau \simeq_{\surd} \omega \to \omega$ has been already considered, we can assume $\nu_i \not\simeq \omega$ without loss of generality. Therefore $B[\sigma/x, \mu_i/z] \vdash_{\surd} (\lambda x.A_{n-1}) : \pi_i \to \nu_i$ and $B[\sigma/x, \mu_i/z] \vdash_{\surd} xz : \pi_i$ for some $\pi_i \not\simeq_{\surd} \omega$, by Lemma 10.1.7.(vii). So, by Lemma 10.1.7.(iv), $B[\sigma/x, \mu_i/z, \pi_i/y] \vdash_{\surd} A_{n-1}[y/x] : \nu_i$, and by induction $\pi_i \leq_{\surd} \nu_i$.
Moreover, by Lemma 10.1.7.(vii) and (ii), $B[\sigma/x, \mu_i/z] \vdash_{\surd} xz : \pi_i$ implies $\sigma \leq_{\surd} \mu_i \to \pi_i$. Thus $\sigma \leq \tau_i$, and the proof follows. □

Now we are ready to prove the next theorem.

Theorem 12.1.22. *Let Z be a call-by-value recursion operator. If $B \equiv \lambda xyz.x(yz)$ then $I \sim_{\mathcal{V}} ZB$.*

Proof. It is easy, but boring, to check that $\mathcal{A}^{\surd}(ZB) = \{\Omega, \lambda x.\Omega, \lambda xy.\Omega\} \cup \{\lambda x.A_n \mid n \geq 1\}$, where $A_n \equiv \lambda z.(\lambda x.A_{n-1})(xz)$. Then the proof follows from Lemma 12.1.21, since $B \vdash I : \sigma$ if and only if either $\sigma \simeq_{\surd} \omega$ or $\sigma \simeq_{\surd} \omega \to \omega$ or $\sigma \simeq_{\surd} \sigma_1 \wedge \sigma_2$ or $\sigma \simeq_{\surd} \mu \to \nu$, where $\mu \leq_{\surd} \nu$. □

Corollary 12.1.23. *The* **V***-operational semantics is not semisensible.*

Proof. Since ZB is a Γ-unsolvable term of infinite order, the Γ-theory $\mathcal{V}$ is not semisensible, by Theorem 12.1.22. Then, by correctness, **V** is also not semisensible. □

The fact that **V** is not semisensible was first proved in [73], using syntactical tools. The model $\mathcal{V}$ is not fully abstract with respect to **V**-operational semantics. In fact, the next theorem shows its incompleteness.

Theorem 12.1.24 ($\mathcal{V}$-Incompleteness).
The model $\mathcal{V}$ is incomplete with respect to the **V***-operational semantics.*

Proof. The proof is in Sect. 12.1.4 □

The next theorem proves that there is not a filter $\lambda\Gamma$-model that is fully abstract with respect to the **V**-operational semantics. The proof is similar to the corresponding one for the **L**-operational semantics, given in Sect. 11.3.2. Namely, we will show that there are two terms that are $\approx_{\mathbf{V}}$, but they cannot be equated in every filter model correct with respect to the **V**-operational semantics.

Theorem 12.1.25. *There is not a filter $\lambda\Gamma$-model that is fully abstract with respect to the* **V***-operational semantics.*

Proof. The proof is in Sect. 12.1.4. □

12.1.1 The $\leq_{\surd}$-Intersection Relation

In order to prove that the $\leq_{\surd}$-intersection relation is well posed, i.e. it does not identify all types, we need to characterize the types $\simeq_{\surd} \omega$.

Theorem 12.1.26. *$\sigma \simeq_{\surd} \omega$ if and only if $\sigma \notin I(C_{\surd})$, for all $\sigma \in T(C_{\surd})$.*

Proof. Note that $\omega \simeq_{\surd} \sigma$ if and only if $\omega \leq_{\surd} \sigma$, by the rule (a).

($\Leftarrow$) By induction on σ. The case $\sigma \equiv \omega$ is obvious. The case $\sigma \equiv \tau \to \pi$ is not possible. The case of intersection follows by induction.

($\Rightarrow$) Let us first prove that if $\sigma \notin I(C_{\surd})$ and $\sigma \leq_{\surd} \tau$ then $\tau \notin I(C_{\surd})$, by induction on the rule of $\leq_{\surd}$.
(a),(b),(c),(c'),(r) Trivial.
(d),(f),(g),(v) Not possible.
(e),(t) By induction.
Since $\omega \notin I(C_{\surd})$, if $\sigma \simeq_{\surd} \omega$ then $\omega \leq_{\surd} \sigma$, so the proof is done. □

Corollary 12.1.27. *Let* $\Omega^{\mathcal{V}} = \{\sigma_0 \wedge ..... \wedge \sigma_n \ (n \in \mathbb{N}) \mid \forall i \leq n \quad \sigma_i \equiv \omega\}$. $\Omega^{\mathcal{V}} = T(C_{\surd}) - I(C_{\surd})$.

The previous theorem implies that $I(C_{\surd})$ is a well-defined set of input types, since it satisfies the conditions of Definition 10.1.1.(iv). Moreover, not all types in $T(C_{\surd})$ are equated by $\simeq_{\surd}$; in fact, $\omega \to \omega \not\simeq_{\surd} \omega$.

Property 12.1.28. (i) $\omega \to \sigma \simeq_{\surd} (\omega \to \omega) \to \sigma$.
(ii) $\sigma \to \tau \leq_{\surd} \omega \to \omega$, for all $\sigma, \tau \in T(C_{\surd})$.
(iii) If $\sigma \not\simeq_{\surd} \omega$ then $\sigma \simeq_{\surd} \sigma_0 \wedge ... \wedge \sigma_n \ (n \in \mathbb{N})$ where

$$\sigma_i \simeq_{\surd} \tau_1^i \to ... \to \tau_{m_i}^i \to \omega \to \omega \qquad (m_i \in \mathbb{N}, i \leq n).$$

Proof. (i) By rule (a), (f) and (v).
(ii) Clearly $\sigma \to \tau \not\simeq_{\surd} \omega$ and $\sigma \to \tau \leq_{\surd} \sigma \to \omega \leq_{\surd} \omega \to \omega$ by rules (a), (f) and (g).
(iii) By induction on σ. The case $\sigma \equiv \omega$ is against the hypothesis. If $\sigma \equiv \pi \wedge \tau$ and $\pi, \tau \not\simeq_{\surd} \omega$ then the proof follows by induction. If $\sigma \equiv \pi \wedge \tau$ and $\pi \not\simeq_{\surd} \omega$ but $\tau \simeq_{\surd} \omega$ the proof follows by induction on π, since $\sigma \simeq_{\surd} \pi$. If $\sigma \equiv \pi \wedge \tau$ and $\tau \not\simeq_{\surd} \omega$ but $\pi \simeq_{\surd} \omega$ the proof follows by induction on τ, since $\sigma \simeq_{\surd} \tau$.
Let $\sigma \equiv \tau \to \pi$. If $\pi \simeq_{\surd} \omega$ then $\sigma \simeq_{\surd} \omega \to \omega$ by rules (g), (f) and (a); so, let $\pi \not\simeq_{\surd} \omega$. By induction $\pi \simeq_{\surd} \pi_0 \wedge ... \wedge \pi_k \ (k \in \mathbb{N})$, where

$$\pi_i \simeq_{\surd} \pi_1^i \to ... \to \pi_{k_i}^i \to \omega \to \omega \qquad (k_i \in \mathbb{N}, i \leq k).$$

Hence, $\sigma \simeq_{\surd} (\tau \to \pi_0) \wedge ... \wedge (\tau \to \pi_k)$ by Lemma 10.1.4.(iii). □

It is easy to check that a type in $T(C_{\surd})$ has the following shape:

$$(\sigma_1 \to \tau_1) \wedge ... \wedge (\sigma_n \to \tau_n) \wedge \underbrace{\omega \wedge ... \wedge \omega}_{m}$$

where $n, m \geq 0, m + n \geq 1$. The following lemma implies the legality of $\surd$.

Property 12.1.29. Let $n, m, p, q \in \mathbb{N}$ be such that $n, p \geq 1$ and

$$(\sigma_1 \to \tau_1) \wedge ... \wedge (\sigma_n \to \tau_n) \wedge \underbrace{\omega \wedge ... \wedge \omega}_{m} \leq_{\surd} (\sigma'_1 \to \tau'_1) \wedge ... \wedge (\sigma'_p \to \tau'_p) \wedge \underbrace{\omega \wedge ... \wedge \omega}_{q}.$$

Let $h \leq p$; if $\tau'_h \not\simeq_{\surd} \omega$ then $\exists \{i_1, ..., i_k\} \subseteq \{1, ..., n\}$, for some $k \in \mathbb{N}$, such that $\sigma_{i_1} \wedge ... \wedge \sigma_{i_k} \geq_{\surd} \sigma'_h \wedge (\omega \to \omega)$ and $\tau_{i_1} \wedge ... \wedge \tau_{i_k} \leq_{\nabla} \tau'_h$, where $\tau_{i_j} \not\simeq_{\surd} \omega$ $(i_j \in \{i_1, ..., i_k\})$.

Proof. By induction on the definition of $\leq_{\surd}$.

(a), (g), (b), (c), (c'), (e), (f), (r) Easy.

(*d*) Let $(\sigma \to \tau) \wedge (\sigma \to \pi) \leq_{\surd} \sigma \to (\tau \wedge \pi)$. The case $\tau, \pi \not\simeq_{\surd} \omega$ is easy. If $\tau \simeq_{\surd} \omega$ and $\pi \not\simeq_{\surd} \omega$ then it is easy to see that $\tau \wedge \pi \simeq_{\surd} \pi$, so the proof is immediate. The remaining case is similar to the previous one.

(*v*) It is sufficient to check that $\omega \to \omega \geq_{\surd} \omega \wedge (\omega \to \omega)$; the proof follows by rule (c').

(*t*) Let ρ be such that $(\sigma_1 \to \tau_1) \wedge ... \wedge (\sigma_n \to \tau_n) \wedge \underbrace{\omega \wedge ... \wedge \omega}_{m} \leq_{\surd} \rho$ and

$\rho \leq_{\surd} (\sigma'_1 \to \tau'_1) \wedge ... \wedge (\sigma'_p \to \tau'_p) \wedge \underbrace{\omega \wedge ... \wedge \omega}_{q}$.

If $\rho \simeq_{\surd} \omega$ then $\omega \leq_{\surd} (\sigma'_1 \to \tau'_1) \wedge ... \wedge (\sigma'_p \to \tau'_p) \wedge \underbrace{\omega \wedge ... \wedge \omega}_{q} \leq_{\surd} \sigma'_j \to \tau'_j$

implies $\omega \simeq_{\surd} \sigma'_j \to \tau'_j$ $(j \leq p)$ and so $p = 0$, by Theorem 12.1.26. Let $\rho \equiv (\mu_1 \to \nu_1) \wedge ... \wedge (\mu_r \to \nu_r) \wedge \underbrace{\omega \wedge ... \wedge \omega}_{s}$, for some $r, s \in \mathbb{N}$

such that $r \geq 1$. If $\tau'_h \not\simeq \omega$ $(h \leq p)$ then, by inductive hypothesis, $\exists\{i_1, ..., i_h\} \subseteq \{1, ..., r\}$ such that $\mu_{i_1} \wedge ... \wedge \mu_{i_h} \geq_{\surd} \sigma'_h \wedge (\omega \to \omega)$ and $\nu_{i_1} \wedge ... \wedge \nu_{i_h} \leq_{\surd} \tau'_h$. Since $\nu_{i_j} \not\simeq_{\surd} \omega$ $(i_j \in \{i_1, ..., i_h\})$, the proof follows by applying the inductive hypothesis to each arrow $\mu_{i_j} \to \nu_{i_j}$. □

▼ **Proof of Theorem 12.1.3.(i)** (pag. 183).

The legality of the type system $\surd$ is a particular case of Property 12.1.29. ■

12.1.2 Proof of Theorem 12.1.6

In order to show that the type assignment system $\vdash_{\surd}$ is closed under $\Xi\ell$-reduction, first we prove that the type assignment system $\vdash_{\surd}$ is closed under Λ-reduction.

Lemma 12.1.30 (Λ-Subject reduction).
If $M \to_{\Lambda} N$ and $B \vdash_{\surd} M : \sigma$ then $B \vdash_{\surd} N : \sigma$.

Proof. Let $M \equiv (\lambda x.P)Q$ and $N \equiv P[Q/x]$. If $\sigma \simeq_{\surd} \omega$ then the proof is trivial, so let $\sigma \not\simeq_{\surd} \omega$. $B \vdash_{\surd} (\lambda x.P)Q : \sigma$ implies, by Lemma 10.1.7.(vii), both $B \vdash_{\surd} (\lambda x.P) : \tau \to \sigma$ and $B \vdash_{\surd} Q : \tau$, for some $\tau \in I(C_{\surd})$. By Lemma 10.1.7.(vi), $B \vdash_{\surd} \lambda x.P : \tau \to \sigma$ if and only if $B[\tau/x] \vdash_{\surd} P : \sigma$.
Without loss of generality, we can assume that there is a derivation d proving $B[\tau/x] \vdash_{\surd} P : \sigma$ such that all typings occurring in it have the same basis $B[\tau/x]$. Indeed, the only rule that can change the basis is $(\to I)$, and we can assume that free and bound variables have different names in P. Derivation d can be transformed into a derivation d' proving $B \vdash_{\surd} P[Q/x] : \sigma$, by performing the following operations:

1. replace each subderivation of d of the shape

$$\frac{}{B[\tau/x] \vdash_{\surd} x : \tau} (var)$$

by a copy of a derivation proving $B \vdash_{\surd} Q : \tau$;
2. replace each typing $B[\tau/x] \vdash_{\surd} P^* : \mu$ in d by $B \vdash_{\surd} P^*[Q/x] : \mu$.

By induction on the derivation d, it is easy to check that d' is well defined. Let $M \equiv C[(\lambda x.P)Q]$ and $N \equiv C[P[Q/x]]$. If an occurrence of $(\lambda x.P)Q$ in M is inside a subterm of M typed by the rule (ω), then just replace Q to each free occurrence of x in term being subject of typings in d. Otherwise, replace each subderivation d proving a typing for $(\lambda x.P)Q$ by a subderivation d' built as described before. □

Obviously, $\vdash_{\surd}$ cannot be closed under Λ-expansion, i.e. $B \vdash_{\surd} P[Q/x]$ cannot imply $B \vdash_{\surd} (\lambda x.P)Q$, since in this case the model would be incorrect. But a restricted form of Λ-expansion can be proved. First, we will prove a property.

Property 12.1.31. Let d be a derivation proving $B \vdash_{\surd} M : \sigma$ where $\sigma \not\simeq_{\surd} \omega$. If N is a subterm of M not occurring under the scope of a λ-abstraction, then in d there is a subderivation d' proving $B \vdash_{\surd} N : \tau$, where $\tau \not\simeq_{\surd} \omega$.

Proof. We will prove that all subterms S of M not occurring under the scope of a λ-abstraction are typed in d by a subderivation d_S proving a typing $B_S \vdash_{\surd} S : \tau_s$ where $\tau_s \not\simeq_{\surd} \omega$. The proof is given by induction on M.
The proof is obvious for $M \equiv x$; so let $M \equiv M_1 M_2$.
$B \vdash_{\surd} M_1 M_2 : \sigma$ and $\sigma \not\simeq_{\surd} \omega$ imply that there is $\tau \in I(C_{\surd})$ such that $B \vdash_{\surd} M_1 : \tau \to \sigma$ and $B \vdash_{\surd} M_2 : \tau$, by Lemma 10.1.7.(vii). Moreover, by induction the property is true for all subterms S of M_i not occurring under the scope of a λ-abstraction $(1 \leq i \leq 2)$.
The case $M \equiv \lambda x.N$ and S occur in N is against the hypothesis. □

Lemma 12.1.32. *Let* d *be a derivation proving* $B \vdash_{\surd} C[M] : \sigma$, *where* $\sigma \not\simeq_{\surd} \omega$. *If* M *occurs in* $C[M]$, *and there is at least one subderivation of* d *assigning to* M *a type* $\not\simeq_{\surd} \omega$, *then* $B \vdash_{\surd} (\lambda x.C[x])M : \sigma$.

Proof. Without loss of generality, let each typing in d have the same basis B; indeed, the only rule that can change the basis is $(\to I)$, and we can assume that free and bound variables have different names in $C[M]$.
Let M occur in $C[M]$ and let there exist $n \geq 1$ subderivations d_i in d proving $B \vdash_{\surd} M : \tau_i$, where $\tau_i \not\simeq_{\surd} \omega$ $(1 \leq i \leq n)$.
Let x be a fresh variable, so d can be transformed into a derivation d' proving $B[\tau_1 \wedge ... \wedge \tau_n/x] \vdash_{\surd} (\lambda x.C[x])M : \sigma$ by performing the following operations. First,

- replace d_i by

$$\frac{\frac{}{B[\tau_1 \wedge ... \wedge \tau_n/x] \vdash_{\surd} x : \tau_1 \wedge ... \wedge \tau_n} (var)}{B[\tau_1 \wedge ... \wedge \tau_n/x] \vdash_{\surd} x : \tau_i} (\leq_{\surd})$$

- replace each typing $B \vdash_{\surd} P[M/x] : \mu$ occurring in the derivation d by the typing $B[\tau_1 \wedge ... \wedge \tau_n/x] \vdash_{\surd} P : \mu$.

It is easy to check that d' is well defined, by induction on d.
So, by rule $(\rightarrow I)$, $B \vdash_{\surd} \lambda x.C[x] : (\tau_1 \wedge ... \wedge \tau_n) \rightarrow \sigma$ and, by rule $(\wedge I)$, $B \vdash_{\surd} M : \tau_1 \wedge ... \wedge \tau_n$, so the proof follows by rule $(\rightarrow E)$, since $\tau_1 \wedge ... \wedge \tau_n \in I(C_{\surd})$. □

Lemma 12.1.33 (Weak $\Xi\ell$-subject expansion).
$M \rightarrow_{\Xi\ell} N$ and $B \vdash_{\surd} N : \sigma$ imply that there is B' such that $B' \vdash_{\surd} M : \sigma$ and $B'(x) \leq_{\surd} B(x)$, for each $x \in$ Var.

Proof. Let $M \equiv C[(\lambda x.P)Q]$, let $N \equiv C[P[Q/x]]$ and, let d be the derivation proving $B \vdash_{\surd} C[P[Q/x]] : \sigma$. If $\sigma \simeq_{\surd} \omega$ then the proof is trivial, so let $\sigma \in I(C_{\surd})$. The proof is given by induction on $C[.]$.
Without loss of generality, let each typing in d have the same basis B. Indeed, the only rule that can change the basis is $(\rightarrow I)$, and we can assume that free and bound variables have different names in P. Let $C[.] \equiv [.]$. There are two cases.

(i) Either Q does not occur in $P[Q/x]$, so $x \notin \mathrm{FV}(P)$ and $N \equiv P$, or Q occurs in subterms of P that are subjects of an application of the rule (ω). Since Q is a $\Xi\ell$-nf, then there is a basis B^* such that $B^* \vdash_{\surd} Q : \omega \rightarrow \omega$ by Lemma 12.1.5. Let $B'(y) = B(y) \wedge B^*(y)$, for each $y \in$ Var; so, by Lemma 10.1.7.(ii) $B'[\omega \rightarrow \omega/x] \vdash_{\surd} N : \sigma$. Thus, by rule $(\rightarrow I)$, $B' \vdash_{\surd} \lambda x.P : (\omega \rightarrow \omega) \rightarrow \sigma$. Hence $B' \vdash_{\surd} (\lambda x.P)Q : \sigma$ by rule $(\rightarrow E)$.
(ii) In the case where Q occurs in $P[Q/x]$ and there is at least one subderivation of d assigning to Q a type $\not\simeq_{\surd} \omega$, the proof follows by Lemma 12.1.32.

In the general case either $C[.] \equiv M'C'[.]$ or $C[.] \equiv C'[.]M'$, since the reduction is lazy. Let us consider the first case. By Lemma 10.1.7.(vi), there are subderivations d_0 and d_1 of d proving respectively $B \vdash_{\surd} C'[P[Q/x]] : \sigma_0$ and $B \vdash_{\surd} M' : \sigma_0 \rightarrow \sigma$, for $\sigma_0 \not\simeq_{\surd} \omega$. By induction there is a derivation d^* proving $B^* \vdash_{\surd} C'[(\lambda x.P)Q] : \sigma_0$ where $B^*(x) \leq_{\surd} B(x)$, for each $x \in$ Var. By Lemma 10.1.7.(iii) $B^* \vdash_{\surd} M' : \sigma_0 \rightarrow \sigma$; hence it is easy to build a derivation proving $B^* \vdash_{\surd} C[(\lambda x.P)Q] : \sigma$. The second case is similar. □

Now we are able to prove the theorem.

▼ **Proof of Theorem 12.1.6** (pag.184).

(i) By the Λ-subject reduction lemma, taking into account that $\Xi\ell$-reduction is a special case of Λ-reduction.
(ii) By the weak $\Xi\ell$-subject expansion lemma. ■

12.1.3 Proof of the $\mathcal{V}$-Approximation Theorem

The proof follows the same lines as the corresponding proof in the previous models. In order to prove the ($\Rightarrow$) implication both parts of the theorem, we need to define a computability predicate.

A basis B is *finite* if and only if $B(y) \simeq_{\surd} \omega \to \omega$ except in a finite number of variables. We will use $[\sigma_1/x_1, ..., \sigma_n/x_n]$ to denote a finite basis. By Lemma 10.1.7.(i), in this section we limit ourselves to consider only such a kind of basis.

Let B and B' be two basis. $B \cup B'$ denotes the basis such that, for every x, $B \cup B'(x) = B(x) \wedge B'(x)$ (remember that $\sigma \wedge (\omega \to \omega) \simeq_{\surd} \sigma$, for every type $\sigma \not\simeq_{\surd} \omega$).

Definition 12.1.34. (i) *$App_{\mathcal{V}}(B, \sigma, M)$ if and only if there is $A \in \mathcal{A}^{\surd}(M)$ such that $B \vdash_{\surd} A : \sigma$.*
(ii) *The predicate $Comp_{\mathcal{V}}$ is defined by induction on types as follows:*
- *$Comp_{\mathcal{V}}(B, \omega, M)$ is true;*
- *$Comp_{\mathcal{V}}(B, \sigma \to \tau, M)$ where $\tau \simeq_{\surd} \omega$, if and only if $App_{\mathcal{V}}(B, \omega \to \omega, M)$;*
- *$Comp_{\mathcal{V}}(B, \sigma \to \tau, M)$ where $\tau \not\simeq_{\surd} \omega$, if and only if*
  $\forall N \in \Gamma$, *$Comp_{\mathcal{V}}(B', \sigma, N)$ implies $Comp_{\mathcal{V}}(B \cup B', \tau, MN)$;*
- *$Comp_{\mathcal{L}}(B, \sigma \wedge \tau, M)$ if and only if $Comp_{\mathcal{V}}(B, \sigma, M)$ and $Comp_{\mathcal{V}}(B, \tau, M)$.*

In the usual way, we prove that $B \vdash_{\surd} M : \sigma$ implies $Comp_{\mathcal{V}}(B, \sigma, M)$, which in turn implies $App_{\mathcal{V}}(B, \sigma, M)$.

Lemma 12.1.35. *$Comp_{\mathcal{V}}(B, \sigma, M)$ and $M =_{\Gamma} M'$ imply $Comp_{\mathcal{V}}(B, \sigma, M')$.*

Proof. The proof is given by induction on σ. The case $\sigma \equiv \omega$ is obvious.
If $\sigma \equiv \sigma \to \tau$ where $\tau \simeq_{\surd} \omega$, then the proof follows from the definition of $App_{\mathcal{V}}$, since $App_{\mathcal{V}}$ is closed under $=_{\Gamma}$.
The other cases follow by the inductive hypothesis. □

Hence, $Comp_{\mathcal{V}}$ is defined modulo $=_{\Gamma}$ on terms. The following property holds.

Property 12.1.36. Let B be a basis, M be a term and τ a type.
$Comp_{\mathcal{V}}(B, \omega \to \tau, M)$ if and only if $Comp_{\mathcal{V}}(B, (\omega \to \omega) \to \tau, M)$.

Proof. The proof is easy by induction on the definition of $Comp_{\mathcal{V}}$, since $N \in \Gamma$ implies $B' \vdash N : \omega \to \omega$ for some B', by Theorem 12.1.3.(ii). □

In order to prove that $Comp_{\mathcal{V}}(B, \sigma, M)$ implies $App_{\mathcal{V}}(B, \sigma, M)$, we need the following property.

Property 12.1.37. Let A be an approximant such that $A \equiv \lambda z.\zeta A_1...A_m z$, where ζ is either a variable or a head block, and $z \notin \mathrm{FV}(\zeta A_1...A_m)$.
If $B \vdash_{\surd} A : \sigma \to \tau$ where $\sigma, \tau \not\simeq_{\surd} \omega$ then $B \vdash_{\surd} \zeta A_1...A_m : \sigma \to \tau$.

Proof. $\sigma \in I(C_{\surd})$ and $B \vdash_{\surd} A : \sigma \to \tau$ imply $B[\sigma/z] \vdash_{\surd} \zeta A_1...A_m z : \tau$, by Lemma 10.1.7.(vi). Since $\tau \not\simeq_{\surd} \omega$, by Lemma 10.1.7.(vii) there is $\epsilon \in I(C_{\surd})$ such that $B[\sigma/z] \vdash_{\surd} z : \epsilon$ and $B[\sigma/z] \vdash_{\surd} \zeta A_1...A_m : \epsilon \to \tau$. By Lemma 10.1.7.(ii) $\sigma \leq_{\surd} \epsilon$; hence $\epsilon \to \tau \leq_{\surd} \sigma \to \tau$. So $B[\sigma/z] \vdash_{\surd} \zeta A_1...A_m : \sigma \to \tau$ by rule $(\leq_{\surd})$. Clearly $B \vdash_{\surd} \zeta A_1...A_m : \sigma \to \tau$, since $z \notin \mathrm{FV}(\zeta A_1...A_m)$. □

Lemma 12.1.38. (i) *$App_{\mathcal{V}}(B, \sigma, x\vec{M})$ implies $Comp_{\mathcal{V}}(B, \sigma, x\vec{M})$.*
(ii) *$Comp_{\mathcal{V}}(B, \sigma, M)$ implies $App_{\mathcal{V}}(B, \sigma, M)$.*

Proof. The proof is done by mutual induction on σ.
The only nonobvious case is when $\sigma \equiv \tau \to \rho$, where $\rho \not\simeq_{\surd} \omega$.

(i) We will prove that $N \in \Gamma$ and $Comp_{\mathcal{V}}(B', \tau, N)$ imply $Comp_{\mathcal{V}}(B \cup B', \rho, x\vec{M}N)$, thus $Comp_{\mathcal{V}}(B, \tau \to \rho, x\vec{M})$ follows by definition.
$Comp_{\mathcal{V}}(B', \tau, N)$ implies $App_{\mathcal{V}}(B', \tau, N)$, by induction on (ii).
By hypothesis $App_{\mathcal{V}}(B, \tau \to \rho, x\vec{M})$; thus $B \cup B' \vdash_{\surd} A^* : \rho$, for some $A^* \in \mathcal{A}^{\surd}(x\vec{M}N)$ by rule $(\to E)$, since $x\vec{A} \in \mathcal{A}^{\surd}(x\vec{M})$ and $A' \in \mathcal{A}^{\surd}(N)$ imply $x\vec{A}A' \in \mathcal{A}^{\surd}(x\vec{M}N)$. Thus $App_{\mathcal{V}}(B \cup B', \rho, x\vec{M}N)$ and by induction, $Comp_{\mathcal{V}}(B \cup B', \rho, x\vec{M}N)$.

(ii) Let $z \notin \mathrm{FV}(M)$ and $B(z) \simeq_{\surd} \omega \to \omega$. Note that both $z \in \mathcal{A}^{\surd}$ and $[\tau/x] \vdash_{\surd} x : \tau$, thus $App_{\mathcal{V}}([\tau/z], \tau, z)$. Hence, $Comp_{\mathcal{V}}([\tau/z], \tau, z)$ by induction on (i).
$Comp_{\mathcal{V}}(B, \tau \to \rho, M)$ and $Comp_{\mathcal{V}}([\tau/z], \tau, z)$ imply $Comp_{\mathcal{V}}(B[\tau/z], \rho, Mz)$ and this implies $App_{\mathcal{V}}(B[\tau/z], \rho, Mz)$, by induction; which means there is $A \in \mathcal{A}^{\surd}(Mz)$ such that $B[\tau/z] \vdash_{\surd} A : \rho$.
The case $A \equiv \Omega$ is not possible, since by the hypothesis $\rho \not\simeq_{\surd} \omega$. Hence $B \vdash_{\surd} \lambda z.A : \tau \to \rho$, by rule $(\to I)$. By definition of the $\surd$-approximants of a term, $A \in \mathcal{A}(Mz)$ implies $\lambda z.A \in \mathcal{A}(\lambda z.Mz)$. Now there are two cases.
  1. M is of order 0, so A is of the shape $A'z$, where either $A' \equiv xA_1...A_m$ or $A' \equiv (\lambda x.A')A''A_1...A_m z$ and $z \notin \mathrm{FV}(A')$.
     In both cases $A' \in \mathcal{A}^{\surd}(M)$. By Property 12.1.37, $B \vdash A' : \tau \to \rho$, and so $App_{\mathcal{V}}(B, \tau \to \rho, M)$.
  2. Otherwise $M =_{\Gamma} \lambda y.M'$, so $\lambda z.Mz =_{\Gamma} \lambda z.M'[z/y] =_{\alpha} \lambda y.M'$, which implies $\lambda z.A \in \mathcal{A}^{\surd}(M)$ and the proof is given. □

Lemma 12.1.39. *$Comp_{\mathcal{V}}(B, \sigma, M)$ and $\sigma \leq_{\surd} \tau$ implies $Comp_{\mathcal{V}}(B, \tau, M)$.*

Proof. By induction on the definition of $\leq_{\surd}$. The more complex case is that of rule (f), so let $\pi' \leq_{\surd} \pi$, $\tau \leq_{\surd} \tau'$ and $Comp_{\mathcal{V}}(B, \pi \to \tau, M)$.
If $\tau \simeq_{\surd} \omega$ then $\tau' \simeq_{\surd} \omega$, so the proof is immediate. If $\tau, \pi \not\simeq_{\surd} \omega$ then the proof follows by induction. If $\tau \not\simeq_{\surd} \omega$ but $\tau' \simeq_{\surd} \omega$ then, by Lemma 12.1.38.(ii), $App_{\mathcal{V}}(B, \pi \to \tau, M)$. By definition of $App_{\mathcal{V}}$, there is an $A \in \mathcal{A}^{\angle}(M)$ such that $B \vdash_{\surd} A : \pi \to \tau$; thus $B \vdash_{\surd} A : \omega \to \omega$, and the proof follows by definition of $Comp_{\mathcal{V}}$. □

Lemma 12.1.40. *Let* $\mathrm{FV}(M) \subseteq \{x_1, ..., x_n\}$ *and* $B = [\sigma_1/x_1, ..., \sigma_n/x_n]$. *If* $N_i \in \Gamma$, $Comp_{\mathcal{V}}(B_i, \sigma_i, N_i)$ $(1 \leq i \leq n)$ *and* $B \vdash_{\surd} M : \tau$, *then*

$$Comp_{\mathcal{V}}(B_1 \cup ... \cup B_n, \tau, M[N_1/x_1, ..., N_n/x_n]).$$

Proof. The proof is given by induction on the derivation of $B \vdash_{\surd} M : \tau$. The most interesting case is when the last applied rule is $(\rightarrow I)$.
Let $M \equiv \lambda x.M'$, $\tau \equiv \mu \rightarrow \nu$, $\mu \in I(C_{\surd})$ and

$$\frac{B[\mu/x] \vdash_{\surd} M' : \nu}{B \vdash_{\surd} \lambda x.M' : \mu \rightarrow \nu} (\rightarrow I).$$

If $N \in \Gamma$ and $Comp_{\mathcal{V}}(B', \mu, N)$, then by induction

$$Comp_{\mathcal{V}}(B' \cup B_1 \cup ... \cup B_n, \nu, M'[N_1/x_1, ..., N_n/x_n, N/x])$$

which implies

$$Comp_{\mathcal{V}}(B' \cup B_1 \cup ... \cup B_n, \nu, (\lambda x.M'[N_1/x_1, ..., N_n/x_n])N)$$

by Lemma 12.1.35. So $Comp_{\mathcal{V}}(B_1 \cup ... \cup B_n, \tau, M[N_1/x_1, ..., N_n/x_n])$ by definition of $Comp_{\mathcal{V}}$. All other cases follow directly from the inductive hypothesis. □

Moreover, the following property holds.

Property 12.1.41. Let $M, N \in \Lambda\Omega$ such that $M \rightarrow_{\Xi\ell\Omega} N$.

(i) If $B \vdash M : \sigma$ then $B \vdash N : \sigma$.
(ii) If $B \vdash N : \sigma$ then $B' \vdash M : \sigma$, for some B' such that $\forall x \in \mathrm{Var}$, $B'(x) \leq_{\surd} B(x)$.

Proof. Easy, by Theorem 12.1.6, and by the fact that to the term Ω only the type ω can be assigned. □

▼ **Proof of the $\mathcal{V}$-Approximation Theorem** (Theorem 12.1.15 pag. 186).

(i)($\Rightarrow$) Clearly $Comp_{\mathcal{V}}([\tau/x], \tau, x)$ by Lemma 12.1.38.(i).
Let $\mathrm{FV}(M) \subseteq \{x_1, ..., x_n\}$; if $B \vdash_{\surd} M : \sigma$, $B = [\sigma_1/x_1, ..., \sigma_n/x_n]$ and $Comp_{\mathcal{V}}([\sigma_i/x_i], \sigma_i, x_i)$ $(1 \leq i \leq n)$ then $Comp_{\mathcal{V}}(B, \sigma, M)$ by Lemma 12.1.40. By Lemma 12.1.38.(ii) and definition of $App_{\mathcal{V}}$ the proof is done.
($\Leftarrow$) By definition, there is M' such that $M =_{\Gamma} M'$ and A matches M' except at occurrences of Ω. A derivation of $B \vdash_{\surd} A : \sigma$ can be transformed into a derivation of $B \vdash_{\surd} M' : \sigma$, simply by replacing every subderivation

$$\frac{}{B \vdash_{\surd} \Omega : \omega} (\omega) \quad \text{by} \quad \frac{}{B \vdash_{\surd} N : \omega} (\omega),$$

where N is the subterm replaced by Ω in M'. $B \vdash_{\surd} M' : \sigma$ implies $B \vdash_{\surd} M : \sigma$, since the type assignment system is closed under $=_{\Gamma}$ on terms as consequence of the fact that it induces a $\lambda\Gamma$-model, so the proof is given.

(ii) It follows by point (i) of the $\mathcal{V}$-approximation theorem and Property 12.1.41. ■

12.1.4 Proof of Theorems 12.1.24 and 12.1.25

Let $V_0 \equiv \lambda x.(\lambda x_1 x_2.DD)(x(\lambda x_1.DD)(\lambda x_1.DD))$,
$V_1 \equiv \lambda x.(\lambda x_1 x_2 x_3.DD)(x(\lambda x_1.DD)(\lambda x_1 x_2.DD))(x(\lambda x_1 x_2.DD)(\lambda x_1.DD))$.

In Sect. 7.1.1 we proved that $V_0 \approx_{\mathbf{V}} V_1$; now we will prove $V_0 \not\approx_{\mathcal{V}} V_1$. Note that both V_0 and V_1 are Γ-unsolvable of order 2, so the model does not equate all Γ-unsolvable terms of the same order.

▼ **Proof of V-Incompleteness Theorem** (Theorem 12.1.24 pag. 190).

Let $\sigma \equiv \sigma_0 \wedge \sigma_1$, where
$$\sigma_0 \equiv (\omega \to \omega) \to (\omega \to \omega \to \omega) \to \omega \to \omega,$$
$$\sigma_1 \equiv (\omega \to \omega \to \omega) \to (\omega \to \omega) \to \omega \to \omega,$$

and let $D^1 \equiv \lambda x_1.DD$, $D^2 \equiv \lambda x_1 x_2.DD$ and $D^3 \equiv \lambda x_1 x_2 x_3.DD$.
We will show that $B \vdash_{\surd} V_1 : \sigma \to \omega \to \omega$ while $B \not\vdash_{\surd} V_0 : \sigma \to \omega \to \omega$, for all bases B. Let d_{12} be the derivation

$$\dfrac{\dfrac{\dfrac{\dfrac{}{B[\sigma/x] \vdash_{\surd} x : \sigma}\,(var)}{B[\sigma/x] \vdash_{\surd} x : \sigma_0}\,(\wedge E) \quad \dfrac{\dfrac{\dfrac{}{B[\sigma/x, \omega \to \omega/x_1] \vdash_{\surd} DD : \omega}\,(\omega)}{B[\sigma/x] \vdash_{\surd} D^1 : (\omega \to \omega) \to \omega}\,(\to I)}{B[\sigma/x] \vdash_{\surd} D^1 : \omega \to \omega}\,(\leq_{\surd})}{B[\sigma/x] \vdash_{\surd} xD^1 : (\omega \to \omega \to \omega) \to \omega \to \omega}\,(\to E) \quad \dfrac{\dfrac{\dfrac{\dfrac{\dfrac{}{B[\sigma/x, \omega \to \omega/x_1, \omega \to \omega/x_2] \vdash_{\surd} DD : \omega}\,(\omega)}{B[\sigma/x, \omega \to \omega/x_1] \vdash_{\surd} D^1 : (\omega \to \omega) \to \omega}\,(\to I)}{B[\sigma/x, \omega \to \omega/x_1] \vdash_{\surd} D^1 : \omega \to \omega}\,(\leq_{\surd})}{B[\sigma/x] \vdash_{\surd} D^2 : (\omega \to \omega) \to \omega \to \omega}\,(\to I)}{B[\sigma/x] \vdash_{\surd} D^2 : \omega \to \omega \to \omega}\,(\leq_{\surd})}{B[\sigma/x] \vdash_{\surd} xD^1D^2 : \omega \to \omega}\,(\to E)$$

and let d_{21} be the derivation

$$\dfrac{\dfrac{\dfrac{\dfrac{}{B[\sigma/x] \vdash_{\surd} x : \sigma}\,(var)}{B[\sigma/x] \vdash_{\surd} x : \sigma_1}\,(\wedge E) \quad \dfrac{\dfrac{\dfrac{\dfrac{\dfrac{}{B[\sigma/x, \omega \to \omega/x_1, \omega \to \omega/x_2] \vdash_{\surd} DD : \omega}\,(\omega)}{B[\sigma/x, \omega \to \omega/x_1] \vdash_{\surd} D^1 : (\omega \to \omega) \to \omega}\,(\to I)}{B[\sigma/x, \omega \to \omega/x_1] \vdash_{\surd} D^1 : \omega \to \omega}\,(\leq_{\surd})}{B[\sigma/x] \vdash_{\surd} D^2 : (\omega \to \omega) \to \omega \to \omega}\,(\to I)}{B[\sigma/x] \vdash_{\surd} D^2 : \omega \to \omega \to \omega}\,(\leq_{\surd})}{B[\sigma/x] \vdash_{\surd} xD^2 : (\omega \to \omega) \to \omega \to \omega}\,(\to E) \quad \dfrac{\dfrac{\dfrac{}{B[\sigma/x, \omega \to \omega/x_1] \vdash_{\surd} DD : \omega}\,(\omega)}{B[\sigma/x] \vdash_{\surd} D^1 : (\omega \to \omega) \to \omega}\,(\to I)}{B[\sigma/x] \vdash_{\surd} D^1 : \omega \to \omega}\,(\leq_{\surd})}{B[\sigma/x] \vdash_{\surd} xD^2D^1 : \omega \to \omega}\,(\to E)$$

It is easy to build a derivation d_3 proving the typing

$$B[\sigma/x] \vdash_{\surd} D^3 : (\omega \to \omega) \to (\omega \to \omega) \to \omega \to \omega.$$

Now, we can build the following derivation:

$$\dfrac{\dfrac{\dfrac{d_3}{B[\sigma/x] \vdash_{\surd} D^3 : (\omega \to \omega) \to (\omega \to \omega) \to \omega \to \omega} \cdots \quad \dfrac{d_{12}}{B[\sigma/x] \vdash_{\surd} xD^1D^2 : \omega \to \omega} \cdots}{B[\sigma/x] \vdash_{\surd} D^3(xD^1D^2) : (\omega \to \omega) \to \omega \to \omega} (\to E) \quad \dfrac{d_{21}}{B[\sigma/x] \vdash_{\surd} xD^2D^1 : \omega \to \omega} (\ldots)}{\dfrac{B[\sigma/x] \vdash_{\surd} (\lambda x_1x_2x_3.DD)(x(\lambda x_1.DD)(\lambda x_1x_2.DD))(x(\lambda x_1x_2.DD)(\lambda x_1.DD)) : \omega \to \omega}{B \vdash_{\surd} V_1 : \sigma \to \omega \to \omega} (\to I)} (\to E)$$

Note that we can apply the rule $(\to E)$, since $\omega \to \omega \in I(C_{\surd})$; moreover, each type considered in the basis is an input type.

On the other hand, $B \vdash_{\surd} V_0 : \sigma \to \omega \to \omega$ implies, by Lemma 10.1.7.(vi), $B[\sigma/x] \vdash_{\surd} (\lambda x_1x_2.DD)(x(\lambda x_1.DD)(\lambda x_1.DD)) : \omega \to \omega$.
Therefore there is $\mu \in I(C_{\surd})$ such that $B[\sigma/x] \vdash_{\surd} \lambda x_1x_2.DD : \mu \to \omega \to \omega$ and $B[\sigma/x] \vdash_{\surd} x(\lambda x_1.DD)(\lambda x_1.DD) : \mu$, by Lemma 10.1.7.(vii).
Since $\mu \not\simeq_{\surd} \omega$, again by Lemma 10.1.7.(vii), there is $\tau \in I(C_{\surd})$ such that $B[\sigma/x] \vdash_{\surd} x(\lambda x_1.DD) : \tau \to \mu$ and

$$B[\sigma/x] \vdash_{\surd} \lambda x_1.DD : \tau. \tag{12.1}$$

Since $\tau \to \mu \not\simeq_{\surd} \omega$, again by Lemma 10.1.7.(vii), there is $\pi \in I(C_{\surd})$ such that $B[\sigma/x] \vdash_{\surd} x : \pi \to \tau \to \mu$ and

$$B[\sigma/x] \vdash_{\surd} \lambda x_1.DD : \pi. \tag{12.2}$$

By Lemma 10.1.7.(ii) $\sigma \leq_{\surd} \pi \to \tau \to \mu$, so, since $\surd$ is legal, there are 3 possible cases.

1. $\pi \leq_{\surd} \omega \to \omega \to \omega$ is not possible; in fact, it is easy to see that the typing given in Eq. 12.2 implies $\omega \to \omega \leq_{\surd} \pi$, by Theorem 12.1.19.(ii). So, by Property 12.1.28.(ii), this would imply $\omega \to \omega \simeq_{\surd} \omega \to \omega \to \omega$, which is an absurd.
2. $\pi \leq_{\surd} (\omega \to \omega) \wedge (\omega \to \omega \to \omega)$ is not possible. In fact, by rules (c), $(\omega \to \omega) \wedge (\omega \to \omega \to \omega) \leq_{\surd} \omega \to \omega \to \omega$, so we can reason as in the previous case.
3. $\pi \leq_{\surd} \omega \to \omega$ and $(\omega \to \omega \to \omega) \to \omega \to \omega \leq_{\surd} \tau \to \mu$, therefore $\tau \leq_{\surd} \omega \to \omega \to \omega$ by Property 10.1.6. Yet an absurd, by the typing given in Eq. 12.1 and by Theorem 12.1.19.(ii).

■

In order to prove Theorem 12.1.25, we need Lemma 12.1.42.

Lemma 12.1.42. *Let U be a closed Γ-unsolvable term of order* 0 *and let ∇ be a type system $< C, \leq_{\nabla}, I(C) >$ inducing a $\lambda\Gamma$-model $\mathcal{M}$ that is fully abstract with respect to the* **V***-operational semantics.*

(i) *All closed Γ-unsolvable terms of the same finite order n are equated in $\mathcal{M}$.*
(ii) $U \sqsubset_{\mathcal{M}} \lambda x.U$.
(iii) *There exists $\theta \in I(C)$ such that $B \vdash_\nabla \lambda x.U : \theta$ and $B \vdash_\nabla \lambda xy.U : \theta \to \theta$, while $B \not\vdash_\nabla U : \theta$ and $B \not\vdash_\nabla \lambda x.U : \theta \to \theta$, for all bases B; moreover, $\theta \to \theta \in I(C)$.*
(iv) *If θ is the type considered in the previous point then $(\theta \to \theta) \to \theta \to \theta, \theta \to (\theta \to \theta) \to \theta \in I(C)$.*

Proof. (i) By the fact that all closed Γ-unsolvable terms of the same finite order are equated in **V** (see Corollary 7.1.9) and by definition of full abstraction.
(ii) By the fact that $U \prec_{\mathbf{V}} \lambda x.U$ and by definition of full abstraction.
(iii) Note that $\lambda x.U$ is an input value for the $\lambda\Gamma$-calculus, so by the definition of the $\lambda\Gamma$-model and by the previous point of this lemma, there is $\theta \in I(C)$ such that $B \vdash_\nabla \lambda x.U : \theta$ and $B \not\vdash_\nabla U : \theta$, for all bases B (since U is closed). It is easy to build a derivation proving $B \vdash_\nabla \lambda xy.U : \theta \to \theta$. If $B \vdash_\nabla \lambda x.U : \theta \to \theta$ then $B \vdash_\nabla (\lambda x.U)\lambda x.U : \theta$, but $(\lambda x.U)\lambda x.U$ is a Λ-unsolvable term of order 0, so $B \not\vdash_\nabla \lambda x.U : \theta \to \theta$.
If $\theta \to \theta \notin I(C)$ then $\theta \leq_\nabla \theta \to \theta$, so $B \vdash_\nabla \lambda x.U : \theta \to \theta$; hence, $\theta \to \theta \in I(C)$.
(iv) If $(\theta \to \theta) \to \theta \to \theta \notin I(C)$ then $\theta \leq_\nabla (\theta \to \theta) \to \theta \to \theta$ and $B \vdash_\nabla \lambda x.U : (\theta \to \theta) \to \theta \to \theta$, so $B \vdash_\nabla (\lambda x.U)(\lambda xy.U)(\lambda x.U) : \theta$ which is an absurdum; hence, $(\theta \to \theta) \to \theta \to \theta \in I(C)$. In a similar way, $\theta \to (\theta \to \theta) \to \theta \in I(C)$. □

▼ **Proof of Theorem 12.1.25** (pag.190).

We prove that every Γ-model fully abstract with respect to the **V**-operational semantics would equate the two terms V_0 and V_1.
Let $< C, \leq_\nabla, I(C) >$ be a legal type system inducing a filter $\lambda\Gamma$-model which is fully abstract with respect to the **V**-operational semantics, and let θ be the input type considered in the Lemma 12.1.42. Note that $\theta \not\simeq_\nabla \omega$.
Let $\sigma \equiv \sigma_0 \wedge \sigma_1$ where $\sigma_0 \equiv \theta \to (\theta \to \theta) \to \theta$ and $\sigma_1 \equiv (\theta \to \theta) \to \theta \to \theta$; moreover let $D^1 \equiv (\lambda x_1.DD)$, $D^2 \equiv (\lambda x_1 x_2.DD)$ and $D^3 \equiv (\lambda x_1 x_2 x_3.DD)$. We will show that $B \vdash_\nabla V_1 : \sigma \to \theta$ while $B \not\vdash_\nabla V_0 : \sigma \to \theta$, for all basis B.
It is easy to build derivations proving the typings

$$B[\sigma/x] \vdash_\nabla xD^1D^2 : \theta$$
$$B[\sigma/x] \vdash_\nabla xD^2D^1 : \theta$$
$$B[\sigma/x] \vdash_\nabla D^3 : \theta \to \theta \to \theta$$

thus can build the following derivation

$$\dfrac{\dfrac{\dfrac{\vdots}{B[\sigma/x] \vdash_\nabla D^3 : \theta \to \theta \to \theta}(\ldots) \quad \dfrac{\vdots}{B[\sigma/x] \vdash_\nabla xD^1D^2 : \theta}(\ldots)}{B[\sigma/x] \vdash_\nabla D^3(xD^1D^2) : \theta \to \theta}(\to E) \quad \dfrac{\vdots}{B[\sigma/x] \vdash_\nabla xD^2D^1 : \theta}(\ldots)}{\dfrac{B[\sigma/x] \vdash_\nabla (\lambda x_1x_2x_3.DD)(x(\lambda x_1.DD)(\lambda x_1x_2.DD))(x(\lambda x_1x_2.DD)(\lambda x_1.DD)) : \theta}{B \vdash_\nabla V_1 : \sigma \to \theta}(\to I)}(\to E)$$

Note that we can apply the rule $(\to E)$, since $\theta \in I(C)$; moreover each type considered in the basis is an input type.

Since in Sect. 7.1.1 we proved that $V_0 \approx_{\mathbf{V}} V_1$, by the full abstraction hypothesis it follows that $B \vdash_\nabla V_0 : \sigma \to \theta$; so, by Lemma 10.1.7.(vi), $B[\sigma/x] \vdash_\nabla (\lambda x_1x_2.DD)(x(\lambda x_1.DD)(\lambda x_1.DD)) : \theta$.
So there is $\mu \in I(C)$ such that $B[\sigma/x] \vdash_\nabla \lambda x_1x_2.DD : \mu \to \theta$ and $B[\sigma/x] \vdash_\nabla x(\lambda x_1.DD)(\lambda x_1.DD) : \mu$, by Lemma 10.1.7.(vii).
If $\mu \simeq_{\surd} \omega$ then $\omega \in I(C)$, so $I(C) = T(C)$ and $B[\sigma/x] \vdash_\nabla \lambda x_1x_2.DD : \omega \to \theta$; so $B[\sigma/x] \vdash_\nabla (\lambda x_1x_2.DD)(DD) : \theta$ by rule $(\to E)$, against Lemma 12.1.42.(iii), since $(\lambda x_1x_2.DD)(DD)$ is a Γ-unsolvable term of order 0.
Let $\mu \not\simeq_{\surd} \omega$; again by Lemma 10.1.7.(vii), there exists $\tau \in I(C)$ such that $B[\sigma/x] \vdash_\nabla x(\lambda x_1.DD) : \tau \to \mu$ and,

$$B[\sigma/x] \vdash_\nabla \lambda x_1.DD : \tau. \tag{12.3}$$

If $\tau \to \mu \simeq_{\surd} \omega$ then $\tau \to \omega \leq_{\surd} \omega \leq_{\surd} \tau \to \mu$, so by Property 10.1.6 $\omega \leq_{\surd} \mu$ and thus $\mu \simeq_{\surd} \omega$, which is not possible; hence, $\tau \to \mu \not\simeq_{\surd} \omega$.
Since $\tau \to \mu \not\simeq_{\surd} \omega$, again by Lemma 10.1.7.(vii), there is $\pi \in I(C)$ such that $B[\sigma/x] \vdash_\nabla x : \pi \to \tau \to \mu$ and,

$$B[\sigma/x] \vdash_\nabla \lambda x_1.DD : \pi. \tag{12.4}$$

By Lemma 10.1.7.(ii) $\sigma \leq_{\surd} \pi \to \tau \to \mu$, so, since ∇ is legal, there are three possible cases:

1. $\pi \leq_\nabla \theta \to \theta$ is not possible; otherwise the typing given in Eq. 12.4 would imply $B[\sigma/x] \vdash_\nabla \lambda x_1.DD : \theta \to \theta$ against Lemma 12.1.42.
2. $\pi \leq_\nabla \theta\wedge(\theta \to \theta)$ is not possible. In fact, by rule (c), $\theta\wedge(\theta \to \theta) \leq_\nabla \theta \to \theta$, so we can reason as in the previous case.
3. $\pi \leq_\nabla \theta$ and $(\theta \to \theta) \to \theta \leq_\nabla \tau \to \mu$, so by Property 10.1.6, $\tau \leq_{\surd} \theta \to \theta$; yet an absurdum, by the typing given in Eq. 12.3. ■

12.2 A Fully Abstract Model for the V-Operational Semantics

It was proved in Theorem 12.1.25 that there is not a filter $\lambda\Gamma$-model that is fully abstract with respect to the **V**-operational semantics. But we will show that it is possible to build a fully abstract model starting from the model $\mathcal{V}$, in a way similar to that presented in the Sect. 11.4. We start by defining a preorder relation on terms.

Definition 12.2.1. (i) $\trianglelefteq_\sigma$ *is a relation on* Λ^0 *defined as follows:*

- $M \trianglelefteq_\omega N$ *is true;*
- $M \trianglelefteq_{\sigma\to\tau} N$ *where* $\tau \simeq_{\surd} \omega$, *if and only if*
  $B \vdash_{\surd} M : \omega \to \omega$ *implies* $B \vdash_{\surd} N : \omega \to \omega$*, for all basis* B*;*
- $M \trianglelefteq_{\sigma\to\tau} N$ *where* $\tau \not\simeq_{\surd} \omega$, *if and only if*
  $\forall P$ *closed* Γ*-valuable term,* $B \vdash_{\surd} P : \sigma$ *implies* $MP \trianglelefteq_\tau NP$*;*
- $M \trianglelefteq_{\sigma\wedge\tau} N$ *if and only if both* $M \trianglelefteq_\sigma N$ *and* $M \trianglelefteq_\tau N$*.*

(ii) $M \trianglelefteq N$ *if and only if* $M \trianglelefteq_\sigma N$*, for all* σ*.*

The previous definition is well posed, thanks to the following property.

Property 12.2.2. There is $P \in \Gamma^0$ such that $B \vdash_{\surd} P : \sigma$, for all B and σ.

Proof. By induction on σ, we will prove that there is P of the shape: $\lambda x_1...x_n.DD$, for $n \geq 0$, to which σ can be assigned.
If $\sigma \equiv \omega$ then $B \vdash_{\surd} DD : \omega$, by rule (ω). Let $\sigma \equiv \mu \to \nu$.
If $\nu \simeq_{\surd} \omega$ and $\mu \not\simeq_{\surd} \omega$ then, by rule (ω), $B[\mu/x] \vdash_{\surd} DD : \omega$, and then, by rule $(\to I)$ and $(\leq_{\surd})$, $B \vdash_{\surd} DD : \mu \to \nu$.
If $\mu, \nu \simeq_{\surd} \omega$, then, by rule (ω), $B[\omega \to \omega/x] \vdash_{\surd} DD : \omega$, and the result follows by rules $(\to I)$ and $(\leq_{\surd})$, taking into account Property 12.1.28.(i).
If $\mu, \nu \not\simeq_{\surd} \omega$, then by induction there is $P \in \Gamma^0$ such that $B \vdash_{\surd} P : \nu$ and, since $P \in \Gamma^0$, $B[\mu/x] \vdash_{\surd} P : \nu$, so the proof follows by rule $(\to I)$.
Let $\sigma \equiv \mu \wedge \nu$. By induction, there are $\lambda x_1...x_p.DD$ and $\lambda x_1...x_q.DD$ such that $B \vdash_{\surd} \lambda x_1...x_p.DD : \mu$ and $B \vdash_{\surd} \lambda x_1...x_q.DD : \nu$. Let $n = \max\{p, q\}$, so $\lambda x_1...x_n.DD$ is the desired term. □

Note that although in the model $\mathcal{V}$ all types are inhabited, this does not imply that all filters are inhabited. Indeed, the filter $\uparrow \{\sigma\}$, where σ is $((\omega \to \omega) \to (\omega \to \omega \to \omega) \to \omega \to \omega) \wedge ((\omega \to \omega \to \omega) \to (\omega \to \omega) \to \omega \to \omega)$ is not the interpretation of any term, since the reader can check that every term having type σ has also the type $(\omega \to \omega) \to (\omega \to \omega) \to \omega \to \omega$, which is not in the filter. If this filter were inhabited, then it would be $V_0 \not\approx_{\mathbf{V}} V_1$ (see Sect. 12.1.4).

Property 12.2.3. Let $M, N \in \Lambda^0$.

(i) If $M \sqsubseteq_{\mathcal{V}} N$ then $M \trianglelefteq N$.
(ii) $\trianglelefteq$ is reflexive.
(iii) $\trianglelefteq$ is transitive.

Proof. (i) We will prove that $M \ntrianglelefteq N$ implies $M \not\sqsubseteq_{\mathcal{V}} N$. By definition, $M \ntrianglelefteq N$ means there is σ such that $M \ntrianglelefteq_\sigma N$. The proof is given by induction on σ.

Clearly $\sigma \not\simeq_{\surd} \omega$, since by definition $M \trianglelefteq_{\omega} N$ is true. If $\sigma \equiv \mu \to \nu$, there are two cases. If $\nu \simeq_{\surd} \omega$ then $B \vdash_{\surd} M : \omega \to \omega$ and $B \not\vdash_{\surd} N : \omega \to \omega$, so the proof is immediate by definition of $\sqsubseteq_{\mathcal{V}}$.
If $\nu \not\simeq_{\surd} \omega$ then there is a Γ-valuable $P \in \Lambda^0$ such that $MP \not\trianglelefteq_{\nu} NP$, by definition of $\trianglelefteq$. Hence, $MP \not\sqsubseteq_{\mathcal{L}} NP$ by induction, so $M \not\sqsubseteq_{\mathcal{V}} N$ by Lemma 10.1.13.(i). If $\sigma \equiv \mu \wedge \nu$ then the proof follows by induction.

(ii) We will prove that $M \trianglelefteq_{\sigma} M$, for all σ, by induction on σ. The case ω is obvious. Let $\sigma \equiv \mu \to \nu$; the case $\nu \simeq_{\surd} \omega$ is obvious. Let $\nu \not\simeq_{\surd} \omega$ and let $P \in \Lambda^0$ be a closed Γ-valuable term such that $B \vdash_{\surd} P : \mu$. By induction $MP \trianglelefteq_{\nu} MP$, so the result follows by definition of $\trianglelefteq$.
The case $\sigma \equiv \mu \wedge \nu$ follows by induction.

(iii) By induction on σ we prove that $\trianglelefteq_{\sigma}$ is transitive. The only nontrivial case is $\sigma \equiv \pi \to \tau$, where $\tau \not\simeq_{\surd} \omega$. Let $M_0 \trianglelefteq_{\pi\to\tau} M_1$ and $M_1 \trianglelefteq_{\pi\to\tau} M_2$. If $P \in \Lambda^0$ is a Γ-valuable term and $B \vdash_{\surd} P : \pi$, then $M_0P \trianglelefteq_{\tau} M_1P$ and $M_1P \trianglelefteq_{\tau} M_2P$, by definition of $\trianglelefteq$. So $M_0P \trianglelefteq_{\tau} M_2P$ by induction; hence $M_0 \trianglelefteq_{\pi\to\tau} M_2$ by definition of $\trianglelefteq$. □

Next two lemmas prove that the relation $\trianglelefteq$ grasps exactly the behaviour of the **V**-operational semantics.

Lemma 12.2.4. *Let $M, N \in \Lambda^0$.*
$M \trianglelefteq N$ if and only if $M\vec{P} \trianglelefteq_{\omega\to\omega} N\vec{P}$, for each sequence of closed Γ-valuable terms $\vec{P}$.

Proof. ($\Leftarrow$) We will prove that $M \not\trianglelefteq N$ implies that there is a closed sequence of Γ-valuable terms $\vec{P}$ such that $M\vec{P} \not\trianglelefteq_{\omega\to\omega} N\vec{P}$. By hypothesis there is a type σ such that $M \not\trianglelefteq_{\sigma} N$, so the proof is done by induction on σ.
If $\sigma \simeq_{\surd} \omega$ then $\sigma \equiv \underbrace{\omega \wedge \ldots\ldots \wedge \omega}_{n}$ $(n \geq 1)$, by Theorem 12.1.26; but since $M \trianglelefteq_{\omega} N$ by definition, this is not possible. If $\sigma \equiv \mu \to \nu$ and $\nu \simeq_{\surd} \omega$ then the proof is trivial. If $\sigma \equiv \mu \to \nu$ and $\nu \not\simeq_{\surd} \omega$ then there is a Γ-valuable term $P \in \Lambda^0$ such that $MP \not\trianglelefteq_{\nu} NP$, so the proof follows by induction. If $\sigma \equiv \mu \wedge \nu$ then the proof follows by induction.

($\Rightarrow$) We will prove that, if there is a sequence of closed Γ-valuable terms $\vec{P}$ and a type $\tau \not\simeq_{\surd} \omega$ such that $M\vec{P} \not\trianglelefteq_{\tau} N\vec{P}$, then $M \not\trianglelefteq N$. The proof will be given by induction on $\|\vec{P}\|$.
If $\|\vec{P}\| = 0$ then the proof is trivial, so let $\|\vec{P}\| \geq 1$ and $\vec{P} \equiv \vec{Q}Q'$. Since Q' is a closed Γ-valuable term, $B \vdash_{\surd} Q' : \omega \to \omega$ by Property 12.1.8. This implies $M\vec{Q} \not\trianglelefteq_{(\omega\to\omega)\to\tau} N\vec{Q}$ by definition of $\trianglelefteq$; so the proof follows by induction. □

Note that $M \trianglelefteq_{\sigma} N$ and $\sigma \leq_{\angle} \tau$ do not imply $M \trianglelefteq_{\tau} N$. Nevertheless, $M \trianglelefteq_{\omega\to\sigma} N$ if and only if $M \trianglelefteq_{(\omega\to\omega)\to\sigma} N$.

Lemma 12.2.5. *Let $M, N \in \Lambda^0$.*
$M \preceq_{\mathbf{V}} N$ if and only if $M\vec{P} \trianglelefteq_{\omega\to\omega} N\vec{P}$, for each sequence of closed Γ-valuable terms $\vec{P}$.

Proof. Let Q be a closed Γ-valuable term. Then $Q \Downarrow_{\mathbf{V}}$ if and only if $B \vdash_{\surd} Q : \omega \to \omega$, by Property 12.1.8 and Lemma 12.1.17.

$\Rightarrow$ Let $\vec{P}$ be a sequence of closed Γ-valuable terms, and let B be a basis. If $M \preceq_{\mathbf{V}} N$ then $M\vec{P} \Downarrow_{\mathbf{V}}$ implies $N\vec{P} \Downarrow_{\mathbf{V}}$; thus $B \vdash_{\surd} M\vec{P} : \omega \to \omega$ implies $B \vdash_{\surd} N\vec{P} : \omega \to \omega$. So the proof is done, by definition of $\unlhd_{\omega \to \omega}$.

$\Leftarrow$ Let $M\vec{P} \unlhd_{\omega \to \omega} N\vec{P}$, for each sequence of closed terms $\vec{P}$. Let us recall the notion of weight of a term, defined in Definition 3.1.29 (pag. 43), and the fact, proved in Corollary 3.2.2, that the weight of a term is defined if and only if it has $\Xi\ell$-normal form.

We will prove that, if $C[M], C[N] \in \Lambda^0$ and $\langle C[M] \rangle$ is defined then $\langle C[N] \rangle$ is defined, for all contexts $C[.]$. Hence the result follows from Theorem 7.1.3, taking into consideration that the set of closed Γ-lazy blocked normal forms coincides with the set of closed $\Xi\ell$-normal forms.

The proof will be given by induction on $\langle C[M] \rangle$. There are two cases, according to the possible shape of $C[.]$.

- $C[.] \equiv [.]C_1[.]...C_m[.]$ $(m \in \mathbb{N})$.
  If $m = 0$ then $\langle M \rangle$ defined implies M has $\Xi\ell$-normal form, so $B \vdash_{\surd} M : \omega \to \omega$. But $B \vdash_{\surd} N : \omega \to \omega$ by definition of $\unlhd_{\omega \to \omega}$, and the proof follows by Property 12.1.8 and Lemma 12.1.17.
  Let $m \geq 1$ and let $M \equiv (\lambda x.M_0)M_1...M_p$. Pose $D[.] \equiv MC_1[.]...C_m[.]$, so $D[M] \equiv C[M]$ and $D[.] \equiv (\lambda x.M_0)M_1...M_pC_1[.]...C_m[.]$ $(m \in \mathbb{N})$. If $p > 0$ then let $D^*[.] \equiv M_0[M_1/x]M_2...M_pC_1[.]...C_m[.]$, otherwise let $D^*[.] \equiv M_0[C_1[.]/x]C_2[.]...C_m[.]$; in both cases the weight of $D^*[M]$ is defined, since $\langle C[M] \rangle$ is defined. Moreover, $\langle D^*[M] \rangle < \langle C[M] \rangle$, so by induction $\langle D^*[N] \rangle$ is defined. But $D^*[N] \equiv MC_1[N]C_2[N]...C_m[N]$ has $\Xi\ell$-normal form implies $B \vdash_{\surd} MC_1[N]....C_m[N] : \omega \to \omega$, so by hypothesis $B \vdash_{\surd} NC_1[N]...C_m[N] : \omega \to \omega$. Hence, $NC_1[N]...C_m[N]$ has $\Xi\ell$-normal and the proof follows by Corollary 3.2.2.
- $C[.] \equiv (\lambda y.C_0[.])C_1[.]...C_m[.]$ $(m \in \mathbb{N})$.
  The case $m = 0$ is trivial; otherwise the proof follows by induction on the weight of $C_0[M][C_1[M]/y]C_2[M]...C_m[M]$ and $C_1[M]$. □

So the desired result follows.

Theorem 12.2.6. *$M \unlhd N$ if and only if $M \preceq_{\mathbf{V}} N$, for all $M, N \in \Lambda^0$.*

Proof. By Lemmas 12.2.4 and 12.2.5. □

The next definition overload the meaning of $\unlhd$ on a subset of filters, namely $\unlhd$ induces a preorder on $\mathcal{F}^0(\surd)$, i.e. the set of filters of $\mathcal{F}(\surd)$ that are interpretations of closed terms.

Definition 12.2.7. *Let $f, g \in \mathcal{F}^0(\surd)$ and let ρ be an environment. $f \unlhd g$ if and only if $M, N \in \Lambda^0$ such that $[\![M]\!]_\rho^{\mathcal{F}(\surd)} = f$ and $[\![N]\!]_\rho^{\mathcal{F}(\surd)} = g$ imply $M \unlhd N$. Moreover, $f \triangleq g$ if and only if $f \unlhd g$ and $g \unlhd f$.*

Note that if M is closed then $[\![M]\!]^{\mathcal{V}}_{\rho} = [\![M]\!]^{\mathcal{V}}_{\rho'}$, for all ρ, ρ'; moreover, if M, N are closed then $[\![M]\!]^{\mathcal{V}}_{\rho} = [\![N]\!]^{\mathcal{V}}_{\rho'}$ implies $M \trianglelefteq N$ and $N \trianglelefteq M$, by Property 12.2.3.(i). Note that $\trianglelefteq$ is overloaded, since it denotes both a relation on Λ^0 and a relation on $\mathcal{F}^0(\surd)$.

Now we can define the new $\lambda\Gamma$-model.

Definition 12.2.8. *Let* $f, g \in \mathcal{F}^0(\surd)$.

(i) $[f]$ *is the equivalence class of* f *with respect to the equivalence relation* $\triangleq$, *while* $\mathcal{F}^0_{\triangleq}$ *is the set of of equivalence classes induced from* $\triangleq$ *on* $\mathcal{F}^0(\surd)$. *Moreover, let* $\mathcal{I}^0_{\triangleq} = \{[f] \in \mathcal{F}^0_{\triangleq} \mid \exists M \in \Gamma^0$ *such that* $[\![M]\!]^{\mathcal{F}(\surd)}_{\rho} \in f\}$.

(ii) $\circ_{\triangleq} : \mathcal{F}^0_{\triangleq} \times \mathcal{F}^0_{\triangleq} \to \mathcal{F}^0_{\triangleq}$ *is defined as* $[f] \circ_{\triangleq} [g] = [f \circ_{\surd} g]$, *for all* $[f], [g] \in \mathcal{F}^0_{\triangleq}$.

(iii) *The interpretation function* $[\![.]\!]^{\mathcal{VV}} : \Lambda \times (\mathrm{Var} \to \mathcal{I}^0_{\triangleq}) \to \mathcal{F}^0_{\triangleq}$ *is defined as:* $[\![M]\!]^{\mathcal{VV}}_{\zeta} = [[\![M]\!]^{\mathcal{F}(\surd)}_{\rho}]$, *where* ρ *is such that* $\rho(x) \in \zeta(x)$ *for all* $x \in \mathrm{Var}$.

(iv) *Let* $\mathcal{VV}$ *be the quadruple:* $< \mathcal{F}^0_{\triangleq}, \mathcal{I}^0_{\triangleq}, \circ, [\![.]\!]^{\mathcal{VV}} >$.

Note that the interpretation is defined for open terms too.

Property 12.2.9. Let $M, N, P, Q \in \Lambda^0$.
If $M \trianglelefteq N$ and $P \trianglelefteq Q$ then $MP \trianglelefteq NQ$.

Proof. Clearly $M \preceq_{\mathbf{V}} N$ and $P \preceq_{\mathbf{V}} Q$ imply $MP \preceq_{\mathbf{V}} NQ$, therefore the proof follows by Theorem 12.2.6. □

Note that $\circ_{\triangleq}$ is well defined, by using the previous property. Furthermore, it is easy to see that $[f] \in \mathcal{I}^0_{\triangleq}$ and $f' \in [f]$ imply that $f' \in \mathcal{I}(\surd)$.

Lemma 12.2.10. $\mathcal{VV}$ *is a* $\lambda\Gamma$*-model.*

Proof. We check that $\mathcal{VV}$ satisfies the conditions of Definition 10.0.1.
If $\zeta \in (\mathrm{Var} \to \mathcal{I}^0_{\triangleq})$ then let ρ be such that $\rho(x) \in \zeta(x)$ for all $x \in \mathrm{Var}$.

1. $[\![x]\!]^{\mathcal{VV}}_{\zeta} = [[\![x]\!]^{\mathcal{F}(\surd)}_{\rho}] = [\rho(x)] = \zeta(x)$.
2. $[\![MN]\!]^{\mathcal{VV}}_{\zeta} = [[\![MN]\!]^{\mathcal{F}(\surd)}_{\rho}] = [[\![M]\!]^{\mathcal{F}(\surd)}_{\rho} \circ_{\surd} [\![N]\!]^{\mathcal{F}(\surd)}_{\rho}] = [[\![M]\!]^{\mathcal{F}(\surd)}_{\rho}] \circ_{\triangleq} [[\![N]\!]^{\mathcal{F}(\surd)}_{\rho}] = [\![M]\!]^{\mathcal{VV}}_{\zeta} \circ_{\triangleq} [\![N]\!]^{\mathcal{VV}}_{\zeta}$.
3. $[\![\lambda x.M]\!]^{\mathcal{VV}}_{\zeta} \circ_{\triangleq} d = [[\![\lambda x.M]\!]^{\mathcal{F}(\surd)}_{\rho}] \circ_{\triangleq} d = [[\![\lambda x.M]\!]^{\mathcal{F}(\surd)}_{\rho} \circ_{\surd} f] = [[\![M]\!]^{\mathcal{F}(\surd)}_{\rho[f/x]}] = [\![M]\!]^{\mathcal{VV}}_{\zeta[d/x]}$, for all $d \in \mathcal{I}^0_{\triangleq}$ and $f \in d$.
4. Let $[\![M]\!]^{\mathcal{VV}}_{\zeta[[d]/x]} = [\![N]\!]^{\mathcal{VV}}_{\zeta'[[d']/x']}$, where $d, d' \in \mathcal{I}^0(\surd)$.
Thus $[[\![M]\!]^{\mathcal{F}(\surd)}_{\rho[d/x]}] = [[\![N]\!]^{\mathcal{F}(\surd)}_{\rho[d'/x']}]$, therefore $[[\![\lambda x.M]\!]_{\rho}] = [[\![\lambda x'.N]\!]_{\rho'}]$, and so $[\![\lambda x.M]\!]^{\mathcal{VV}}_{\zeta} = [\![\lambda x'.N]\!]^{\mathcal{VV}}_{\zeta'}$.
5. Trivial.

□

Since $\trianglelefteq$ is a preorder on $\mathcal{F}^0(\surd)$ then it induces a partial order on $\mathcal{F}^0_{\triangleq}$.

Definition 12.2.11. *Let* $M \sqsubseteq_{\mathcal{VV}} N$ *denote* $[\![M]\!]_\zeta^{\mathcal{VV}} \trianglelefteq [\![N]\!]_\zeta^{\mathcal{VV}}$, *for all* $\zeta \in (\mathrm{Var} \to \mathcal{I}^0_{\triangleq})$. *Moreover, let* $M \sim_{\mathcal{VV}} N$ *denote* $M \sqsubseteq_{\mathcal{VV}} N$ *and* $N \sqsubseteq_{\mathcal{VV}} M$.

Consequently, the model $\mathcal{VV}$ induces a partial order on the interpretation of terms (not only closed terms).

Lemma 12.2.12. *Let* $M, N \in \Lambda^0$. $M \sqsubseteq_{\mathcal{VV}} N$ *if and only if* $M \trianglelefteq N$.

Proof. Let $\zeta \in (\mathrm{Var} \to \mathcal{I}^0_{\triangleq})$, and let ρ be such that $\rho(x) \in \zeta(x)$ for all $x \in \mathrm{Var}$. $M \sqsubseteq_{\mathcal{VV}} N$ if and only if $[\![M]\!]_\zeta^{\mathcal{VV}} \trianglelefteq [\![N]\!]_\zeta^{\mathcal{VV}}$ if and only if $[[\![M]\!]_\rho^{\mathcal{F}(\surd)}] \trianglelefteq [[\![N]\!]_\rho^{\mathcal{F}(\surd)}]$ if and only if $[\![M]\!]_\rho^{\mathcal{F}(\surd)} \trianglelefteq [\![N]\!]_\rho^{\mathcal{F}(\surd)}$ if and only if $M \trianglelefteq N$. □

The correctness is easy.

Theorem 12.2.13 ($\mathcal{VV}$-Correctness).
The model $\mathcal{VV}$ *is correct with respect to the* **V**-*operational semantics.*

Proof. $M \sqsubseteq_{\mathcal{VV}} N$ implies $C[M] \sqsubseteq_{\mathcal{VV}} C[N]$, for each closing context $C[.]$, by Property 10.0.2.(v). Hence $C[M] \trianglelefteq C[N]$ by Lemma 12.2.12; in particular, $C[M] \trianglelefteq_{\omega\to\omega} C[N]$, so $B \vdash_\surd C[M] : \omega \to \omega$ implies $B \vdash_\surd C[N] : \omega \to \omega$, for all bases B. Therefore, if $C[M]$ is Γ-valuable then $C[N]$ is Γ-valuable, by Property 12.1.8, since $C[M]$ and $C[N]$ are closed. Hence $M \preceq_{\mathbf{V}} N$. □

The following theorem implies the full abstraction of $\mathcal{VV}$ with respect to the **V**-operational semantics.

Theorem 12.2.14 ($\mathcal{VV}$-Completeness).
The model $\mathcal{VV}$ *is complete with respect to the* **V**-*operational semantics.*

Proof. We will prove $\not\sqsubseteq_{\mathcal{VV}}$ implies $\not\preceq_{\mathbf{V}}$.
$M \not\sqsubseteq_{\mathcal{VV}} N$ means $[\![M]\!]_\zeta^{\mathcal{VV}} \ntrianglelefteq [\![N]\!]_\zeta^{\mathcal{VV}}$, for some $\zeta \in (\mathrm{Var} \to \mathcal{I}^0_{\triangleq})$. Since the codomain of ζ is $\mathcal{I}^0_{\triangleq}$, if $\mathrm{FV}(M)\cup\mathrm{FV}(N) = \{x_1, ..., x_m\}$ then there are $P_i \in \Gamma^0$ such that $\zeta(x_i) = [[\![P_i]\!]_\rho^{\mathcal{F}(\surd)}]$. Thus, let $\mathbf{s}$ be such that $\mathbf{s}(x_i) = P_i$ $(1 \le i \le m)$, hence $\mathbf{s}(M), \mathbf{s}(N) \in \Lambda^0$. By Property 10.0.2.(iv), $[\![\mathbf{s}(M)]\!]_{\zeta'}^{\mathcal{VV}} \ntrianglelefteq [\![\mathbf{s}(N)]\!]_{\zeta'}^{\mathcal{VV}}$, for all $\zeta' \in (\mathrm{Var} \to \mathcal{I}^0_{\triangleq})$, so in particular $\mathbf{s}(M) \not\sqsubseteq_{\mathcal{VV}} \mathbf{s}(N)$.
By Lemma 12.2.12, $\mathbf{s}(M) \ntrianglelefteq \mathbf{s}(N)$, so there is a sequence of closed Γ-valuable terms $\vec{Q}$ such that $\mathbf{s}(M)\vec{Q} \ntrianglelefteq_{\omega\to\omega} \mathbf{s}(N)\vec{Q}$, by Lemma 12.2.4.
Let $C[.] \equiv (\lambda x_1...x_m.[.])\mathbf{s}(x_1)...\mathbf{s}(x_m)\vec{Q}$; clearly $C[M], C[N] \in \Lambda^0$, and moreover $C[M] \Downarrow_{\mathbf{V}}$ and $C[N] \Uparrow_{\mathbf{V}}$, so $M \not\preceq_{\mathbf{V}} N$. □

Corollary 12.2.15. *If* $M \not\preceq_{\mathbf{V}} N$ *then there is a head context separating* M *and* N.

Proof. Immediate, by the proof of Theorem 12.2.14. □

The technique used here for building the fully abstract model of the **V**-operational semantics is similar to that used in [71] and [44], for different calculi. The use of intersection types and filter models allows for the application of such techniques to a wider class of models.

13. Filter $\lambda\Delta$-Models and Domains

13.1 Domains

There is an analogy between $\lambda\Delta$-filter models and $\lambda\Delta$-models that are ω-algebraic lattices, which was first noticed in [28] and further developed in [1] and [3]. This analogy lies in the fact that type symbols in a $\lambda\Delta$-filter model play the role of names for compact elements in the corresponding ω-algebraic lattice. It is out of the aim of this book to give a complete survey of the $\lambda\Delta$-models based on ω-algebraic lattices. In case where $\Delta = \Lambda$, there are some textbooks giving a complete development of this topic, e.g. [5, 81, 87]. Here we will just give some basic informations in order to assure readability to those readers who are not expert in this topic, without developing the proofs for standard properties.
Let us recall the definition of a ω-algebraic complete lattice.

Definition 13.1.1. (i) *A* complete lattice $(\mathbb{L}, \sqsubseteq_{\mathbb{L}})$ *is a set* $\mathbb{L}$, *equipped by a order relation* $\sqsubseteq_{\mathbb{L}}$, *such that for all* $X \subseteq \mathbb{L}$ *both* $\sqcup X$ *(the* least upper bound *of* X*) and* $\sqcap X$ *(the* greatest lower bound *of* X*) exist.*
(ii) $X \subseteq \mathbb{L}$ *is directed if and only if every two elements of* X *have an upper bound in* X.
(iii) $x \in \mathbb{L}$ *is* compact *if and only if every directed* $X \subseteq \mathbb{L}$ *is such that:*
$$x \sqsubseteq_{\mathbb{L}} \sqcup X \textit{ implies } x \sqsubseteq_{\mathbb{L}} y \textit{ for some } y \in X.$$
Let $\mathbf{comp}(\mathbb{L})$ *be the set of compact elements of* $\mathbb{L}$.
(iv) $\mathbb{L}$ *is* ω-algebraic *if and only if* $x = \sqcup\{y \sqsubseteq x \mid y \textit{ compact}\}$ *and* $\mathbf{comp}(\mathbb{L})$ *is countable.*

Let us use the word *domain* in order to denote a ω-*algebraic* complete lattice. It is easy to see that in a domain there is always a bottom (minimum) element, that as usual we denote by $\bot$.

Definition 13.1.2. (i) *A function* $h : \mathbb{L} \to \mathbb{L}'$ *is* monotone *if and only if:*
$$x \sqsubseteq_{\mathbb{L}} y \textit{ implies } h(x) \sqsubseteq_{\mathbb{L}'} h(y).$$
(ii) *A function* $h : \mathbb{L} \to \mathbb{L}'$ *is* continuous *if and only if it is monotone, and moreover:*
$$h(\sqcup X) = \sqcup\{h(x) \mid x \in X\} \textit{ for all sets } X \subseteq \mathbb{L}.$$

(iii) *A continuous function* $h : \mathbb{L} \to \mathbb{L}'$ *is* strict *if and only if:*

$$h(\bot_{\mathbb{L}}) =_{\mathbb{L}'} \bot_{\mathbb{L}'}.$$

(iv) *The* pointwise order *between two continuous functions* $h, k : \mathbb{L} \to \mathbb{L}'$ *is defined in the following way:*

$$h \sqsubseteq_{\mathbb{L}\to\mathbb{L}'} k \text{ if and only if } \forall x \in \mathbb{L}.\ h(x) \sqsubseteq_{\mathbb{L}'} k(x).$$

(v) *Two domains* $(\mathbb{L}, \sqsubseteq_{\mathbb{L}})$ *and* $(\mathbb{L}', \sqsubseteq_{\mathbb{L}'})$ *are* isomorphic *if and only if there are two continuous functions* $h : \mathbb{L} \to \mathbb{L}'$ *and* $k : \mathbb{L}' \to \mathbb{L}$ *such that:*
- $h \circ k = id_{\mathbb{L}'}$,
- $k \circ h = id_{\mathbb{L}}$,

where $id_{\mathbb{L}}$ *and* $id_{\mathbb{L}'}$ *denote the identity function respectively on* $\mathbb{L}$ *and* $\mathbb{L}'$.

The notion of step function will play a key role in the construction of the isomorphism between filter spaces and domains.

Definition 13.1.3. (i) *Let* $a \in \mathbb{L}$ *and* $b \in \mathbb{L}'$.
The step function $s_{a,b} : \mathbb{L} \to \mathbb{L}'$ *is defined as*

$$\boldsymbol{\lambda} x : \mathbb{L}.\ \text{if } a \sqsubseteq_{\mathbb{L}} x \text{ then } b \text{ else } \bot_{\mathbb{L}'} ,$$

where $\boldsymbol{\lambda}$ *denotes the metatheoretic abstraction.*
(ii) *A step function* $s_{a,b}$ *is* strict *if and only if* $s_{a,b}(\bot_{\mathbb{L}}) = \bot_{\mathbb{L}'}$.
(iii) *The partial order between step functions from* $\mathbb{L}$ *to* $\mathbb{L}'$ *is defined as follows:*

$$s_{a,b} \sqsubseteq_{\mathbb{L}\to\mathbb{L}'} s_{c,d} \text{ if and only if } c \sqsubseteq_{\mathbb{L}} a \text{ and } b \sqsubseteq_{\mathbb{L}'} d.$$

Let $[\mathbb{L} \to \mathbb{L}'] = \{f \mid f : \mathbb{L} \to \mathbb{L}' \text{ is continuous }\}$, and let $[\mathbb{L} \to_{\bot} \mathbb{L}'] = \{f \mid f : \mathbb{L} \to \mathbb{L}' \text{ is continuous and strict}\}$. The following result holds.

Lemma 13.1.4. *Let* $\mathbb{L}$ *and* $\mathbb{L}'$ *be domains.*

(i) $([\mathbb{L} \to \mathbb{L}'], \sqsubseteq_{\mathbb{L}\to\mathbb{L}'})$ *is a domain whose compact elements are least upper bounds of finite sets of step functions.*
(ii) $([\mathbb{L} \to_{\bot} \mathbb{L}'], \sqsubseteq_{\mathbb{L}\to\mathbb{L}'})$ *is a domain whose compact elements are least upper bounds of finite sets of strict step functions.*

Proof. Define $\sqcup\{f, g\}(x) = \sqcup\{f(x), g(x)\}$ and $\sqcap\{f, g\}(x) = \sqcap\{f(x), g(x)\}$; then both $([\mathbb{L} \to \mathbb{L}'], \sqsubseteq_{\mathbb{L}\to\mathbb{L}'})$ and $([\mathbb{L} \to_{\bot} \mathbb{L}'], \sqsubseteq_{\mathbb{L}\to\mathbb{L}'})$ are complete lattices since $(\mathbb{L}', \sqsubseteq_{\mathbb{L}'})$ is a complete lattice.
The fact that both constructions give rise to an ω-algebraic lattice is an obvious consequence of the fact that both $\mathbb{L}$ and $\mathbb{L}'$ are ω-algebraic.
Moreover, note that if f is a continuous function from $\mathbb{L}$ to $\mathbb{L}'$, such that $f(a) = b$, then $s_{a,b} \sqsubseteq_{\mathbb{L}\to\mathbb{L}'} f$. Then $f = \sqcup\{s_{a,b} \mid f(a) = b\}$. □

A further operation on domains that will be useful is the *lifting*. Let $(\mathbb{L}, \sqsubseteq_{\mathbb{L}})$ be a domain and let $\mathbb{L}_\perp = \mathbb{L} \cup \{\perp\}$, where $\perp$ is a fresh element not belonging to $\mathbb{L}$. Moreover, let $a \sqsubseteq_{\mathbb{L}_\perp} b$ if and only if either $a = \perp$ or $a \sqsubseteq_{\mathbb{L}} b$. The following lemma holds.

Lemma 13.1.5. *If* $(\mathbb{L}, \sqsubseteq_{\mathbb{L}})$ *is a domain then* $(\mathbb{L}_\perp, \sqsubseteq_{\mathbb{L}_\perp})$ *is a domain (the lifting of* $(\mathbb{L}, \sqsubseteq_{\mathbb{L}})$*).*

Proof. Easy. □

Let us call *domain constructor* an operation on domains. We will consider in this section a restricted set of domain constructors, namely

$$\mathcal{C} = \{[\cdot \to \cdot], [\cdot \to_\perp \cdot], (\cdot)_\perp\}.$$

Let $\mathbf{c}$ denote an element of $\mathcal{C}$. We will use all constructors in $\mathcal{C}$ as being unary. It is possible to compose domain constructors, in order to obtain further domain constructors.

Property 13.1.6. Let $(\mathbb{L}, \sqsubseteq_{\mathbb{L}})$ be a domain.
If $\mathbf{c}_1 \in \{[\cdot \to \cdot], [\cdot \to_\perp \cdot]\}$ and $\mathbf{c}_2 \equiv (\cdot)_\perp$ then $(\mathbf{c}_2(\mathbf{c}_1(\mathbb{L})), \sqsubseteq_{\mathbf{c}_2(\mathbf{c}_1(\mathbb{L}))})$ is a domain.

Proof. Easy. □

Definition 13.1.7. *Let* $(\mathbb{L}, \sqsubseteq_{\mathbb{L}})$ *and* $(\mathbb{L}', \sqsubseteq_{\mathbb{L}'})$ *be domains. A* retraction pair *is a pair of continuous functions* $(i : \mathbb{L} \to \mathbb{L}', j : \mathbb{L}' \to \mathbb{L})$ *such that:*

- $j \circ i = id_{\mathbb{L}}$,
- $i \circ j \sqsubseteq id_{\mathbb{L}'}$.

If (i, j) *is a retraction pair from* $\mathbb{L}$ *to* $\mathbb{L}'$, i *is called the* embedding *and* j *is called the* projection.

Recalling the notion of isomorphism between domains, given in Definition 13.1.2.(v), if there is a retraction pair from $\mathbb{L}$ to $\mathbb{L}'$, then sometimes $\mathbb{L}$ is called a *subdomain* of $\mathbb{L}'$.

Property 13.1.8. (i) Let (i_1, j_1) be a retraction pair from $\mathbb{L}$ to $\mathbb{L}'$ and (i_2, j_2) be a retraction pair from $\mathbb{L}'$ to $\mathbb{L}''$. Then $(i_2 \circ i_1, j_1 \circ j_2)$ is a retraction pair from $\mathbb{L}$ to $\mathbb{L}''$.
(ii) An embedding (projection) function has a unique corresponding projection (embedding).
(iii) If (i, j) is a retraction pair from $\mathbb{L}$ to $\mathbb{L}'$ then both i and j are strict.

Domain constructors can be extended to retraction pairs. Let us show how the extension can be made in the particular cases we are interested in.

- Let (i, j) be a retraction pair between $\mathbb{L}$ and $\mathbf{c}(\mathbb{L})$, where either $\mathbf{c} = [\cdot \to \cdot]$ or $\mathbf{c} = [\cdot \to_\perp \cdot]$. Let

- $\mathbf{c}(i) = \boldsymbol{\lambda} x{:}\mathbf{c}(\mathbb{L}).\, i \circ x \circ j$,
- $\mathbf{c}(j) = \boldsymbol{\lambda} x{:}\mathbf{c}^2(\mathbb{L}).\, j \circ x \circ i$.

It is easy to check that $(\mathbf{c}(i), \mathbf{c}(j))$ is a retraction pair between $\mathbf{c}(\mathbb{L})$ and $\mathbf{c}^2(\mathbb{L})$.

- In case of lifting, let (i, j) be a retraction pair between $\mathbb{L}$ and $\mathbf{c}(\mathbb{L})$, where $\mathbf{c} = (.)_\perp$. Let
  - $\mathbf{c}(i) = \boldsymbol{\lambda} x{:}\mathbf{c}(\mathbb{L}).$ if $x = \perp_{\mathbf{c}(\mathbb{L})}$ then $\perp_{\mathbf{c}^2(\mathbb{L})}$ else $i(x)$,
  - $\mathbf{c}(j) = \boldsymbol{\lambda} x{:}\mathbf{c}^2(\mathbb{L}).$ if $x = \perp_{\mathbf{c}^2(\mathbb{L})}$ then $\perp_{\mathbf{c}(\mathbb{L})}$ else $j(x)$.

  Then $(\mathbf{c}(i), \mathbf{c}(j))$ is a retraction pair between $\mathbf{c}(\mathbb{L})$ and $\mathbf{c}^2(\mathbb{L})$.

In case $\mathbf{c}$ is a compound domain constructor, the extension of $\mathbf{c}$ to retraction pairs can be made starting from the previous defined extension and then using Property 13.1.8.(i).

Definition 13.1.9. *Let $\mathbb{L}_0, \mathbb{L}_1, ..., \mathbb{L}_n, ...$ be domains.*

(i) *A* retraction sequence *is a pair whose first component is the set*

$$\{\mathbb{L}_i \mid i \geq 0\},$$

and whose second component is the set

$$\{(i_i, j_i) \mid (i_i, j_i) \text{ is a retraction pair from } \mathbb{L}_i \text{ to } \mathbb{L}_{i+1}, i \geq 0\}.$$

(ii) *The* inverse limit *of a retraction sequence is the set*

$$\mathbb{L}_\infty = \{(a_0, a_1, ..., a_n, ...) \mid n \geq 0, a_n \in \mathbb{L}_n, a_n = j_n(a_{n+1})\},$$

partially ordered by the relation $\sqsubseteq_{\mathbb{L}_\infty}$, defined as follows:

$$(a_0, a_1, ..., a_n, ...) \sqsubseteq_{\mathbb{L}_\infty} (b_0, b_1, ..., b_n, ...) \text{ if and only if } a_n \sqsubseteq_{\mathbb{L}_n} b_n, \text{ for all } n.$$

The following property holds.

Property 13.1.10. The inverse limit $\mathbb{L}_\infty$ of retraction sequence is a domain.

A domain equation is an equation of the shape

$$\mathbb{X} = \mathbf{c}(\mathbb{X}),$$

where $\mathbf{c}$ is a domain constructor, and $=$ denotes the isomorphism between domains.

Theorem 13.1.11. *Let $(\mathbb{L}, \sqsubseteq_{\mathbb{L}})$ be a domain, and let (i, j) be a retraction pair between $\mathbb{L}$ and $\mathbf{c}(\mathbb{L})$. The inverse limit*

$$\mathbb{L}_\infty = \{(a_0, a_1, ..., a_n, ...) \mid n \geq 0, a_n \in \mathbf{c}^n(\mathbb{L}), a_n = \mathbf{c}^n(j)(a_{n+1})\}$$

is a solution of the domain equation $\mathbb{X} = \mathbf{c}(\mathbb{X})$, i.e. $\mathbb{L}_\infty = \mathbf{c}(\mathbb{L}_\infty)$.

Proof. Let $\mathbf{c}^0(x) = x$ and $\mathbf{c}^{n+1}(x) = \mathbf{c}(\mathbf{c}^n(x))$ for all $n \in \mathbb{N}$.

$$\mathbf{c}^0(\mathbb{L}) \underset{j}{\overset{i}{\rightleftarrows}} \mathbf{c}^1(\mathbb{L}) \underset{\mathbf{c}^1(j)\circ..\circ\mathbf{c}^{k-1}(j)}{\overset{\mathbf{c}^{k-1}(i)\circ..\circ\mathbf{c}^1(i)}{\overset{\cdots}{\rightleftarrows}}} \mathbf{c}^k(\mathbb{L}) \underset{\mathbf{c}^k(j)}{\overset{\mathbf{c}^k(i)}{\rightleftarrows}} \mathbf{c}^{k+1}(\mathbb{L}) \qquad \cdots$$

Let $r_{m,n} : \mathbf{c}^m(\mathbb{L}) \to \mathbf{c}^n(\mathbb{L})$ be the following function:

$$r_{m,n} = \begin{cases} \mathsf{id}_{\mathbf{c}^m(\mathbb{L})} & \text{if } m = n, \\ \mathbf{c}^{n-1}(i) \circ .. \circ \mathbf{c}^m(i) & \text{if } m < n, \\ \mathbf{c}^n(j) \circ .. \circ \mathbf{c}^{m-1}(j) & \text{if } n < m \end{cases}$$

It is easy to check that if $m \leq n$ then $(r_{m,n}, r_{n,m})$ is a retraction pair between $\mathbf{c}^m(\mathbb{L})$ and $\mathbf{c}^n(\mathbb{L})$, by Property 13.1.8.(i). Let

- $i_{n,\infty} : \mathbf{c}^n(\mathbb{L}) \to \mathbb{L}_\infty$ be $\boldsymbol{\lambda} x : \mathbf{c}^n(\mathbb{L}).\,(r_{n,0}(x), r_{n,1}(x), ..., r_{n,n}(x), r_{n,n+1}(x), ...)$;
- $j_{n,\infty} : \mathbb{L}_\infty \to \mathbf{c}^n(\mathbb{L})$ be $\boldsymbol{\lambda} x : \mathbb{L}_\infty.\,(x)_n$,
  where $(.)_n$ denotes the n-th element of a sequence;
- $I : \mathbb{L}_\infty \to \mathbf{c}(\mathbb{L}_\infty)$ be $\bigsqcup_{(n\geq 0)}(i_{n+1,\infty} \circ r_{n,n+1} \circ j_{n,\infty})$;
- $J : \mathbf{c}(\mathbb{L}_\infty) \to \mathbb{L}_\infty$ be $\bigsqcup_{(n\geq 0)}(i_{n,\infty} \circ r_{n+1,n} \circ j_{n+1,\infty})$.

Then (I, J) is a retraction pair between $\mathbb{L}_\infty$ and $\mathbf{c}(\mathbb{L}_\infty)$, such that $I \circ J = id_{\mathbf{c}(\mathbb{L}_\infty)}$, so the two domains are isomorphic. □

By Property 13.1.8.(ii), the solution of a domain equation $\mathbb{X} = \mathbf{c}(\mathbb{X})$ is completely determined by the initial domain $(\mathbb{L}, \sqsubseteq_\mathbb{L})$ and the embedding function i between $\mathbb{L}$ and $\mathbf{c}(\mathbb{L})$.

Definition 13.1.12. *A solution of a domain equation is* minimal, *if the initial domain* $(\mathbb{L}, \sqsubseteq_\mathbb{L})$ *is isomorphic to the domain* $(\{\bot\}, id_{\{\bot\}})$.

Now we have all the ingredients in order to show the correspondence between a $\lambda\Delta$-model that is an inverse limit solution of a domain equation of a given shape and a filter model.

Let us assume that the initial domain $\mathbb{L}$ always has a finite number of elements, which implies that all the elements of $\mathbb{L}$ are compact. Let us define the following procedure, in order to build from $\mathbb{L}_\infty$ the filter space $\mathcal{F}(\mathbb{L}_\infty)$.

> Let $\mathbb{L}_\infty$ be a solution of the domain equation $\mathbb{X} = \mathbf{c}(\mathbb{X})$, where $\mathbf{c} \in \{[. \to .], [. \to .]_\bot, [. \to_\bot .]_\bot\}$, starting from the initial domain $(\mathbb{L}, \sqsubseteq_\mathbb{L})$ and from the embedding function i between $\mathbb{L}$ and $\mathbf{c}(\mathbb{L})$. Note that by the particular set of constructors we chosen and by the fact that domain equations are defined modulo isomorphisms, by Lemma 13.1.4 we can consider the compact elements of $\mathbb{L}_\infty$ to be either $\bot_{\mathbb{L}_\infty}$ or least upper bounds of finite sets of step functions from $\mathbb{L}_\infty$ to $\mathbb{L}_\infty$. Moreover, $\mathbf{comp}(\mathbb{L}_\infty) = \bigsqcup_{n\geq 0} i_{n,\infty}(\mathbf{comp}(\mathbf{c}^n(\mathbb{L})))$. The set of type constants $C_\mathbb{L}$ and the inclusion relation $\leq_{\nabla_\mathbb{L}}$ can be built according to the following procedure **compact-as-types**(.).

Procedure compact-as-types($\mathbb{L}_\infty$)

1. Choose a set of type constants $C_\mathbb{L}$ such that there is a bijection $(.)^+$ between $C_\mathbb{L}$ and the compact elements of $\mathbb{L}$, such that $(\omega)^+ = \bot_\mathbb{L}$.
2. Define an intersection relation $\leq_{\nabla_\mathbb{L}}$ such that $\sigma \leq_{\nabla_\mathbb{L}} \tau$ if and only if $(\tau)^+ \sqsubseteq_\mathbb{L} (\sigma)^+$, for all $\sigma, \tau \in C_\mathbb{L}$.
3. Let $T(C_\mathbb{L})$ be the set of types built from the set of constants $C_\mathbb{L}$.
   Let $(.)^*$ be the function from $T(C_\mathbb{L})$ to $\mathbf{comp}(\mathbb{L}_\infty)$ defined as follows.
   (3.1) If $\sigma \in C_\mathbb{L}$ then $(\sigma)^* = i_{0,\infty}((\sigma)^+)$.
   (3.2) If $\mathbf{c} \in \{[. \to .], [. \to .]_\bot, \}$ then $(\sigma \to \tau)^* = s_{(\sigma)^*,(\tau)^*}$.
   Otherwise, in case $\mathbf{c} = [. \to_\bot .]_\bot$, if $(\sigma)^* \neq \bot_{\mathbb{L}_\infty}$ then $(\sigma \to \tau)^* = s_{(\sigma)^*,(\tau)^*}$, while if $(\sigma)^* = \bot_{\mathbb{L}_\infty}$ then $(\sigma \to \tau)^* = s_{a,(\tau)^*}$ where $a = s_{\bot_{\mathbb{L}_\infty},\bot_{\mathbb{L}_\infty}}$.
   (3.3) $(\sigma \wedge \tau)^* = (\sigma)^* \sqcup (\tau)^*$.
4. Extend the intersection relation $\leq_{\nabla_\mathbb{L}}$ as follows.
   (4.1) If $\sigma, \mu_i, \nu_i \in C_\mathbb{L}$ $(1 \leq i \leq n)$ and $i((\sigma)^+) = \bigsqcup_{1\leq i\leq n} s_{(\mu_i)^+,(\nu_i)^+}$ where $n \in \mathbb{N}$, then both $\sigma \leq_{\nabla_\mathbb{L}} (\mu_1 \to \nu_1) \wedge ... \wedge (\mu_n \to \nu_n)$ and $(\mu_1 \to \nu_1) \wedge ... \wedge (\mu_n \to \nu_n) \leq_{\nabla_\mathbb{L}} \sigma$.
   (4.2) If $\mathbf{c} = [. \to_\bot .]_\bot$ then $(\omega \to \omega) \to \tau \leq_{\nabla_\mathbb{L}} \omega \to \tau$,

The definition of the mapping $(.)^*$ need some comments. Point (3.2) maps every arrow type into a step function. Note that if $\mathbf{c} = [. \to_\bot .]_\bot$, by definition of strict function, $s_{a,b} \in \mathbb{L}_\infty$ implies either $a \neq \bot$ or $a, b = \bot$. So types of the shape $\omega \to \tau$, where $(\tau)^* \neq \bot$, are in some sense redundant, and they are mapped into a step function that is the maximum one less to the step function $s_{\bot,(\tau)^*}$.

Point (4.1) takes into account the initial retraction pair. Point (4.2) reflects point (3.2). Moreover, if $C_\mathbb{L}$ is finite (so $\mathbb{L}$ and $\mathbf{c}(\mathbb{L})$ are finite) then the number of rules to be joined to the intersection relation $\leq_{\nabla_\mathbb{L}}$ can be transformed in a finite number.

The following lemma proves that the procedure is correct, in the sense that $(.)^*$ is a surjection, and the inclusion relation between types respects the order relation of the domain.

Lemma 13.1.13. *Let $\mathbb{L}_\infty$ be a solution of the domain equation $\mathbb{X} = \mathbf{c}(\mathbb{X})$, where $\mathbf{c} \in \{[. \to .], [. \to .]_\bot, [. \to_\bot .]_\bot\}$, starting from the initial domain $(\mathbb{L}, \sqsubseteq_\mathbb{L})$ and from the embedding function i between $\mathbb{L}$ and $\mathbf{c}(\mathbb{L})$.*

(i) $(.)^* : T(C_\mathbb{L}) \to \mathbf{comp}(\mathbb{L}_\infty)$ *is a surjection.*
(ii) *For all $\sigma, \tau \in T(C_\mathbb{L})$, $\sigma \leq_{\nabla_\mathbb{L}} \tau$ if and only if $(\tau)^* \sqsubseteq_{\mathbb{L}_\infty} (\sigma)^*$.*

Proof. (i) Let $(\mathbb{L}_\infty, \sqsubseteq_{\mathbb{L}_\infty})$ be the solution of the given domain equation. The set $\mathbf{comp}(\mathbb{L}_\infty)$ can be viewed as $\bigsqcup_{n\geq 0} i_{n,\infty}(\mathbf{comp}(\mathbf{c}^n(\mathbb{L})))$. We will prove that, if $a \in i_{n,\infty}(\mathbf{comp}(\mathbf{c}^n(\mathbb{L})))$ then there is σ such that $(\sigma)^* = a$, by induction on n. If $n = 0$, then the proof follows by construction.
Let $n \geq 1$. Since the set of constructors $\mathbf{c}$ we are taking into consideration

is restricted, a compact element d of $\mathbf{c}^n(\mathbb{L})$ meets one of the following constraints.

- d is the bottom, then $\bot = (\omega)^*$, by construction.
- d is a step function $s_{a,b} : \mathbf{c}^{n-1}(\mathbb{L}) \to \mathbf{c}^{n-1}(\mathbb{L})$, where $a, b \in \mathbf{c}^{n-1}(\mathbb{L})$. Note that $\mathbf{c}^{n-1}(\mathbb{L})$ has a finite number of elements, for all $n \geq 1$; hence, if $a \in \mathbf{c}^{n-1}(\mathbb{L})$ then $i_{n-1,\infty}(a)$ is compact. So, by induction, there are σ, τ such that $i_{n-1,\infty}(a) = (\sigma)^*$ and $i_{n-1,\infty}(b) = (\tau)^*$, and therefore $s_{i_{n-1,\infty}(a), i_{n-1,\infty}(b)} = (\sigma \to \tau)^*$.
- d is the least upper bound of a finite set of step functions, namely $\sqcup\{s_{a_i,b_i} \mid 1 \leq i \leq m\}$; then by induction there are σ_i, τ_i such that $i_{n-1,\infty}(a_i) = (\sigma_i)^*$ and $i_{n-1,\infty}(b_i) = (\tau_i)^*$, and

$$\sqcup\{s_{a_i,b_i} \mid 1 \leq i \leq m\} = ((\sigma_1 \to \tau_1) \wedge \ldots \wedge (\sigma_m \to \tau_m))^*.$$

(ii) ($\Rightarrow$) By induction on the definition of $\leq_{\nabla_{\mathbb{L}}}$, taking into account the last used rule.
(a) Then the proof follows since $\bot \sqsubseteq_{\mathbb{L}_\infty} a$, for all a, and since $(\omega)^* = \bot$.
$(b), (c), (c'), (e)$ By the definition of least upper bound.
$(f), (d)$ By the Definition 13.1.3.iii.
(g) By the fact that $(\omega \to \omega)^* = \lambda x.\bot$, that is the smallest step function.
(r) Obvious.
(t) By induction.
Let $\sigma \leq_{\nabla_{\mathbb{L}}} (\mu_1 \to \nu_1) \wedge \ldots \wedge (\mu_n \to \nu_n)$ be the conclusion of a rule added by the point (4.1) of the procedure compact-as-types. Clearly also the rule $(\mu_1 \to \nu_1) \wedge \ldots \wedge (\mu_n \to \nu_n) \leq_{\nabla_{\mathbb{L}}} \sigma$ was added by construction, i.e. $\sigma \simeq_{\nabla_{\mathbb{L}}} (\mu_1 \to \nu_1) \wedge \ldots \wedge (\mu_n \to \nu_n)$. Hence $(\sigma)^*$ and $\bigsqcup_{1 \leq i \leq n} s_{(\mu_i)^*,(\nu_i)^*}$ denote the same compact element.
The case $(\mu_1 \to \nu_1) \wedge \ldots \wedge (\mu_n \to \nu_n) \leq_{\nabla_{\mathbb{L}}} \sigma$ is symmetric.
In both cases, $\sigma, \mu_i, \nu_i \in C_{\mathbb{L}}$ $(1 \leq i \leq n)$, $i((\sigma)^+) = \bigsqcup_{1 \leq i \leq n} s_{(\mu_i)^+,(\nu_i)^+}$ and $(\sigma)^+ = j\big(\bigsqcup_{1 \leq i \leq n} s_{(\mu_i)^+,(\nu_i)^+}\big)$.
Let $(\omega \to \omega) \to \tau \leq_{\nabla_{\mathbb{L}}} \omega \to \tau$ be the conclusion of a rule added by the point (4.2) of the procedure compact-as-types. Hence $\mathbf{c} = [. \to_\bot .]_\bot$ and if $(\tau)^* \neq (\omega)^*$ then $s_{(\omega)^*,(\tau)^*}$ does not belong to $\mathbb{L}_\infty$. In such case, by the point (3.2) of the procedure compact-as-types,

$$((\omega \to \omega) \to \tau)^* = (\omega \to \tau)^*.$$

($\Leftarrow$) Let $(\tau)^* \sqsubseteq_{\mathbb{L}_\infty} (\sigma)^*$. The proof is given by induction on the total number of symbols of σ and τ.
If $(\tau)^* = \bot$ then the proof is trivial, since $\bot = (\omega)^*$, and ω is the biggest type. If τ or σ are type constants then the proof follows by construction. If $\sigma \equiv \sigma_1 \to \sigma_2$ and $\tau \equiv \tau_1 \to \tau_2$, then by the Definition 13.1.3.(iii) and the definition of $(.)^*$, $(\tau_2)^* \sqsubseteq_{\mathbb{L}_\infty} (\sigma_2)^*$ and $(\sigma_1)^* \sqsubseteq_{\mathbb{L}_\infty} (\tau_1)^*$. Thus by induction $\tau_1 \leq_{\nabla_{\mathbb{L}}} \sigma_1$ and $\sigma_2 \leq_{\nabla_{\mathbb{L}}} \tau_2$, and the proof follows by rule (f) of Definition 10.1.1.(ii).

If $\tau \equiv \tau_1 \to \tau_2$ and $(\sigma)^* \equiv \sqcup\{(\mu_i \to \nu_i)^* \mid 1 \leq i \leq m\}$, then there is $\{i_1, ..., i_k\} \subseteq \{1, ..., m\}$ such that $(\mu_{i_1} \wedge ... \wedge \mu_{i_k})^* \sqsubseteq_{\mathbb{L}_\infty} (\tau_1)^*$ and $(\tau_2)^* \sqsubseteq_{\mathbb{L}_\infty} (\nu_{i_1} \wedge ... \wedge \nu_{i_k})^*$, and so $(\tau_1 \to \tau_2)^* \sqsubseteq_{\mathbb{L}_\infty} ((\mu_{i_1} \wedge ... \wedge \mu_{i_k}) \to (\nu_{i_1} \wedge ... \wedge \nu_{i_k}))^*$. So, by induction, $\tau_1 \leq_{\nabla_\mathbb{L}} \mu_{i_1} \wedge ... \wedge \mu_{i_k}$ and $\nu_{i_1} \wedge ... \wedge \nu_{i_k} \leq_{\nabla_\mathbb{L}} \tau_2$, and, by rule (f) of Definition 10.1.1.(ii), $\tau_1 \to \tau_2 \geq_{\nabla_\mathbb{L}} (\mu_{i_1} \wedge ... \wedge \mu_{i_k}) \to \nu_{i_1} \wedge ... \wedge \nu_{i_k} \simeq_{\nabla_\mathbb{L}} ((\mu_{i_1} \wedge ... \wedge \mu_{i_k}) \to \nu_{i_1}) \wedge ... \wedge (\mu_{i_1} \wedge ... \wedge \mu_{i_k}) \to \nu_{i_k} \geq_{\nabla_\mathbb{L}} (\mu_{i_1} \to \nu_{i_1}) \wedge ... \wedge (\mu_{i_k} \to \nu_{i_k}) \geq_{\nabla_\mathbb{L}} \sigma$, by applying respectively Lemma 10.1.4.(iii), and the rules (c) and (f) of Definition 10.1.1.(ii).
If $(\tau)^* \equiv \sqcup\{(\mu_i \to \nu_i)^* \mid 1 \leq i \leq m\}$, then $(\mu_i \to \nu_i)^* \sqsubseteq_{\mathbb{L}_\infty} (\sigma)^*$, for all i $(1 \leq i \leq m)$, and the proof follows the same lines as the previous point.

□

Note that the proof of part $(\Leftarrow)$ of the Lemma 13.1.13 gives a justification of the legality condition (Definition 10.1.5) on a type system in order to induce a $\lambda\Delta$-model. In fact, this condition reflects a semantic property on step functions.

Theorem 13.1.14. *Let $\mathbb{L}_\infty$ be a solution of the domain equation $\mathbb{X} = \mathbf{c}(\mathbb{X})$, where $\mathbf{c} \in \{[. \to .], [. \to .]_\perp, [. \to_\perp .]_\perp\}$, starting from the initial domain $(\mathbb{L}, \sqsubseteq_\mathbb{L})$ and from the embedding function i between $\mathbb{L}$ and $\mathbf{c}(\mathbb{L})$.*
Let ∇ be a type system such that $C(\mathbb{L})$ and $\leq_{\nabla_\mathbb{L}}$ are built according to the procedure **compact-as-types**$(\mathbb{L}_\infty)$.
Then $\mathbb{L}_\infty$ is isomorphic to the space of filters $\mathcal{F}(\nabla_\mathbb{L})$, ordered by set inclusion.

Proof. Let us define the functions $h : \mathbb{L}_\infty \to \mathcal{F}(\nabla_\mathbb{L})$ and $k : \mathcal{F}(\nabla_\mathbb{L}) \to \mathbb{L}_\infty$ in the following way:

- $h(a) = \uparrow \{\sigma \mid (\sigma)^* \sqsubseteq a\}$,
- $k(f) = \sqcup\{(\sigma)^* \mid \sigma \in f\}$.

By Lemma 13.1.13, h and k realize the desired isomorphism. □

Now we are ready to prove that each of the filter models we presented is isomorphic to a model built as the inverse limit solution of a domain equation.

13.1.1 $\mathcal{H}$ as Domain

Take the domain equation:

$$\mathbb{X} = [\mathbb{X} \to \mathbb{X}].$$

Note that the minimal solution of this equation is the domain $(\{\perp\}, id_{\{\perp\}})$, since there is just one continuous function from $\{\perp\}$ to $\{\perp\}$, namely the function $\lambda x.\perp$. So take as initial domain $(\mathbb{I}, \sqsubseteq_\mathbb{I})$, where $\mathbb{I} = \{\perp_\mathbb{I}, \top_\mathbb{I}\}$, and $\sqsubseteq_\mathbb{I}$ is defined as $\perp_\mathbb{I} \sqsubseteq_\mathbb{I} \top_\mathbb{I}$. Take the domain $([\mathbb{I} \to \mathbb{I}], \sqsubseteq_{\mathbb{I}\to\mathbb{I}})$, which is a domain by Lemma 13.1.4.(i), and choose, as embedding function between $(\mathbb{I}, \sqsubseteq_\mathbb{I})$ and $([\mathbb{I} \to \mathbb{I}], \sqsubseteq_{\mathbb{I}\to\mathbb{I}})$, the function i so defined:

- $i(\bot_{\mathbb{I}}) = \bot_{[\mathbb{I}\to\mathbb{I}]}$,
- $i(\top_{\mathbb{I}}) = s_{\bot_{\mathbb{I}},\top_{\mathbb{I}}}$,

and let $\mathbb{I}_\infty$ be the inverse limit solution so obtained.
The correspondence between C_∞ and **comp**($\mathbb{I}$) is defined in the following way:

- $(\omega)^+ = \bot_{\mathbb{I}}$,
- $(\phi)^+ = \top_{\mathbb{I}}$.

Let the function $(.)^* : T(C_\infty) \to$ **comp**($\mathbb{I}_\infty$) be the function made according to the procedure **compact-as-types**.
The procedure **compact-as-types**($\mathbb{I}_\infty$) generates, at point (4.1), the rules $(h1), (h2)$ and $(h3)$ of Fig. 11.1 (pag. 120) of the intersection relation $\leq_\infty$. Note that point (4.2) is not applied in this case.

Let $\rho : \text{Var} \to \mathbb{I}_\infty$; the interpretation function $[\![.]\!]^{\mathbb{I}_\infty}$ (see [67]) is:

- $[\![x]\!]_\rho^{\mathbb{I}_\infty} = \rho(x)$,
- $[\![MN]\!]_\rho^{\mathbb{I}_\infty} = I([\![M]\!]_\rho^{\mathbb{I}_\infty})([\![N]\!]_\rho^{\mathbb{I}_\infty})$,
- $[\![\lambda x.M]\!]_\rho^{\mathbb{I}_\infty} = J(\lambda d.[\![M]\!]_{\rho[d/x]}^{\mathbb{I}_\infty})$,

where I and J are defined in the proof of Theorem 13.1.11.

Theorem 13.1.15. *Let* $\mathcal{I} =< \mathbb{I}_\infty, \mathbb{I}_\infty, \circ_{\mathbb{I}_\infty}, [\![.]\!]^{\mathbb{I}_\infty} >$ *where* $\circ_{\mathbb{I}_\infty} \equiv \lambda xy.I(x)(y)$. $\mathcal{I}$ *is a* $\lambda\Lambda$*-calculus model, and it is isomorphic to* $\mathcal{H}$.

Proof. It is easy to check that $\mathcal{I}$ satisfies the conditions of Definition 10.0.1, so it is a $\lambda\Delta$-model. By Theorem 13.1.14, $(\mathbb{I}_\infty, \sqsubseteq_{\mathbb{I}_\infty})$ is isomorphic to $\mathcal{F}(\infty)$, ordered by set inclusion. Now we will prove that, if $\zeta : \text{Var} \to \mathbb{I}_\infty$ and $\rho : \text{Var} \to \mathcal{F}(\infty)$ are such that $\forall x \in \text{Var}$, $\zeta(x)$ and $\rho(x)$ are isomorphic elements, then $\forall M \in \Lambda$ $[\![M]\!]_\zeta^{\mathbb{I}_\infty}$ is isomorphic to $[\![M]\!]_\rho^{\mathcal{F}(\infty)}$. Since an element in a domain is completely determined by the set of compact elements less equal to it, we need only to prove that:

$$B_\rho \vdash_\infty M : \sigma \quad \text{if and only if} \quad (\sigma)^* \sqsubseteq_{\mathcal{I}} [\![M]\!]_\zeta^{\mathbb{I}_\infty},$$

where $B_\rho(x) \in \rho(x)$, for all $x \in \text{Var}$.

$(\Rightarrow)$ The proof follows by induction on M.
If $M \equiv x$ then $B_\rho(x) \leq_\infty \sigma$, so $(\sigma)^* \sqsubseteq_{\mathcal{I}} \zeta(x)$ and the proof is given.
Let $M \equiv PQ$. So $B_\rho \vdash_\infty PQ : \sigma$ implies that there is a type τ such that $B_\rho \vdash_\infty P : \tau \to \sigma$ and $B_\rho \vdash_\infty Q : \tau$ by Lemma 10.1.7.(vii). By induction both $(\tau \to \sigma)^* \sqsubseteq_{\mathcal{I}} [\![P]\!]_\zeta^{\mathbb{I}_\infty}$ and $(\tau)^* \sqsubseteq_{\mathcal{I}} [\![Q]\!]_\zeta^{\mathbb{I}_\infty}$; thus $s_{(\tau)^*,(\sigma)^*} \sqsubseteq_{\mathcal{I}} [\![P]\!]_\zeta^{\mathbb{I}_\infty} = I([\![P]\!]_\zeta^{\mathbb{I}_\infty})$ by isomorphism. Hence $(\sigma)^* \sqsubseteq_{\mathcal{I}} I([\![P]\!]_\zeta^{\mathbb{I}_\infty})([\![Q]\!]_\zeta^{\mathbb{I}_\infty}) = ([\![PQ]\!]_\zeta^{\mathbb{I}_\infty})$.
Let $M \equiv \lambda x.N$. By Lemma 10.1.7.(iv), $B_\rho \vdash_\infty \lambda x.N : \sigma$ implies that there are μ_i, ν_i $(1 \leq i \leq n)$ for some $n \in \mathbb{N}$, such that $(\mu_1 \to \nu_1) \wedge \ldots \wedge (\mu_n \to \nu_n) \leq_\infty \sigma$, $B_\rho[\mu_i/x] \vdash_\infty N : \nu_i$ $(1 \leq i \leq n)$. By induction $(\nu_i)^* \sqsubseteq_{\mathcal{I}} [\![N]\!]_{\zeta[d/x]}^{\mathbb{I}_\infty}$ where $(\mu_i)^* \sqsubseteq_{\mathcal{I}} d$. Therefore

$$(\sigma)^* \sqsubseteq_{\mathcal{I}} \bigsqcup_{1 \le i \le n} s_{(\mu_i)^*,(\nu_i)^*} \sqsubseteq_{\mathcal{I}} \bigsqcup_{\substack{1 \le i \le n \\ d \in \mathbf{comp}(\mathbb{I}_\infty)}} s_{d,(\nu_i)^*} \sqsubseteq_{\mathcal{I}} J(\lambda\!\!\lambda d.[\![N]\!]^{\mathbb{I}_\infty}_{\varsigma[d/x]}).$$

($\Leftarrow$) The proof follows by induction on M. If $M \equiv x$ then the proof is easy. Let $M \equiv PQ$. So $(\sigma)^* \sqsubseteq_{\mathcal{I}} [\![PQ]\!]^{\mathbb{I}_\infty}_\varsigma = I([\![P]\!]^{\mathbb{I}_\infty}_\varsigma)([\![Q]\!]^{\mathbb{I}_\infty}_\varsigma)$. Thus there is τ such that $s_{(\tau)^*,(\sigma)^*} \sqsubseteq_{\mathcal{I}} [\![P]\!]^{\mathbb{I}_\infty}_\varsigma$ and $(\tau)^* \sqsubseteq_{\mathcal{I}} [\![Q]\!]^{\mathbb{I}_\infty}_\varsigma$. The proof follows by induction. Let $M \equiv \lambda x.N$. So $(\sigma)^* \sqsubseteq_{\mathcal{I}} [\![M]\!]^{\mathbb{I}_\infty}_\varsigma = J(\lambda\!\!\lambda d.[\![N]\!]^{\mathbb{I}_\infty}_{\varsigma[d/x]})$ implies that there are μ_i, ν_i $(1 \le i \le n)$ for some $n \in \mathbb{N}$, such that both $(\nu_i)^* \sqsubseteq_{\mathcal{I}} [\![N]\!]^{\mathbb{I}_\infty}_{\varsigma[d/x]}$ and $(\sigma)^* \sqsubseteq_{\mathcal{I}} \bigsqcup_{1 \le i \le n} s_{(\mu_i)^*,(\nu_i)^*}$ and $(\mu_i)^* \sqsubseteq_{\mathcal{I}} d$. Hence $\sigma \ge_\infty \bigwedge_{1 \le i \le n}(\mu_i \to \nu_i)$ where $B_\rho[\mu_i/x] \vdash_\infty N : \nu_i$ $(1 \le i \le n)$, so $B_\rho \vdash_\infty \lambda x.N : \mu_i \to \nu_i$ $(1 \le i \le n)$ by rule $(\to I)$, and $B_\rho \vdash_\infty \lambda x.N : \sigma$ by rule $(\le_\infty)$. □

The model $\mathcal{I}$ was the first denotational $\lambda\Lambda$-calculus model. It was built by Scott [89], and the induced Λ-theory was extensively studied in [54, 97], where the approximation theorem is proved by using the technique of indexed reductions. An analysis of the characterization of term in the $\mathcal{I}$-model can be found in [32]. By the structure of $\mathbb{I}$, it is possible to have a different inverse limit solution, by choosing as initial embedding function the function i', such that $i'(\bot_{\mathbb{I}}) = \bot_{[\mathbb{I}\to\mathbb{I}]}$ and $i'(\top_{\mathbb{I}}) = s_{\top_{\mathbb{I}},\top_{\mathbb{I}}}$. This model was first defined by Park [75], and it induce a Λ-theory quite different from **H**, which was extensively studied in [53].

13.1.2 $\mathcal{N}$ as Domain

$\mathcal{N}$ is isomorphic to a $\lambda\Lambda$-model, which arises from an inverse limit solution of the same domain equation as $\mathcal{H}$, i.e.

$$\mathbb{X} = [\mathbb{X} \to \mathbb{X}].$$

Take as initial domain $(\mathbb{N}, \sqsubseteq_{\mathbb{N}})$, where $\mathbb{N} = \{\bot_{\mathbb{N}}, \varkappa, \top_{\mathbb{N}}\}$, and $\sqsubseteq_{\mathbb{N}}$ is defined as $\bot_{\mathbb{N}} \sqsubseteq_{\mathbb{N}} b$, for all $b \in \mathbb{N}$ and $\varkappa \sqsubseteq_{\mathbb{N}} \top_{\mathbb{N}}$. Take the domain $([\mathbb{N} \to \mathbb{N}], \sqsubseteq_{\mathbb{N}\to\mathbb{N}})$, which is a domain by Lemma 13.1.4.(i), and choose, as embedding function between $(\mathbb{N}, \sqsubseteq_{\mathbb{N}})$ and $([\mathbb{N} \to \mathbb{N}], \sqsubseteq_{\mathbb{N}\to\mathbb{N}})$, the function i so defined:

- $i(\bot_{\mathbb{N}}) = \bot_{[\mathbb{N}\to\mathbb{N}]}$,
- $i(\varkappa) = s_{\top_{\mathbb{N}},\varkappa}$,
- $i(\top_{\mathbb{N}}) = s_{\varkappa,\top_{\mathbb{N}}}$,

and let $\mathbb{N}_\infty$ be the inverse limit solution so obtained.
The correspondence between $C_{\bowtie}$ and $\mathbf{comp}(\mathbb{N})$ is defined in the following way:

- $(\omega)^+ = \bot_{\mathbb{N}}$,
- $(\psi)^+ = \varkappa$,
- $(\phi)^+ = \top_{\mathbb{N}}$.

Let the function $(.)^* : T(C_{\bowtie}) \to \mathbf{comp}(\mathbb{N}_\infty)$ be the function made according to the procedure **compact-as-types**.

The procedure **compact-as-types**($\mathbb{N}_\infty$) generates, by point 2, the rule $(n0)$ of Fig. 11.3 (pag. 145). Furthermore, point (4.1) generates the rules $(n1), (n2), (n3), (n4)$ and $(n5)$ of Fig. 11.3 of the intersection relation $\leq_{\bowtie}$. Note that point (4.2) is not applied in this case.

Let the interpretation function $[\![.]\!]^{\mathbb{N}_\infty}$ and the composition $\circ_{\mathbb{N}_\infty}$ be defined as for the the model $\mathbb{I}_\infty$.

Theorem 13.1.16. *Let* $\mathcal{J} = <\mathbb{N}_\infty, \mathbb{N}_\infty, \circ_{\mathbb{N}_\infty}, [\![.]\!]^{\mathbb{N}_\infty}>$ *and* $\circ_{\mathbb{N}_\infty} \equiv \lambda\!\!\lambda xy.I(x)(y)$. $\mathcal{J}$ *is a* $\lambda\Lambda$*-calculus model, and it is isomorphic to* $\mathcal{N}$.

Proof. Similar to the proof of Theorem 13.1.15. □

The model $\mathcal{J}$ was first presented and studied in [30], as filter model. There the approximation theorem was proved using the computability technique we also used here. The notion of Λ-persistent normal form, on which the construction of the model is based, was first introduced in [17].

13.1.3 $\mathcal{L}$ as Domain

Take the following domain equation:

$$\mathbb{X} = [\mathbb{X} \to \mathbb{X}]_\perp.$$

By the presence of the lifting domain constructor, this equation admits a minimal solution. In fact, take as initial domain $(\mathbb{E}, \sqsubseteq_{\mathbb{E}})$, where $\mathbb{E} = \{\perp_{\mathbb{E}}\}$, and $\sqsubseteq_{\mathbb{E}}$ is the identity relation. Take the domain $([\mathbb{E} \to \mathbb{E}]_\perp, \sqsubseteq_{[\mathbb{E}\to\mathbb{E}]_\perp})$, which is a domain by Lemma 13.1.4.(i) and Property 13.1.6, and choose, as embedding function between $(\mathbb{E}, \sqsubseteq_{\mathbb{E}})$ and $([\mathbb{E} \to \mathbb{E}], \sqsubseteq_{[\mathbb{E}\to\mathbb{E}]})$, the function i so defined:

- $i(\perp_{\mathbb{E}}) = \perp_{[\mathbb{E}\to\mathbb{E}]_\perp}$,

and let $\mathbb{E}_\infty$ be the inverse limit solution so obtained.
The correspondence between $C_\angle$ and $\mathbf{comp}(\mathbb{E})$ is defined in the following way:

- $(\omega)^+ = \perp_{\mathbb{E}}$.

Let the function $(.)^* : T(C_{\bowtie}) \to \mathbf{comp}(\mathbb{E}_\infty)$ be the function made according to the procedure **compact-as-types**.

The procedure **compact-as-types**($\mathbb{E}_\infty$) generates, at point (4.1), the trivial rule $\omega \leq \omega$ of the intersection relation $\leq_\angle$. Note that point (4.2) is not applied in this case.

Let the interpretation function $[\![.]\!]^{\mathbb{E}_\infty}$ and the composition $\circ_{\mathbb{E}_\infty}$ be defined as follows:

- $[\![x]\!]^{\mathbb{E}_\infty}_\rho = \rho(x)$,

- $[\![MN]\!]_\rho^{\mathbb{E}_\infty} = J'(I([\![M]\!]_\rho^{\mathbb{E}_\infty}))([\![N]\!]_\rho^{\mathbb{E}_\infty})$,
- $[\![\lambda x.M]\!]_\rho^{\mathbb{E}_\infty} = J(I'(\boldsymbol{\lambda} d.[\![M]\!]_{\rho[d/x]}^{\mathbb{E}_\infty}))$,

where I and J are defined in the proof of Theorem 13.1.11, I' and J' are the isomorphism pair between $[\mathbb{E}_\infty \to \mathbb{E}_\infty]$ and $[\mathbb{E}_\infty \to \mathbb{E}_\infty]_\perp$.

Theorem 13.1.17. *Let* $\mathcal{E} =< \mathbb{E}_\infty, \mathbb{E}_\infty, \circ_{\mathbb{E}_\infty}, [\![.]\!]^{\mathbb{E}_\infty} >$ *where* $\circ_{\mathbb{E}_\infty}$ *is defined as* $\boldsymbol{\lambda} xy.I(x)(y)$.
$\mathcal{E}$ *is a* $\lambda\Lambda$*-calculus model, and it is isomorphic to* $\mathcal{L}$.

Proof. Similar to the proof of Theorem 13.1.15, taking into account the different definition of interpretation. □

The model $\mathcal{E}$ was first presented and studied in [2].

13.1.4 $\mathcal{V}$ as Domain

Take the following domain equation:

$$\mathbb{X} = [\mathbb{X} \to_\perp \mathbb{X}]_\perp.$$

This equation, like that one showed in the previous subsection, also admits a minimal solution. In fact, take as initial domain $(\mathbb{U}, \sqsubseteq_\mathbb{U})$, where $\mathbb{U} = \{\perp_\mathbb{U}\}$, and $\sqsubseteq_\mathbb{U}$ is the identity relation. Take the domain $([\mathbb{U} \to_\perp \mathbb{U}]_\perp, \sqsubseteq_{[\mathbb{U}\to_\perp\mathbb{U}]_\perp})$, which is a domain by Lemma 13.1.4.(i) and Property 13.1.6, and choose, as embedding function between $(\mathbb{U}, \sqsubseteq_\mathbb{U})$ and $([\mathbb{U} \to_\perp \mathbb{U}], \sqsubseteq_{[\mathbb{U}\to_\perp\mathbb{U}]})$, the function i so defined:

- $i(\perp_\mathbb{U}) = \perp_{[\mathbb{U}\to_\perp\mathbb{U}]_\perp}$

and let $\mathbb{U}_\infty$ be the inverse limit solution so obtained.

The function $(.)^*$, generated by the procedure **compact-as-types**$(\mathbb{U}_\infty)$, has the following behaviour:

$$(\sigma \to \tau)^* = s_{(\sigma)^*,(\tau)^*} \text{ if either } (\sigma)^*, (\tau)^* = \perp_{\mathbb{U}_\infty} \text{ or } (\sigma)^* \neq \perp_{\mathbb{U}_\infty},$$
$$(\sigma \to \tau)^* = s_{(\omega\to\omega)^*,(\tau)^*} \text{ if } (\sigma)^* = \perp_{\mathbb{U}_\infty}.$$

Moreover, the intersection relation built by the procedure generates, at point (4.1), the trivial rule $\omega \leq \omega$, while at point (4.2) rule (v) of Fig. 12.1 (pag. 182) is generated.

Let $\rho : \text{Var} \to \mathbb{U}_\infty/\perp_{\mathbb{U}_\infty}$; the interpretation function $[\![.]\!]^{\mathbb{U}_\infty}$ (see [44]) is:

- $[\![x]\!]_\rho^{\mathbb{U}_\infty} = \rho(x)$,
- $[\![MN]\!]_\rho^{\mathbb{U}_\infty} = J'(I([\![M]\!]_\rho^{\mathbb{U}_\infty}))([\![N]\!]_\rho^{\mathbb{U}_\infty})$,
- $[\![\lambda x.M]\!]_\rho^{\mathbb{U}_\infty} = J(I'(\mathsf{strict}(\boldsymbol{\lambda} d.[\![M]\!]_{\rho[d/x]}^{\mathbb{U}_\infty})))$,

where I and J are defined in the proof of Theorem 13.1.11, I' and J' are the isomorphism pair between $[\mathbb{U}_\infty \rightarrow_\perp \mathbb{U}_\infty]$ and $[\mathbb{U}_\infty \rightarrow_\perp \mathbb{U}_\infty]_\perp$, and strict is a function such that

$$\mathsf{strict}(f)(x) = \begin{cases} \perp_{\mathbb{U}_\infty} & x = \perp_{\mathbb{U}_\infty}, \\ f(x) & \text{otherwise.} \end{cases}$$

Theorem 13.1.18. *Let* $\mathcal{U} =< \mathbb{U}_\infty, \mathbb{U}_\infty/\perp_{\mathbb{U}_\infty}, \circ_{\mathbb{U}_\infty}, [\![.]\!]^{\mathbb{U}_\infty} >$ *where* $\circ_{\mathbb{U}_\infty}$ *is defined as* $\boldsymbol{\lambda} xy.I(x)(y)$.
$\mathcal{U}$ *is a* $\lambda\Gamma$*-calculus model, and it is isomorphic to* $\mathcal{V}$.

Proof. Similar to the proof of Theorem 13.1.15, taking into account the fact that only the strict functions are present in the domain.

By Theorem 13.1.14, $(\mathbb{U}_\infty, \sqsubseteq_{\mathbb{U}_\infty})$ is isomorphic to $\mathcal{F}(\surd)$, ordered by set inclusion. Then the proof follows from the definition of $[\![.]\!]^{\mathbb{U}_\infty}$. □

The model $\mathcal{V}$ was first presented and studied in [44], both as an inverse limit solution of the previous domain equation and as a filter model. There the approximation model was proved using the indexed reduction technique.

13.1.5 Another Domain

Every solution of the domain equation

$$\mathbb{X} = [\mathbb{X} \rightarrow_\perp \mathbb{X}]$$

is a model for the Λ-NF^0-calculus.

This fact was first noticed in [39]. We did not develop the study of such a calculus, since it does not seem to have interesting operational properties.

14. Further Reading

Other filters $\lambda\Lambda$-models. In [37] two filter $\lambda\Lambda$-models are designed which completely characterise sets of terms with similar computational behaviours. Moreover, in [4, 40] filter $\lambda\Lambda$-models characterizing the easiness property of terms are proposed. Shortly, a term is easy when it can be consistently equated to every other term.

Other classes of $\lambda\Lambda$-models. Berry [14] proposed a different class of domains based on the notion of stable functions. Starting from this notion, Girard [47] proposed qualitative domains as $\lambda\Lambda$-models. Later qualitative domains were that were later refined in the coherence domains. The first denotational semantics of linear logics is based on this kind of domains[48]. The definition of intersection relation can be modified in order to describe this class of models using intersection types, as was proved in [52]. The notions of strongly stable functions and hypercoherence spaces, on which another class of $\lambda\Lambda$-models is based, were introduced in [23]. Models based on the notion of bidomain, which is a space endowed with two notions of order (continuous order and stable order), were introduced in [99]. In this setting, a model correct with respect to the **L**-operational semantics was constructed in [61]. A quite complete presentation of the "webbed" $\lambda\Lambda$-models, i.e. those whose domains are subdomains of some $(\mathbf{P}(D), \subseteq)$, can be found in [13]; clearly all the $\lambda\Lambda$-models presented in this book belong to this class. $\lambda\Lambda$-models based on game sematics were presented in [43]. Categorical presentations of $\lambda\Lambda$-models in a typed setting can be found [8, 31, 62, 59, 90].

Incompleteness. The first incompleteness result for the $\lambda\Lambda$-calculus semantics was proved in [53], where a λ-theory was shown for which there do not exist a correct and complete model in the class of Scott's models built by an inverse limit construction. Further investigations on this topic, using topological tools, was made in [86]. Incompleteness results for the class of $\lambda\Lambda$-models based on stable functions was proved in [11].

Lazy semantics. A general characterization of models that are correct with respect to the **L**-operational semantics was given in [12].

$\lambda\Gamma$-Models. A general characterization of models that are correct with respect to the **V**-operational semantics was given in [82].

Semiseparability. An extension of the semiseparability algorithm to a finite set of approximants was introduced in [29].

Part IV

Computational Power

15. Preliminaries

In the Introduction we claimed that both the $\lambda\Lambda$-calculus and the $\lambda\Gamma$-calculus can be seen as paradigms for programming languages in the call-by-name and call-by-value settings respectively. In this chapter this claim will be justified. In fact, we will show that both the call-by-name and the call-by-value λ-calculi have the computational power of Turing machines, or equivalently, they are *computationally complete.* The completeness can be achieved without adding special constants to the language, but all data structures needed for computing, in particular booleans, natural numbers and functions, can be coded into Λ.

We will show how to code useful data structures. Moreover, we will prove that all the call-by-name and call-by-value reduction machines presented in Part II of this book can be effectively used for computing. In fact, computational completeness can be achieved by using each one of them.

15.1 Kleene's Recursive Functions

It is well known that not all the partial functions from natural numbers to natural numbers can be effectively computed. The most famous definition of the class of computable functions was given by Turing, by using the *Turing machines.* But Kleene's definition of partial recursive functions [56] makes it easier to define the coding of functions in a $\lambda\Delta$-calculus [57].

The class of computable functions, or partial recursive functions, is given in two stages. First, the class of *primitive recursive* functions is defined. They are generated from a set of initial functions by closure under particular constructions (composition and primitive recursion). The primitive recursive functions include most functions ever encountered in practical mathematics and computer science. However, all primitive recursive functions are total, and hence the class must necessarily fall short of the full class of computable functions.

The second stage of Kleene's characterization extends the class of primitive recursive functions by adding an additional operator of minimalization that introduces unbounded and possibly nonterminating searches. Therefore the class of all partial recursive functions is obtained.

Definition 15.1.1 (Primitive recursive functions).

(i) *The following functions are* primitive recursive functions*:*
   1. *The function* $Z : \mathbb{N} \to \mathbb{N}$ *such that* $Z(n) = 0$*;*
   2. *The* successor $S : \mathbb{N} \to \mathbb{N}$ *such that* $S(n) = n + 1$*;*
   3. *The* projection *functions* $\pi_i^m(x_1, ..., x_m) = x_i \ \ (1 \leq i \leq m \in \mathbb{N})$.

(ii) *If* $h : \mathbb{N}^n \to \mathbb{N}$ *and* $g_1, ..., g_n : \mathbb{N}^m \to \mathbb{N}$ *are primitive recursive functions then the function* f*, defined as their* composition *in the following way:*

$$f(x_1, ..., x_m) = h(g_1(x_1, ..., x_m), ..., g_n(x_1, ..., x_m)),$$

is primitive recursive too $(n, m \in \mathbb{N})$.

(iii) *If* $h : \mathbb{N}^{m+2} \to \mathbb{N}$ *and* $g : \mathbb{N}^m \to \mathbb{N}$ *are primitive recursive functions then* f *defined by* primitive recursion *in the following way:*

$$f(k, x_1, ..., x_m) = \begin{cases} g(x_1, ..., x_m) & \textit{if } k = 0, \\ h(f(k-1, x_1, ..., x_m), k-1, x_1, ..., x_m) & \textit{otherwise.} \end{cases}$$

is primitive recursive too $(m \in \mathbb{N})$.

By induction on the depth of nested instances of composition and primitive recursion, it is easy to check that each primitive recursive function is total. The next definition allows the construction of partial functions.

Definition 15.1.2. *Let* $h : \mathbb{N}^2 \to \mathbb{N}$ *be a total function, and let* $x \in \mathbb{N}$*. Then a function* $f : \mathbb{N} \to \mathbb{N}$ *can be defined by* minimalization *from* h *in the following way:*

$$f(x) = \mu y\ [h(x, y) = 0] = \begin{cases} \min\{k \in \mathbb{N} \mid h(x, k) = 0\} & \textit{if such a } k \in \mathbb{N} \textit{ exists,} \\ \textit{undefined} & \textit{otherwise.} \end{cases}$$

Note that, in the previous definition, the function h is defined on all natural numbers, by hypothesis. Now, the full class of partial recursive functions can be defined as follows.

Definition 15.1.3 (Partial recursive functions).
A function $f : \mathbb{N}^m \to \mathbb{N}$ $(m \in \mathbb{N})$ *is* partial recursive *if and only if one of the following conditions holds:*

(i) f *is a primitive recursive function;*
(ii) f *is defined by composition of partial recursive functions;*
(iii) f *is defined by minimalization starting from a total recursive function.*

It is important to notice that, in the previous definition, the minimalization construction must be applied only to *total* recursive functions, which form a nonrecursive class. By Church's thesis, the class of partial recursive functions coincides with the whole class of computable functions.

15.2 Representing Data Structures

The starting point for transforming a $\lambda\Delta$-calculus in a programming language is to code some fundamental data structures in it, namely booleans and natural numbers.

Let us study the problem of representing booleans in a $\lambda\Delta$-calculus whose operational behaviour is described by an evaluation relation $\mathbf{O} \in \mathcal{E}(\Delta, \Theta)$. In order to represent the truth values *True* and *False*, we need to define two terms having a suitable behaviour; in particular, they must be the basis for the definition of a further term having the behaviour of a conditional operator.

Definition 15.2.1. *Let $\mathbf{O} \in \mathcal{E}(\Delta, \Theta)$ be a evaluation relation. An $\mathbf{O}$-representation of booleans is any set $\{T, F\}$ such that:*

(i) $T, F \in \Delta \cap \Theta$;
(ii) *there is a term Cond such that, for every $M, N \in \Delta \cap \Theta$:*

$$Cond\, TMN \Downarrow_{\mathbf{O}} M; \qquad\qquad Cond\, FMN \Downarrow_{\mathbf{O}} N.$$

The next lemma shows that in order to represent the booleans, the choice of taking two $\Lambda\eta$-different normal forms is correct in each one of the reduction machines we defined.

Lemma 15.2.2. *Let $\mathbf{O} \in \{\mathbf{H}, \mathbf{N}, \mathbf{L}, \mathbf{V}\}$, and let $M, N \in \Lambda^0$ be two different $\Lambda\eta$-normal forms. $\{M, N\}$ is an $\mathbf{O}$-representation of booleans.*

Proof. If $\mathbf{O} \in \{\mathbf{H}, \mathbf{N}, \mathbf{L}\}$ then let $C[.]$ be such that $M, N \Rrightarrow_{\Lambda} C[.]$ (see Fig. 2.1 pag. 32), otherwise let $C[.]$ be such that $M, N \Rrightarrow_{\Gamma} C[.]$ (see Fig. 3.1 pag. 52). Then $C[M] \Downarrow_{\mathbf{O}} x$ and $C[M] \Downarrow_{\mathbf{O}} y$, for two different variables x and y. The term $Cond \equiv \lambda uxy.C[u]$ plays the desired role in all the reduction machines under consideration. □

It is a standard choice to define $T \equiv \lambda xy.x$ and $F \equiv \lambda xy.y$. In this case *Cond* can be taken as the identity term I, or simply omitted. In fact, if $M, N \in \Lambda$-NF then $TMN \Downarrow_{\mathbf{O}} M$ and $FMN \Downarrow_{\mathbf{O}} N$, for all $\mathbf{O} \in \{\mathbf{H}, \mathbf{N}, \mathbf{L}, \mathbf{V}\}$. A boolean expression is every term B, such that $B \Downarrow_{\mathbf{O}}$ implies either $B \Downarrow_{\mathbf{O}} T$ or $B \Downarrow_{\mathbf{O}} F$.

Through the coding of the booleans it is possible to code more complex data structures, for example, the pairs. Let M, N be two Λ-normal forms; the pair $[M, N]$ can be coded as $\lambda x.xMN$. So projections can be built using booleans, by defining $\lambda x.xT$ and $\lambda x.xF$ as respectively the first and the second projections. In fact, $(\lambda x.xT)[M, N] \Downarrow_{\mathbf{O}} M$ and $(\lambda x.xF)[M, N] \Downarrow_{\mathbf{O}} N$. We will denote $\lambda x.xMN$ by $[M, N]$.

The coding of the natural numbers can be based on Peano's axioms, recalled in the following definition.

Definition 15.2.3 (Peano's natural numbers).

1. *There is a natural number, called zero.*
2. *Given a natural number n, there is a unique natural number m that is its successor.*
3. *Two different natural numbers have different successors.*
4. *If n is a natural number then its successor is different from zero.*
5. *If A is a subset of natural numbers satisfying:*
   - *zero belongs to A,*
   - *if n belongs to A then it successor belongs to A too,*

   then A is the set of all natural numbers.

The notion of an **O**-numeral system, given in the next definition, gives the conditions for building the coding of natural numbers with respect to an evaluation relation **O**. Following the lines of Peano's axioms, the infinite set of natural numbers can be generated by two suitable terms playing the role of zero and successor. The other conditions assure that all the terms generated by iterating the application of successor to zero are different in the operational setting we are considering.

Definition 15.2.4. *Let $\mathbf{O} \in \mathcal{E}(\Delta, \Theta)$ be a deterministic evaluation relation. An* **O**-numeral system *is a 5-tuple $\langle \mathbb{B}, Zero, Succ, Test, Pred \rangle$, where:*

(i) *$\mathbb{B}$ is an **O**-representation of booleans.*

(ii) *$Zero, Succ, Test, Pred \in \Delta \cap \Theta$ are such that, for all $n \in \mathbb{N}$:*

1. $\underbrace{Succ\ (\ ...\ (Succ}_{n}\ Zero)...) \Downarrow_{\mathbf{O}}$.

   Moreover, if $\underbrace{Succ\ (\ ...\ (Succ}_{n}\ Zero)...) \Downarrow_{\mathbf{O}} \ulcorner n \urcorner$ *then* $\ulcorner n \urcorner \in \Delta \cap \Theta$*;*

   we will say that $\ulcorner n \urcorner$ *is the* numeral representation *of* n.
2. $P \Downarrow_{\mathbf{O}} \ulcorner n \urcorner$ *implies* $Succ\ P \Downarrow_{\mathbf{O}} \ulcorner n+1 \urcorner$.
3. $P \Downarrow_{\mathbf{O}} Zero$ *implies* $Test\ P \Downarrow_{\mathbf{O}} T$.
4. $Q \Downarrow_{\mathbf{O}} \ulcorner n+1 \urcorner$ *implies* $Test\ Q \Downarrow_{\mathbf{O}} F$.
5. $P \Downarrow_{\mathbf{O}} \ulcorner n+1 \urcorner$ *implies* $Pred\ P \Downarrow_{\mathbf{O}} \ulcorner n \urcorner$.

This definition is well posed. According to it, the number $n \in \mathbb{N}$ is represented by the numerals $\ulcorner n \urcorner$.

Property 15.2.5. Definition 15.2.4 respects the Peano constraints.

Proof. All points of Peano's definition are satisfied.

1. Immediate, by the definition.
2. Immediate, by the definition and since **O** is deterministic.

3. Let $\ulcorner n \urcorner \not\approx_O \ulcorner m \urcorner$. Assume, by absurd, $Succ \ulcorner m \urcorner \approx_O Succ \ulcorner n \urcorner$. Then, by the contextual closure of $\approx_O$ and point (ii).5 of Definition 15.2.4, $\ulcorner m \urcorner \approx_O Pred(Succ \ulcorner m \urcorner) \approx_O Pred(Succ \ulcorner n \urcorner) \approx_O \ulcorner n \urcorner$, against the hypothesis.
4. Trivial, by using $Test$.
5. $\left\{ \ulcorner n \urcorner \in \Lambda \;\middle|\; \underbrace{Succ\,(\ldots\ldots (Succ}_{n}\, Zero) \ldots) \Downarrow_{\mathbf{O}} \ulcorner n \urcorner \text{ for some } n \in \mathbb{N} \right\}$ is the set of the numerals.

□

In the next definition a numeral system is presented that plays the desired role in all the operational semantics we have studied.

Definition 15.2.6. *Let* $\mathfrak{N} \equiv \langle \{\mathsf{T}, \mathsf{F}\}, \mathsf{Zero}, \mathsf{Succ}, \mathsf{Test}, \mathsf{Pred} \rangle$, *where*

- $\mathsf{T} \equiv \lambda xy.x$ *and* $\mathsf{F} \equiv \lambda xy.y$;
- $\mathsf{Zero} \equiv [\mathsf{T}, \mathsf{T}]$;
- $\mathsf{Succ} \equiv \lambda t.t(\lambda uvx.x\mathsf{F}(\lambda y.yuv))$;
- $\mathsf{Test} \equiv \lambda x.x\mathsf{T}$;
- $\mathsf{Pred} \equiv \lambda x.x\mathsf{F}$.

We will check that $\mathfrak{N}$ is a numeral system in the sense of Definition 15.2.4, for all $\mathbf{O} \in \{\mathbf{H}, \mathbf{N}, \mathbf{L}, \mathbf{V}\}$. It is easy to see that $\forall n \in \mathbb{N}$, the numeral $\ulcorner n \urcorner$ in $\mathfrak{N}$ is the same term for all $\mathbf{O}$-numeral system; in particular, $\ulcorner n \urcorner \equiv \underbrace{[\mathsf{F}, [\mathsf{F} \ldots . [\mathsf{F}}_{n}, \mathsf{Zero}] \ldots]]$ and $\ulcorner n+1 \urcorner \equiv \lambda x.x\mathsf{F} \ulcorner n \urcorner$.

Note that basic elements of $\mathfrak{N}$ are Λ-normal forms, so they are both input and output values for all the machines. The proof can be done in the same manner for all the call-by-name reduction machines.

In all the formal systems presented in order to induce evaluation relations $\mathbf{H}, \mathbf{N}, \mathbf{L}, \mathbf{V}$, there is a rule named (*head*). It is easy to see that, for each one of the given operational machines, (*head*) is reversible in the sense that in all considered cases when the conclusion is derivable then its premises are derivable. In order to simplify the proofs, in the call-by-name setting, i.e. when $\mathbf{O} \in \{\mathbf{H}, \mathbf{N}, \mathbf{L}\}$, we will denote by $(head)^+$ a sequence of $n \geq 1$ applications of rule (*head*) in a derivation.

We need some properties of the operational semantics.

Property 15.2.7. Let $M, N, P, Q \in \Lambda^0$.

(i) Let $\mathbf{O} \in \{\mathbf{H}, \mathbf{N}, \mathbf{L}, \mathbf{V}\}$. $M \Downarrow_{\mathbf{O}} N$ and $NP \Downarrow_{\mathbf{O}} Q$ if and only if $MP \Downarrow_{\mathbf{O}} Q$.
(ii) If $M \Uparrow_{\mathbf{H}}$, $M \Uparrow_{\mathbf{N}}$, $M \Uparrow_{\mathbf{L}}$ and $M \Uparrow_{\mathbf{V}}$ then $MN \Uparrow_{\mathbf{O}}$, for all $N \in \Lambda$ and $\mathbf{O} \in \{\mathbf{H}, \mathbf{N}, \mathbf{L}, \mathbf{V}\}$.

Proof. (i) In case $\mathbf{O} \in \{\mathbf{H}, \mathbf{N}, \mathbf{L}\}$, the proof follows respectively by Properties 6.1.4, 6.2.4 and 6.3.4. Let $\mathbf{O} \equiv \mathbf{V}$. By the confluence theorem we can assume $MP \Downarrow_{\mathbf{V}} R$ if and only if $NP \Downarrow_{\mathbf{V}} Q$. We show that $R \equiv Q$ by induction on the last applied rule in $M \Downarrow_{\mathbf{V}} N$. Rules (*var*) and (*block*) are not possible, since $M \in \Lambda^0$; while (*abs*) and (*head*) are trivial.

(ii) Since $M \Uparrow_{\mathbf{O}}$, for all $\mathbf{O}$, implies that M is both a Λ and a Γ-unsolvable of order 0, then MN is an unsolvable of order 0 too. □

Note that point (i) of the previous property is just a consequence of the fact that every reduction machine reduces at every step the head redex.

Theorem 15.2.8. *If* $\mathbf{O} \in \{\mathbf{H}, \mathbf{N}, \mathbf{L}\}$ *then* $\mathfrak{N}$ *is an* $\mathbf{O}$*-numeral system.*

Proof. We prove that $\mathfrak{N}$ satisfies all the conditions given in Definition 15.2.4.

(i) $\{\mathsf{T}, \mathsf{F}\}$ is an $\mathbf{O}$-representation of booleans, by Lemma 15.2.2 and since in all cases $\approx_{\mathbf{O}}$ is a Λ-theory.

(ii) $\mathsf{Zero}, \mathsf{Succ}, \mathsf{Test}, \mathsf{Pred}$ are both input and output values.

1. $\mathsf{Zero} \Downarrow_{\mathbf{O}} \ulcorner 0 \urcorner$, where $\ulcorner 0 \urcorner \equiv \mathsf{Zero}$. By induction on n we will prove that the numeral $\ulcorner n+1 \urcorner$ is the term $\lambda x.x\mathsf{F}\ulcorner n \urcorner \in \Delta \cap \Theta$.
   Let $\underbrace{\mathsf{Succ}(...(\mathsf{Succ}\,\mathsf{Zero})...)}_{n} \Downarrow_{\mathbf{O}} \ulcorner n \urcorner$; thus
   $$\underbrace{\mathsf{Succ}(\mathsf{Succ}(...(\mathsf{Succ}\,\mathsf{Zero})...))}_{n+1} \Downarrow_{\mathbf{O}} R,$$
   if and only if, since $\mathsf{Succ} \equiv \lambda t.t(\lambda uvx.x\mathsf{F}(\lambda y.yuv))$,
   $$\underbrace{\mathsf{Succ}(...(\mathsf{Succ}\,\mathsf{Zero})...)}_{n}(\lambda uvx.x\mathsf{F}(\lambda y.yuv)) \Downarrow_{\mathbf{O}} R$$
   by (*head*), if and only if $\ulcorner n \urcorner(\lambda uvx.x\mathsf{F}(\lambda y.yuv)) \Downarrow_{\mathbf{O}} R$, by Property 15.2.7.(i).
   If $n = 0$, then this happens only if $(\lambda uvx.x\mathsf{F}(\lambda y.yuv))\mathsf{TT} \Downarrow_{\mathbf{O}} R$, by (*head*), so $R \equiv \lambda x.x\mathsf{F}(\lambda y.y\mathsf{TT}) \equiv \ulcorner 1 \urcorner$.
   Otherwise, $\ulcorner n \urcorner(\lambda uvx.x\mathsf{F}(\lambda y.yuv)) \Downarrow_{\mathbf{O}} R$ if and only if
   $$(\lambda uvx.x\mathsf{F}(\lambda y.yuv))\mathsf{F}\ulcorner n-1 \urcorner \Downarrow_{\mathbf{O}} R$$
   by (*head*), if and only if $\lambda x.x\mathsf{F}(\lambda y.y\mathsf{F}\ulcorner n-1 \urcorner) \Downarrow_{\mathbf{O}} R$ by $(\textit{head})^+$. But $\lambda x.x\mathsf{F}(\lambda y.y\mathsf{F}\ulcorner n-1 \urcorner) \in \Lambda$-NF, so $R \equiv \lambda x.x\mathsf{F}(\lambda y.y\mathsf{F}\ulcorner n-1 \urcorner) \equiv \ulcorner n+1 \urcorner$.
2. If $P \in \Lambda$ is such that $P \Downarrow_{\mathbf{O}} \ulcorner n \urcorner$ then $\mathsf{Succ}\,\ulcorner n \urcorner \Downarrow_{\mathbf{O}} \ulcorner n+1 \urcorner$, reasoning as in the previous point.
3. Let $P \Downarrow_{\mathbf{O}} \mathsf{Zero}$, so $\mathsf{Test} \equiv \lambda x.x\mathsf{T}$ implies $\mathsf{Test}\,P \Downarrow_{\mathbf{O}} R$ if and only if $P\mathsf{T} \Downarrow_{\mathbf{O}} R$ by (*head*), if and only if $\mathsf{Zero}\mathsf{T} \Downarrow_{\mathbf{O}} R$ by Property 15.2.7.(i), if and only if $\mathsf{T} \Downarrow_{\mathbf{O}} R$ (by $(\textit{head})^+$ again); thus $R \equiv \mathsf{T}$.
4. The case $P \Downarrow_{\mathbf{O}} \ulcorner n+1 \urcorner$ is similar to the previous point.
5. Let $P \Downarrow_{\mathbf{O}} \ulcorner n+1 \urcorner$. Then $\mathsf{Pred}\,P \Downarrow_{\mathbf{O}} R$ if and only if $P\mathsf{F} \Downarrow_{\mathbf{O}} R$ by (*head*), if and only if $\ulcorner n+1 \urcorner\mathsf{F} \Downarrow_{\mathbf{O}} R$ (by Property 15.2.7.(i)) if and only if $\ulcorner n \urcorner \Downarrow_{\mathbf{O}} R$ (by $(\textit{head})^+$). So it must be $R \equiv \ulcorner n \urcorner$. □

Now we check that $\mathfrak{N}$ is also a numeral system in a call-by value setting.

Theorem 15.2.9. *$\mathfrak{N}$ is a* **V***-numeral system.*

Proof. (i) $\{\mathsf{T}, \mathsf{F}\}$ is an **V**-representation of booleans, by Lemma 15.2.2.
(ii) $\mathsf{Zero}, \mathsf{Succ}, \mathsf{Test}, \mathsf{Pred} \in \Gamma \cap \Gamma$-LBNF.

$\mathsf{Zero} \Downarrow_{\mathbf{V}} \ulcorner 0 \urcorner$, where $\ulcorner 0 \urcorner \equiv \mathsf{Zero}$. By induction on n we will prove that the numeral $\ulcorner n+1 \urcorner$ is the term $\lambda x.x\mathsf{F}\ulcorner n \urcorner \in \Gamma \cap \Gamma$-LBNF.
Let us assume that $\underbrace{\mathsf{Succ}(...(\mathsf{Succ}}_{n}\,\mathsf{Zero})...) \Downarrow_{\mathbf{V}} \ulcorner n \urcorner$, by induction; therefore $\underbrace{\mathsf{Succ}(\mathsf{Succ}(...(\mathsf{Succ}}_{n+1}\,\mathsf{Zero})...)) \Downarrow_{\mathbf{V}} R$ if and only if $\mathsf{Succ}\ulcorner n \urcorner \Downarrow_{\mathbf{V}} R$ by induction and (*head*), if and only if $\ulcorner n \urcorner(\lambda uvx.x\mathsf{F}(\lambda y.yuv)) \Downarrow_{\mathbf{V}} R$ by rules (*head*) and (*lazy*), since $\mathsf{Succ} \equiv \lambda t.t(\lambda uvx.x\mathsf{F}(\lambda y.yuv))$.
If $n = 0$, then this happens only if $(\lambda uvx.x\mathsf{F}(\lambda y.yuv))\mathsf{TT} \Downarrow_{\mathbf{V}} R$ by rules (*head*) and (*lazy*), so it is easy to see that $R \equiv \lambda x.x\mathsf{F}(\lambda y.y\mathsf{TT}) \equiv \ulcorner 1 \urcorner$.
Otherwise, $\ulcorner n \urcorner(\lambda uvx.x\mathsf{F}(\lambda y.yuv)) \Downarrow_{\mathbf{V}} R$ if and only if

$$(\lambda uvx.x\mathsf{F}(\lambda y.yuv))\mathsf{F}\ulcorner n-1 \urcorner \Downarrow_{\mathbf{V}} R,$$

if and only if $\lambda x.x\mathsf{F}(\lambda y.y\mathsf{F}\ulcorner n-1 \urcorner) \Downarrow_{\mathbf{V}} R$. But $\lambda x.x\mathsf{F}(\lambda y.y\mathsf{F}\ulcorner n-1 \urcorner) \in \Lambda$-NF, so $R \equiv \lambda x.x\mathsf{F}(\lambda y.y\mathsf{F}\ulcorner n-1 \urcorner) \equiv \ulcorner n+1 \urcorner$.

It is easy to check the remaining constraints given in Definition 15.2.4, since the proof follows the same lines as the Theorem 15.2.8. □

16. Representing Functions

In order to represent a numerical function with respect to an evaluation relation $\mathbf{O}$, it is necessary to exhibit a term mimicking the behaviour of the function itself. More precisely the reduction machine, taken as input this term applied to a sequence of terms representing natural numbers, gives as output the term representing the result, if it exists, and does not stop otherwise.

The term representing the function ϕ will be denoted by $\ulcorner\phi\urcorner$, extending the same notation used for natural numbers. The notion is defined in a formal way in the next definition.

Definition 16.0.10. *Let $\mathbf{O} \in \mathcal{E}(\Delta, \Theta)$ be an evaluation relation, and let ϕ be a partial recursive function with arity $p \in \mathbb{N}$; let $\ulcorner n\urcorner$ be the numeral representation of $n \in \mathbb{N}$ in an $\mathbf{O}$-numeral system.*
ϕ is $\mathbf{O}$-representable *if and only if there is a term $\ulcorner\phi\urcorner \in \Lambda^0$ such that, for all terms N_i such that $N_i \Downarrow_{\mathbf{O}} \ulcorner n_i\urcorner$ $(1 \leq i \leq p; n_1, ..., n_p \in \mathbb{N})$:*

- *if $\phi(n_1, ..., n_p)$ is defined then $\ulcorner\phi\urcorner N_1...N_p \Downarrow_{\mathbf{O}} \ulcorner\phi(n_1, ..., n_p)\urcorner$;*
- *if $\phi(n_1, ..., n_p)$ is undefined then $\ulcorner\phi\urcorner N_1...N_p \Uparrow_{\mathbf{O}}$.*

We will prove that in all given reduction machines all partial recursive functions are representable. To do so, we need to consider separately the call-by-name cases and the call-by-value one.

16.1 Call-by-Name Computational Completeness

We will prove that each one of the studied call-by-name reduction machines can be used for computing all partial recursive functions. In all this section, $\mathbf{O}$ will range over the set $\{\mathbf{H}, \mathbf{N}, \mathbf{L}\}$, i.e. it denotes each one of the call-by-name evaluation relations, and $\mathfrak{N}$ is the $\mathbf{O}$-numeral system given in Definition 15.2.8.

As a first step, it must be checked that primitive recursive functions are $\mathbf{O}$-representable.

Lemma 16.1.1. (i) *Z is $\mathbf{O}$-representable,*
(ii) *S is $\mathbf{O}$-representable,*
(iii) *Projections are $\mathbf{O}$-representable.*

Proof. (i) $\ulcorner Z \urcorner \equiv \lambda x.\mathsf{Zero}$,
(ii) $\ulcorner S \urcorner \equiv \mathsf{Succ}$,
(iii) $\ulcorner \pi_m^i \urcorner \equiv \lambda x_1 ... x_m . x_i$ $(1 \leq i \leq m \in \mathbb{N})$. □

Now let us consider the composition between primitive recursive functions.

Lemma 16.1.2. *Let* $h : \mathbb{N}^m \to \mathbb{N}$ *and* $g_1, ..., g_m : \mathbb{N}^p \to \mathbb{N}$ *be* **O***-representable primitive recursive functions; so their composition:*

$$f(x_1, ..., x_p) = h(g_1(x_1, ..., x_p), ..., g_m(x_1, ..., x_p)),$$

is **O***-representable.*

Proof. By hypothesis there are terms $\ulcorner h \urcorner, \ulcorner g_1 \urcorner, ..., \ulcorner g_m \urcorner$ **O**-representing functions $h, g_1, ..., g_m$. Let

$$\ulcorner f \urcorner \equiv \lambda x_1 ... x_p . \ulcorner h \urcorner (\ulcorner g_1 \urcorner x_1 ... x_p) ... (\ulcorner g_m \urcorner x_1 ... x_p).$$

Let $N_i \in \Lambda$ be such that $N_i \Downarrow_{\mathbf{O}} \ulcorner n_i \urcorner$, for some $n_i \in \mathbb{N}$ $(1 \leq i \leq p)$; by hypothesis $\ulcorner g_j \urcorner N_1 ... N_p \Downarrow_{\mathbf{O}} \ulcorner g_j(n_1, ..., n_p) \urcorner$ $(1 \leq j \leq m)$.
Let $R_i \in \Lambda$ be such that $R_i \Downarrow_{\mathbf{O}} \ulcorner n_i \urcorner$, for some $n_i \in \mathbb{N}$ $(1 \leq i \leq m)$; by hypothesis $\ulcorner h \urcorner R_1 ... R_m \Downarrow_{\mathbf{O}} \ulcorner h(n_1, ..., n_m) \urcorner$, so in particular

$$\ulcorner h \urcorner (\ulcorner g_1 \urcorner N_1 ... N_p) ... (\ulcorner g_m \urcorner N_1 ... N_p) \Downarrow_{\mathbf{O}} \ulcorner h(g_1(n_1, ..., n_p), ..., g_m(n_1, ..., n_p)) \urcorner.$$

But $\ulcorner f \urcorner N_1 ... N_p \Downarrow_{\mathbf{O}} R$ if and only if $\ulcorner h \urcorner (\ulcorner g_1 \urcorner N_1 ... N_p) ... (\ulcorner g_m \urcorner N_1 ... N_p) \Downarrow_{\mathbf{O}} R$ (by $(head)^{+}$), so the proof is done since $h, g_1, ..., g_m$ are total. □

In order to represent the functions built by primitive recursion and by minimalization, a "fixed-point operator" is needed, that work well in all the call-by-name reduction machines. We already proved that in the $\lambda\Lambda$-calculus every term has a fixed-point, and we showed, in the proof of Theorem 2.1.8, an operator building it, namely the term Y.

But, while $YM =_{\Lambda} M(YM)$, it does not hold that $YM \to_{\Lambda}^{*} M(YM)$, which is a necessary condition for using it as recursion operator in a call-by-name reduction machine.

So in the next theorem a further fixed-point operator, suitable for our purposes, is defined.

Theorem 16.1.3. *Let* $Y_{\Lambda} \equiv (\lambda xy.y(xxy))(\lambda xy.y(xxy))$*. If* $M \in \Lambda$ *then* $Y_{\Lambda} M \to_{\Lambda}^{*} M(Y_{\Lambda} M)$*; moreover,* $Y_{\Lambda} M \Downarrow_{\mathbf{O}} R$ *if and only if* $M(Y_{\Lambda} M) \Downarrow_{\mathbf{O}} R$*.*

Proof. Trivial. □

The following lemma shows how Y_{Λ} can be used for mimicking primitive recursion.

Lemma 16.1.4. *Let $h : \mathbb{N}^{m+2} \to \mathbb{N}$ and $g : \mathbb{N}^m \to \mathbb{N}$ be* **O**-*representable primitive recursive functions. The following function is* **O**-*representable:*

$$f(k, x_1, ..., x_m) = \begin{cases} g(x_1, ..., x_m) & \textit{if } k = 0, \\ h(f(k-1, x_1, ..., x_m), k-1, x_1, ..., x_m) & \textit{otherwise.} \end{cases}$$

Proof. By hypothesis there are terms $\ulcorner h \urcorner$ and $\ulcorner g \urcorner$ representing h and g. We will prove that $\ulcorner f \urcorner$ is **O**-represented by $Y_\Lambda P$, where P is:

$$\lambda tyx_1...x_m.\mathsf{Test}\ y(\ulcorner g \urcorner x_1...x_m)(\ulcorner h \urcorner(t(\mathsf{Pred}\ y)x_1...x_m)(\mathsf{Pred}\ y)x_1...x_m).$$

Let $N_i \Downarrow_{\mathbf{O}} \ulcorner n_i \urcorner$, $Q \Downarrow_{\mathbf{O}} \ulcorner k \urcorner$ for some $k, n_i \in \mathbb{N}$ $(1 \leq i \leq m)$; the proof will be given by induction on k.

Let $k = 0$. $Y_\Lambda PQN_1...N_m \Downarrow_{\mathbf{O}} R$ if and only if $P(Y_\Lambda P)QN_1...N_m \Downarrow_{\mathbf{O}} R$ (by $(head)^+$), if and only if
$\mathsf{Test}\ Q(\ulcorner g \urcorner x_1...x_m)(\ulcorner h \urcorner(Y_\Lambda P(\mathsf{Pred}\ Q)x_1...x_m)(\mathsf{Pred}\ Q)x_1...x_m) \Downarrow_{\mathbf{O}} R$
(by $(head)^+$), if and only if
$\mathsf{T}(\ulcorner g \urcorner x_1...x_m)(\ulcorner h \urcorner(Y_\Lambda P(\mathsf{Pred}\ Q)N_1...N_m)(\mathsf{Pred}\ Q)N_1...N_m) \Downarrow_{\mathbf{O}} R$
(by Property 15.2.7.(i), since $\mathsf{Test}\ Q \Downarrow_{\mathbf{O}} \mathsf{T}$), if and only if $\ulcorner g \urcorner N_1...N_m \Downarrow_{\mathbf{O}} R$.
But, by hypothesis $\ulcorner g \urcorner N_1...N_m \Downarrow_{\mathbf{O}} \ulcorner g(n_1, ..., n_m) \urcorner$.

Let $k > 0$. $Y_\Lambda PQN_1...N_m \Downarrow_{\mathbf{O}} R$ if and only if $P(Y_\Lambda P)QN_1...N_m \Downarrow_{\mathbf{O}} R$ (by $(head)^+$), if and only if
$\mathsf{Test}\ Q(\ulcorner g \urcorner N_1...N_m)(\ulcorner h \urcorner(Y_\Lambda P(\mathsf{Pred}\ Q)N_1...N_m)(\mathsf{Pred}\ Q)N_1...N_m) \Downarrow_{\mathbf{O}} R$
(by $(head)^+$), if and only if
$\mathsf{F}(\ulcorner g \urcorner N_1...N_m)(\ulcorner h \urcorner(Y_\Lambda P(\mathsf{Pred}\ Q)N_1...N_m)(\mathsf{Pred}\ Q)N_1...N_m) \Downarrow_{\mathbf{O}} R$
(by Property 15.2.7.(i), since $\mathsf{Test}\ Q \Downarrow_{\mathbf{O}} \mathsf{F}$), if and only if
$\ulcorner h \urcorner(Y_\Lambda P(\mathsf{Pred}\ Q)N_1...N_m)(\mathsf{Pred}\ Q)N_1...N_m \Downarrow_{\mathbf{O}} R$ (by $(head)^+$).
But, by induction $(Y_\Lambda P)(\mathsf{Pred}\ Q)N_1...N_m \Downarrow_{\mathbf{O}} \ulcorner f(k-1, n_1, ..., n_m) \urcorner$; thus $R \equiv \ulcorner h(f(k-1, x_1, ..., x_m), k-1, x_1, ..., x_m) \urcorner$, since by hypothesis $\ulcorner h \urcorner$ is an **O**-representation of h. □

Thus all primitive recursive functions are representable in the considered settings.

In order to represent the composition of partial functions in a call-by-name setting, the main problem is to make the representation "strict"; namely, when a function is applied to an undefined argument then its evaluation must diverge. The proposed solution takes into account the fact that terms representing natural numbers are in Λ-head normal form and so are Λ-solvable.

Lemma 16.1.5. *If $M \Downarrow_{\mathbf{O}} \ulcorner n \urcorner$ then $MKII \Downarrow_{\mathbf{O}} I$.*

Proof. By Property 15.2.7.(i), $MKII \Downarrow_{\mathbf{O}} R$ if and only if $\ulcorner n \urcorner KII \Downarrow_{\mathbf{O}} R$. Thus by $(head)^+$ the proof follows by observing the shape of $\ulcorner n \urcorner$. □

Then the representation of a function built by composition of partial functions is a term with the following operational behaviour: first it checks if all its arguments are defined and, in case at least one is undefined then it diverges; otherwise it computes the result.

Lemma 16.1.6. *Let $h : \mathbb{N}^n \to \mathbb{N}$ and $g_1, ..., g_n : \mathbb{N}^m \to \mathbb{N}$ be* **O***-representable partial recursive functions. The function defined from them by composition, namely*

$$f(x_1, ..., x_m) = h(g_1(x_1, ..., x_m), ..., g_n(x_1, ..., x_m)),$$

is **O***-representable.*

Proof. Let

$$F \equiv \lambda x_1...x_m.\ulcorner h \urcorner(\ulcorner g_1 \urcorner x_1...x_m)...(\ulcorner g_n \urcorner x_1...x_m) \qquad \text{and}$$
$$\ulcorner f \urcorner \equiv \lambda x_1...x_m.(\ulcorner g_1 \urcorner x_1...x_m KII)...(\ulcorner g_n \urcorner x_1...x_m KII)(F x_1...x_m).$$

Let $N_i \Downarrow_{\mathbf{O}} \ulcorner n_i \urcorner$ $(1 \leq i \leq m)$. $\ulcorner f \urcorner N_1...N_m \Downarrow_{\mathbf{O}} R$ if and only if $(\ulcorner g_1 \urcorner N_1...N_m KII)...(\ulcorner g_n \urcorner N_1...N_m KII)(F N_1...N_m) \Downarrow_{\mathbf{O}} R$ (by $(head)^+$). Let j be the minimum integer such that $\ulcorner g_j \urcorner N_1...N_m \Uparrow_{\mathbf{O}}$ $(1 \leq j \leq m)$. $(\ulcorner g_1 \urcorner N_1...N_m KII)...(\ulcorner g_n \urcorner N_1...N_m KII)(F N_1...N_m) \Downarrow_{\mathbf{O}} R$ if and only if $I(\ulcorner g_2 \urcorner N_1...N_m KII)...(\ulcorner g_n \urcorner N_1...N_m KII)(F N_1...N_m) \Downarrow_{\mathbf{O}} R$ (by $(head)^+$) if and only if $(\ulcorner g_2 \urcorner N_1...N_m KII)...(\ulcorner g_n \urcorner N_1...N_m KII)(F N_1...N_m) \Downarrow_{\mathbf{O}} R$ (by (head) if and only if $(\ulcorner g_j \urcorner N_1...N_m KII)...(\ulcorner g_n \urcorner N_1...N_m KII)(F N_1...N_m) \Downarrow_{\mathbf{O}} R$. But $\ulcorner g_j \urcorner N_1...N_m \Uparrow_{\mathbf{O}}$ implies $\ulcorner g_j \urcorner N_1...N_m$ is a Λ-unsolvable of order 0 in case $\mathbf{O} \equiv \mathbf{L}$. So $(\ulcorner g_j \urcorner N_1...N_m KII)...(\ulcorner g_n \urcorner N_1...N_m KII)(F N_1...N_m)$ is a Λ-unsolvable of order 0, respectively, and so in all cases,

$$(\ulcorner g_j \urcorner N_1...N_m KII)...(\ulcorner g_n \urcorner N_1...N_m KII)(F N_1...N_m) \Uparrow_{\mathbf{O}} .$$

In case $\ulcorner g_i \urcorner N_1...N_m \Downarrow_{\mathbf{O}}$, for all i $(1 \leq i \leq n)$, $\ulcorner f \urcorner N_1...N_m \Downarrow_{\mathbf{O}} R$ if and only if $(F N_1...N_m) \Downarrow_{\mathbf{O}} R$, and the proof follows the same line as that of Lemma 16.1.2. □

Finally, we check the computability of functions defined by minimalization. Let $P \equiv \lambda thxy.\mathsf{Test}(hxy)y(thx(\mathsf{Succ}\ y))$.

Lemma 16.1.7. *Let $h : \mathbb{N}^2 \to \mathbb{N}$ be an* **O***-representable total recursive function. Let N and Q be such that $N \Downarrow_{\mathbf{O}} \ulcorner n \urcorner$ and $Q \Downarrow_{\mathbf{O}} \ulcorner k \urcorner$.*

(i) *If $h(n, k) = 0$ then $Y_\Lambda P \ulcorner h \urcorner N Q \Downarrow_{\mathbf{O}} \ulcorner k \urcorner$.*

(ii) *Let $h(n, k) \neq 0$; so*

$$Y_\Lambda P \ulcorner h \urcorner N Q \Downarrow_{\mathbf{O}} R \textit{ if and only if } Y_\Lambda P \ulcorner h \urcorner N(\mathsf{Succ}\, Q) \Downarrow_{\mathbf{O}} R.$$

Proof. (i) $Y_\Lambda P \ulcorner h \urcorner N Q \Downarrow_{\mathbf{O}} R$ if and only if $P(Y_\Lambda P)\ulcorner h \urcorner N Q \Downarrow_{\mathbf{O}} R$ (by $(head)^+$) if and only if $\mathsf{Test}(\ulcorner h \urcorner N Q)Q(Y_\Lambda P \ulcorner h \urcorner N(\mathsf{Succ}\ Q)) \Downarrow_{\mathbf{O}} R$, (by $(head)^+$) if and only if $Q \Downarrow_{\mathbf{O}} R$ (since $\ulcorner h \urcorner N Q \Downarrow_{\mathbf{O}} \ulcorner 0 \urcorner$, always by $(head)^+$), but $Q \Downarrow_{\mathbf{O}} \ulcorner k \urcorner$, so the proof is done.

(ii) The proof is similar to that of the previous point, by using the fact that h is a total function. □

Property 16.1.8. Let $h : \mathbb{N}^2 \to \mathbb{N}$ be an **O**-representable total recursive function.

(i) If $f(n) = \mu y[h(n,y) = 0]$ is defined then $Y_\Lambda P \ulcorner h \urcorner N \mathsf{Zero} \Downarrow_{\mathbf{O}} \ulcorner f(n) \urcorner$, for every N such that $N \Downarrow_{\mathbf{O}} \ulcorner n \urcorner$.

(ii) If $f(n) = \mu y[h(n,y) = 0]$ is undefined then $Y_\Lambda P \ulcorner h \urcorner N \mathsf{Zero} \Uparrow_{\mathbf{O}}$, for all N such that $N \Downarrow_{\mathbf{O}} \ulcorner n \urcorner$.

Proof. (i) Let $f(n) = k$, thus k is the minimum integer such that $h(n,k) = 0$. By induction on k, the proof follows by Lemma 16.1.7.

(ii) By Lemma 16.1.7. □

Lemma 16.1.9. *Let $h : \mathbb{N}^2 \to \mathbb{N}$ be an **O**-representable total recursive function. $f(x) = \mu y[h(x,y) = 0]$ is **O**-representable.*

Proof. Let $\ulcorner f \urcorner \equiv \lambda x.Y_\Lambda P \ulcorner h \urcorner x \mathsf{Zero}$, where

$$P \equiv \lambda thxy.\mathsf{Test}(hxy)y\big(thx(\mathsf{Succ}\ y)\big).$$

Then the proof follows directly by Property 16.1.8. □

So the **O**-representability of all partial recursive function follows.

Theorem 16.1.10. *Let $\mathbf{O} \in \{\mathbf{H}, \mathbf{N}, \mathbf{L}\}$.*
*All partial recursive functions are **O**-representable.*

16.2 Call-by-Value Computational Completeness

Now let us prove that also the **V**-reduction machine can compute all partial recursive functions. We will point out just the differences between call-by-name and call-by-value computability. The most interesting difference occurs in the coding of recursion and minimalization, which is done through a fixed-point operator in the call-by-name setting. We have seen that a call-by-value fixed-point operator is such that, when applied to a Γ-valuable term, it is operationally equal to a not Γ-valuable term. In fact, if we think to interpret terms as function, the undefined value is the fixed-point of every function, when parameters are passed by value. So in order to deal with both recursion and minimalization, we will use a call-by-value recursion operator, whose behaviour was been defined at the end of Sect. 3.1. Other quite small differences are in the composition of partial functions.

Let $Y_\Gamma \equiv (\lambda xf.f(\lambda z.xxfz))(\lambda xf.f(\lambda z.xxfz))$; clearly Y_Γ is a recursion operator. The following theorem shows its operational behaviour.

Theorem 16.2.1 (Recursion).
Let $Y_\Gamma \equiv (\lambda xf.f(\lambda z.xxfz))(\lambda xf.f(\lambda z.xxfz))$.
If $M \in \Gamma$ *then* $Y_\Gamma M \Downarrow_{\mathbf{V}}$ *if and only if* $M(\lambda z.Y_\Gamma Mz) \Downarrow_{\mathbf{V}}$*, where* $z \notin FV(M)$.

Proof. Easy. □

Now we can show that the $\lambda\Gamma$-calculus is computationally complete.

Lemma 16.2.2. *Primitive recursive functions are* **V***-representable.*

Proof. The proof follows the same lines as Lemmas 16.1.1, 16.1.2 and 16.1.4, taking into account that Y_Γ must be used instead of Y_Λ, and moreover taking into account the behaviour of the **V**-reduction machine. □

The extension to partial functions is easier than in the call-by-name case. In fact the mathematical functions are naturally "strict", in the sense that a function diverges if one of its arguments diverges, and this behaviour is exactly the behaviour of the **V**-evaluation. In fact the following property holds.

Property 16.2.3. Let $M, N \in \Lambda^0$; $M \Uparrow_{\mathbf{V}}$ implies $NM \Uparrow_{\mathbf{V}}$.

Proof. Trivial, by the rule (*head*). □

Lemma 16.2.4. *Let* $h : \mathbb{N}^n \to \mathbb{N}$ *and* $g_1, ..., g_n : \mathbb{N}^m \to \mathbb{N}$ *be* **V***-representable partial recursive functions. The function* f *defined by composition from them in the following way:*

$$f(x_1, ..., x_m) = h(g_1(x_1, ..., x_m), ..., g_n(x_1, ..., x_m))$$

is **V***-representable.*

Proof. Let $H \equiv \lambda x_1...x_m.\ulcorner h \urcorner(\ulcorner g_1 \urcorner x_1...x_m).....(\ulcorner g_n \urcorner x_1...x_m)$; it is easy to check that H is a **V**-representation of f, by Property 16.2.3. □

Finally, we check the **V**-computability of functions defined by minimalization. Let $F \equiv \lambda thxy.\mathsf{Test}(hIxy)(\lambda v.y)(\lambda u.thx(\mathsf{Succ}\ y))I$.

Lemma 16.2.5. *Let* $h : \mathbb{N}^2 \to \mathbb{N}$ *be a* **V***-representable total recursive function such that* $h(n, k) = m$ *where* $n, k, m \in \mathbb{N}$.

(i) *If* $m = 0$ *then* $(Y_\Gamma F)(\lambda z.\ulcorner h \urcorner)\ulcorner n \urcorner \ulcorner k \urcorner \Downarrow_{\mathbf{V}} \ulcorner k \urcorner$.

(ii) *Let* $m \neq 0$.
$(Y_\Gamma F)(\lambda z.\ulcorner h \urcorner)\ulcorner n \urcorner \ulcorner k \urcorner \Downarrow_{\mathbf{V}}$ *if and only if* $(Y_\Gamma F)(\lambda z.\ulcorner h \urcorner)\ulcorner n \urcorner \ulcorner k+1 \urcorner \Downarrow_{\mathbf{V}}$.

Proof. (i) $(Y_\Gamma F)(\lambda z.\ulcorner h \urcorner)\ulcorner n \urcorner \ulcorner k \urcorner \Downarrow_{\mathbf{V}} R$ if and only if
$F(\lambda z.Y_\Gamma Fz)(\lambda z.\ulcorner h \urcorner)\ulcorner n \urcorner \ulcorner k \urcorner \Downarrow_{\mathbf{V}} R$ if and only if
$\mathsf{Test}((\lambda z.\ulcorner h \urcorner)I\ulcorner n \urcorner \ulcorner k \urcorner)(\lambda v.\ulcorner k \urcorner)(\lambda u.(\lambda z.Y_\Gamma Fz)(\lambda z.\ulcorner h \urcorner)\ulcorner n \urcorner(\mathsf{Succ}\ulcorner k \urcorner))I \Downarrow_{\mathbf{V}}$ R by two applications of rule (*head*), if and only if $(\lambda v.\ulcorner k \urcorner)I \Downarrow_{\mathbf{V}} R$ if and only if $\ulcorner k \urcorner \Downarrow_{\mathbf{V}} R$, always by (*head*), and this implies $R \equiv \ulcorner k \urcorner$.

(ii) $(Y_\Gamma F)(\lambda z.\ulcorner h \urcorner)\ulcorner n \urcorner\ulcorner k \urcorner \Downarrow_{\mathbf{V}}$ if and only if $F(\lambda z.Y_\Gamma F z)(\lambda z.\ulcorner h \urcorner)\ulcorner n \urcorner\ulcorner k \urcorner \Downarrow_{\mathbf{V}}$ if and only if $\mathsf{Test}((\lambda z.\ulcorner h \urcorner)I\ulcorner n \urcorner\ulcorner k \urcorner)(\lambda v.\ulcorner k \urcorner)(\lambda u.(\lambda z.Y_\Gamma F z)(\lambda z.\ulcorner h \urcorner)\ulcorner n \urcorner(\mathsf{Succ}\ulcorner k \urcorner))I \Downarrow_{\mathbf{V}}$ if and only if $\mathsf{F}(\lambda v.\ulcorner k \urcorner)(\lambda u.(\lambda z.Y_\Gamma F z)(\lambda z.\ulcorner h \urcorner)\ulcorner n \urcorner(\mathsf{Succ}\ \ulcorner k \urcorner))I \Downarrow_{\mathbf{V}}$ always by (*head*), if and only if $(\lambda u.(\lambda z.Y_\Gamma F z)(\lambda z.\ulcorner h \urcorner)\ulcorner n \urcorner(\mathsf{Succ}\ulcorner k \urcorner))I \Downarrow_{\mathbf{V}}$ if and only if $(\lambda z.Y_\Gamma F z)(\lambda z.\ulcorner h \urcorner)\ulcorner n \urcorner(\mathsf{Succ}\ulcorner k \urcorner) \Downarrow_{\mathbf{V}}$ And, again by (*head*), if and only if $Y_\Gamma F(\lambda z.\ulcorner h \urcorner)\ulcorner n \urcorner(\mathsf{Succ}\ulcorner k \urcorner) \Downarrow_{\mathbf{V}}$. □

The representation of a function defined by minimalization is different from the call-by-name case, since the term P, performing the iteration in the call-by-name case, does not work correctly in the call-by-value one. In fact, in P, the second argument of the function Test is not a value. The term F is a slight modification of P, playing the desired role in this particular setting.

Lemma 16.2.6. *Let $h : \mathbb{N}^2 \to \mathbb{N}$ be a $\mathbf{V}$-representable total recursive function. $f(x) = \mu y[h(x,y) = 0]$ is $\mathbf{V}$-representable.*

Proof. Let $\ulcorner f \urcorner \equiv \lambda x.Y_\Gamma F(\lambda z.\ulcorner h \urcorner)x\mathsf{Zero}$, where

$$F \equiv \lambda thxy.\mathsf{Test}(hIxy)(\lambda v.y)(\lambda u.t(hx(\mathsf{Succ}\ y)))I.$$

The proof follows by Lemma 16.2.5, in the same way as the proof of Lemma 16.1.9. □

Theorem 16.2.7. *All partial recursive functions are $\mathbf{V}$-representable.*

16.3 Historical Remarks

In the literature, computability, for both the call-by-name and the call-by-value λ-calculi, was defined starting from an approach different from the present one. In fact, the representation of computable functions was developed inside a theory, while we have chosen an operational point of view. The next definition is the classical one for the $\lambda\Lambda$-calculus.

Definition 16.3.1. *Let ϕ be a partial recursive function with arity $p \in \mathbb{N}$. ϕ is λ-definable if and only if there is a term $\ulcorner \phi \urcorner$ such that*

- *$\phi(n_1,...,n_p)$ defined implies $\ulcorner \phi \urcorner\ulcorner n \urcorner_1 ... \ulcorner n \urcorner_p =_\Lambda \ulcorner \phi(n_1,...,n_p) \urcorner$;*
- *$\phi(n_1,...,n_p)$ undefined implies $\ulcorner \phi \urcorner\ulcorner n \urcorner_1 ... \ulcorner n \urcorner_p$ is an unsolvable term.*

Inside the $\lambda\Lambda$-calculus, the first numeral system was introduced by Church. In it, $\ulcorner n \urcorner$ is the term $\lambda fx.\underbrace{f(.....(f}_{n}x)...)$, which represents the n-th iteration of a function f applied to an argument x. This numeral system has been the starting point for the first λ-representation of partial recursive functions, given by Kleene, in the λ-I-calculus [57]. The idea of using fixed-point operators to represent primitive recursion and minimalization comes from Turing [95].

The system we proposed in Section 15.2 is a slight modification of the numeral system of Barendregt [9], recalled in the following definition.

Definition 16.3.2 (Barendregt numeral system).
Let $\mathfrak{B} \equiv \langle \{T, F\}, Zero^{\mathfrak{B}}, Succ^{\mathfrak{B}}, Test^{\mathfrak{B}}, Pred^{\mathfrak{B}} \rangle$, *where*

- $T \equiv \lambda xy.x$ *and* $F \equiv \lambda xy.y$;
- $Zero^{\mathfrak{B}} \equiv \lambda x.x$;
- $Succ^{\mathfrak{B}} \equiv \lambda n.[F, n] \equiv \lambda nx.xFn$;
- $Test^{\mathfrak{B}} \equiv \lambda n.nT$;
- $Pred^{\mathfrak{B}} \equiv \lambda n.nF$.

$\mathfrak{B}$ is a **N**-numeral system but is neither a **L**-numeral system nor a **H**-numeral system. In fact, the Barendregt representation of 1 is $\ulcorner 1 \urcorner \equiv \lambda x.xFZero^{\mathfrak{B}}$, and $Succ^{\mathfrak{B}} Zero^{\mathfrak{B}} \Downarrow_{\mathbf{O}} \lambda x.xFZero^{\mathfrak{B}}$ where $\mathbf{O} \in \{\mathbf{L}, \mathbf{H}\}$; nevertheless

$$Succ^{\mathfrak{B}}(Pred^{\mathfrak{B}} \ulcorner 1 \urcorner) \Downarrow_{\mathbf{H}} \lambda x.x((\lambda n.nF)\ulcorner 1 \urcorner)$$

$$Succ^{\mathfrak{B}}(Pred^{\mathfrak{B}} \ulcorner 1 \urcorner)\ulcorner n \urcorner \Downarrow_{\mathbf{L}} \lambda x.x((\lambda n.nF)\ulcorner 1 \urcorner)$$

but $\lambda x.x((\lambda n.nF)\ulcorner 1 \urcorner) \not\equiv \ulcorner 1 \urcorner$, although

$$\lambda x.x((\lambda n.nF)\ulcorner 1 \urcorner) =_{\Lambda} \ulcorner 1 \urcorner.$$

As far as the call-by-value computability is concerned, Plotkin [78] was the first to point out the difference between call-by-name and call-by-value recursion. He proposed the following recursion operator:

$$\lambda f.(\lambda x.f(\lambda z.xxz))(\lambda x.f(\lambda z.xxz)),$$

which works similarly to Y_{Γ}. The computability in the theory $=_{\Gamma}$ was completely developed by Paolini [72], using Y_{Γ} and the Barendregt numeral system.

Bibliography

1. Samson Abramsky. Domain theory in logical form. *Annals of Pure and Applied Logic*, 51(1-2):1–77, 1991.
2. Samson Abramsky and Luke Ong. Full abstraction in the lazy lambda calculus. *Information and Computation*, 105(2):159–267, 1993.
3. Fabio Alessi. *Strutture di Tipi, Teorie dei Domini e Modelli del Lambda Calcolo*. Tesi di dottorato di ricerca in informatica, Università di Milano e Torino, 1990.
4. Fabio Alessi, Mariangiola Dezani-Ciancaglini, and Furio Honsell. Filter models and easy terms. In Antonio Restivo, Simona Ronchi Della Rocca, and Luca Roversi, editors, *Theoretical Computer Science, 7th Italian Conference, ICTCS 2001, Torino, Italy, October 4-6, 2001*, volume 2202 of *Lecture Notes in Computer Science*, pages 17–37. Springer-Verlag, 2001.
5. Roberto M. Amadio and Pierre-Louis Curien. *Domains and Lambda-Calculi*, volume 46 of *Cambridge Tracts in Theoretical Computer Science*. Cambridge University Press, Cambridge, 1998.
6. Andrea Asperti, Cecilia Giovannetti, and Andrea Naletto. The bologna optimal higher-order machine. Technical Report UBLCS-95-9, University of Bologna, Department of Computer Science, March 1995.
7. Andrea Asperti and Stefano Guerrini. *The Optimal Implementation of Functional Programming Languages*. Cambridge University Press, Cambridge, 1998.
8. Andrea Asperti and Giuseppe Longo. *Categories, Types, and Structures: An Introduction to Category Theory for the Working Computer Scientist*. Foundations of Computing Series. The MIT Press, Cambridge, MA, 1991.
9. Henk Barendregt. *The Lambda Calculus: Its Syntax and Semantics (2nd edition)*. North-Holland, Amsterdam, 1984.
10. Henk Barendregt, Mario Coppo, and Mariangiola Dezani-Ciancaglini. A filter lambda model and the completeness of type assignment. *The Journal of Symbolic Logic*, 48(4):931–940, December 1983.
11. Olivier Bastonero and Xavier Gouy. Strong stability and the incompleteness of stable models for lambda- calculus. *Annals of Pure and Applied Logic*, 100(1-3):247–277, 1999.
12. Olivier Bastonero, Alberto Pravato, and Simona Ronchi Della Rocca. Structures for lazy semantics. In Gries and de Roever, editors, *Programming Concepts and Methods*, pages 30–48. Chaptman & Hall, 1998.
13. Chantal Berline. From computation to foundations via functions and application: The λ-calculus and its webbed models. *Theoretical Computer Science*, 249(1):81–161, October 2000.
14. Gérard Berry. Stable models of typed lambda-calculi. In Giorgio Ausiello and Corrado Böhm, editors, *Automata, Languages and Programming, Fifth Colloquium, ICALP, Udine, Italy, July 17-21, 1978*, volume 62 of *Lecture Notes in Computer Science*, pages 72–89. Springer-Verlag, 1978.

15. Corrado Böhm. Alcune proprieta delle forme $\beta\eta$-normali nel λK-calculus. Pubblicazione n. 696, Instituto per le Applicazioni del Calcolo, Roma, 1968.
16. Corrado Böhm and Mariangiola Dezani-Ciancaglini. A CUCH-machine: the automatic treatment of bound variables. *International Journal of Computer and Information Sciences*, 1(2):171–191, June 1972.
17. Corrado Böhm and Mariangiola Dezani-Ciancaglini. λ-terms as total or partial functions on normal forms. In G. Goos and J. Hartmanis, editors, *λ-Calculus and Computer Science Theory*, volume 37 of *Lecture Notes in Computer Science*, pages 96–121, Berlin, DE, 1975. Springer-Verlag.
18. Corrado Böhm, Mariangiola Dezani-Ciancaglini, P. Peretti, and Simona Ronchi Della Rocca. A discrimination algorithm inside λ-calculus. *Theoretical Computer Science*, 8(3):271–291, 1978.
19. Corrado Böhm and W. Gross. Introduction to the CUCH. In E. R. Caianiello, editor, *Automata Theory*, pages 35–65. Academic Press, New York, 1966.
20. Corrado Böhm and Adolfo Piperno. Characterizing X-separability and one-side invertibility in λ-β-Ω-calculus. In *Proceedings, Third Annual Symposium on Logic in Computer Science - LICS'88*, pages 91–103, Edinburgh, Scotland, 5–8 July 1988. IEEE Computer Society Press.
21. Corrado Böhm, Adolfo Piperno, and Stefano Guerrini. Lambda-definition of function(al)s by normal forms. In Donald Sannella, editor, *Programming Languages and Systems-ESOP'94, 5th European Symposium on Programming*, volume 788 of *Lecture Notes in Computer Science*, pages 135–149. Springer-Verlag, 1994.
22. Corrado Böhm, Adolfo Piperno, and Enrico Tronci. Solving equations in lambda-calculus. In *Logic Colloquium'88*, Amsterdam, 1988. North-Holland.
23. Antonio Bucciarelli and Thomas Ehrhard. A theory of sequentiality. *Theoretical Computer Science*, 113(2):273–291, 7 June 1993.
24. Rod Burstall and Furio Honsell. Operational semantics in a natural deduction setting. In Gérard Huet and Gordon Plotkin, editors, *Logical Frameworks*, pages 185–214, Cambridge, 1991. Cambridge University Press.
25. Alonzo Church. *The Calculi of Lambda Conversion*, volume 6 of *Annals of Mathematical Studies*. Princeton University Press, Princeton, 1941. Reprinted by University Microfilms Inc., Ann Arbor, MI in 1963 and by Klaus Reprint Corp., New York in 1965.
26. Alonzo Church and J. Barkley Rosser. Some properties of conversion. *Transactions of the Aerican Mathematical Society*, 39:472–482, 1936.
27. Mario Coppo and Mariangiola Dezani-Ciancaglini. An extension of the basic functionality theory for the λ-calculus. *Notre Dame Journal of Formal Logic*, 21(4):685–693, October 1980.
28. Mario Coppo, Mariangiola Dezani-Ciancaglini, Furio Honsell, and Giuseppe Longo. Extended type structure and filter lambda models. In G. Lolli, G. Longo, and A. Marcja, editors, *Logic Colloquim'82*, pages 241–262. Elsevier Science Publishers B.V. (North-Holland), Amsterdam, 1984.
29. Mario Coppo, Mariangiola Dezani-Ciancaglini, and Simona Ronchi Della Rocca. (Semi)-separability of finite sets of terms in Scott's D_∞-models of the λ-calculus. In Giorgio Ausiello and Corrado Böhm, editors, *Automata, Languages and Programming, Fifth Colloquium*, volume 62 of *Lecture Notes in Computer Science*, pages 142–164, Udine, Italy, 17–21 July 1978. Berlin, Springer-Verlag.
30. Mario Coppo, Mariangiola Dezani-Ciancaglini, and Maddalena Zacchi. Type theories, normal forms and D_∞ lambda models. *Information and Computation*, 72(2):85–116, 1987.
31. Roy L. Crole. *Categories for Types*. Cambridge University Press, Cambridge, 1993.

32. Pierre Louis Curien. Sur l'eta-expansion infinie. *Comptes Rendus de l'Académie des Sciences*, to appear.
33. Pierre-Louis Curien and Hugo Herbelin. The duality of computation. In *Proceedings of the ACM Sigplan International Conference on Functional Programming (ICFP-00)*, volume 35(9) of *ACM Sigplan Notices*, pages 233–243, Montréal, Canada, September 18–21 2000. ACM Press.
34. Haskell B. Curry and Robert Feys. *Combinatory Logic - Volume 1*. Studies in Logic and the Foundations of Mathematics. Elsevier, North-Holland, (Amsterdam, London, New York), L. E. J. Brouwer, E. W. Beth, A. Heyting editors, edition, 1958. With two sections by William Craig. Second edition, 1968.
35. Haskell B. Curry, J. Roger Hindley, and Jonathan P. Seldin. *Combinatory Logic - Volume 2*, volume 65 of *Studies in Logic and the Foundations of Mathematics*. Elsevier, North-Holland, (Amsterdam, London, New York), A. Heyting, H. J. Keisler, A. Mostowski, A. Robinson, P. Suppes editors, edition, 1972.
36. René David and Karim Nour. A syntactical proof of the operational equivalence of two λ-terms. *Theoretical Computer Science*, 180(1–2):371–375, 10 June 1997.
37. Mariangiola Dezani-Ciancaglini, Silvia Ghilezan, and Silvia Likavec. Behavioural inverse limit models. *Theoretical Computer Science*, 2003. To appear.
38. Mariangiola Dezani-Ciancaglini, Furio Honsell, and Fabio Alessi. A complete characterization of complete intersection-type preorders. *ACM Transactions on Computational Logic*, 4(1):120–147, January 2003.
39. Mariangiola Dezani-Ciancaglini, Furio Honsell, and Simona Ronchi Della Rocca. Models for theories of functions strictly depending on all their arguments. *The Journal of Symbolic Logic*, 51(3):845–846, 1986. (Abstract).
40. Mariangiola Dezani-Ciancaglini and Stefania Lusin. Intersection types and lambda theories. In *Electronic Proceedings of WIT'02 (http://www.irit.fr/zeno/WIT2002/proceedings.shtml)*, 2002.
41. Roberto Di Cosmo. A brief history of rewriting with extensionality. In Fairouz Kamareddine, editor, *International Summer School on Type Theory and Rewriting*, Glasgow, 1996. Kluwer.
42. Pietro Digianantonio. Game semantics for the pure lazy λ-calculus. In Samson Abramsky, editor, *Typed Lambda Calculi and Applications: 5th International Conference, TLCA 2001 Krakow, Poland, May 2-5, 2001*, volume 2044 of *Lecture Notes in Computer Science*, pages 106–120, Berlin, June 2003. Springer-Verlag.
43. Pietro Digianantonio, Gianluca Franco, and Furio Honsell. Game semantics for untyped $\lambda\beta\eta$-calculus. In Jean-Yves Girard, editor, *Typed Lambda Calculi and Applications: 4th International Conference, TLCA'99, L'Aquila, Italy, April 1999*, volume 1581 of *Lecture Notes in Computer Science*, pages 114–128, Berlin, July 2003. Springer-Verlag.
44. Lavinia Egidi, Furio Honsell, and Simona Ronchi Della Rocca. Operational, denotational and logical descriptions: a case study. *Fundamenta Informaticæ*, 16(2):149–170, 1992.
45. Matthias Felleisen and Daniel P. Friedman. A syntactic theory of sequential state. *Theoretical Computer Science*, 69(3):243–287, 1989. Preliminary version in *Proc. 14th ACM Symp. Principles of Programming Languages* 1987, pages 314–325.
46. Matthias Felleisen, Daniel P. Friedman, Eugene E. Kohlbecker, and Bruce F. Duba. A syntactic theory of sequential control. *Theoretical Computer Science*, 52:205–237, 1987.
47. Jean-Yves Girard. The system F of variable types, fifteen years later. *Theoretical Computer Science*, 45(2):159–192, 1986.

48. Jean-Yves Girard. Linear logic. *Theoretical Computer Science*, 50:1–102, 1987.
49. J. Roger Hindley. *Basic Simple Type Theory*, volume 42 of *Cambridge Tracts in Theoretical Computer Science*. Cambridge University Press, Cambridge, UK, 1997.
50. J. Roger Hindley and Giuseppe Longo. Lambda calculus models and extensionality. *Zeitschrift für mathematische Logik und Grundlagen der Mathematik*, 26:289–310, 1980.
51. J. Roger Hindley and Jonathan P. Seldin. *Introduction to Combinators and λ-Calculus*, volume 1 of *London Mathematical Society Student Texts*. Cambridge University Press, Cambridge, UK, 1986.
52. Furio Honsell and Simona Ronchi della Rocca. Reasoning about interpretation in qualitative lambda-models. In M. Broy and C.B. Jones, editors, *Proceeding of IFIP 2.2 Working Conference on Programming Concepts and Methods*, pages 505–521, Sea of Galilee, Israel, 1990. North Holland.
53. Furio Honsell and Simona Ronchi Della Rocca. An approximation theorem for topological lambda models and the topological incompleteness of lambda calculus. *Journal of Computer and System Sciences*, 45(1):49–75, August 1992.
54. J. Martin E. Hyland. A syntactic characterization of the equality in some models of the lambda calculus. *Journal of the London Mathematical Society*, 2(12):361–370, 1976.
55. Gilles Kahn. Natural semantics. In *Symposium on Theoretical Aspects of Computer Science*, volume 247 of *Lecture Notes in Computer Science*, pages 22–39, 1987.
56. A. J. Kfoury, Robert A. Moll, and Michael A. Arbib. *A Programming Approach to Computability*. Texts and Monographs in Computer Science. Springer-Verlag, Berlin, 1986. Second edition.
57. Stephen Cole Kleene. Lambda definability and recursiveness. *Duke Mathematical Journal*, 2:340–353, 1936.
58. Jan Willem Klop. *Combinatory Reduction Systems*, volume 127 of *Mathematical Centre Tracts*. Mathematischen Centrum, 413 Kruislaan, Amsterdam, 1980.
59. C. P. J. Koymans. Models of the lambda calculus. *Information and Computation*, 52(3):306–323, 1982.
60. Jean Louis Krivine. *Lambda-Calculus, Types and Models*. Ellis Horwood Series in Computers and Their Applications. Masson, Paris, and Ellis Horwood, Hemel Hempstead, 1993. Transation from French by René Cori, French orig. ed., Masson, Paris, 1990.
61. James Laird. A fully abstract bidomain model of unary PCF. In Martin Hofmann, editor, *Typed Lambda Calculi and Applications, 6th International Conference, TLCA 2003, Valencia, Spain, June 10-12, 2003, Proceedings*, volume 2701 of *Lecture Notes in Computer Science*, pages 211–225. Springer-Verlag, 2003.
62. Joachim Lambek. From lambda calculus to cartesian closed categories. In *To H.B. Curry: Essays on Combinatory Logic, Lambda Calculus and Formalism*, pages 375–402. Academic Press, 1980.
63. Peter J. Landin. The mechanical evaluation of expressions. *Computer Journal*, 6:308–320, January 1964.
64. Peter J. Landin. A correspondence between ALGOL 60 and Church's lambda-notation: Part I and Part II. *Communications of the ACM*, 8(2-3):89–101,158–165, 1965.
65. Peter J. Landin. The next 700 programming languages. *Communications of the ACM*, 9(3):157–166, March 1966.
66. John McCarthy. *LISP 1.5 Programmer's Manual*. The MIT Press, Cambridge, Mass., 1962. (with Abrahams, Edwards, Hart, and Levin).

67. Albert Meyer. What is a model of the lambda calculus? *Information and Computation*, 52(1):87–122, 1982.
68. Robert Milner. Fully abstract models of typed lambda-calculus. *Theoretical Computer Science*, 4:1–22, 1977.
69. John C. Mitchell. *Foundations of Programming Languages*. The MIT Press, Cambridge, MA, 1996.
70. Eugenio Moggi. *The Partial Lambda-Calculus*. PhD thesis, Edinburgh University, February 1988. Report CST-53-88.
71. C.-H. Luke Ong. Fully abstract models of the lazy lambda calculus. In *29th Annual Symposium on Foundations of Computer Science*, pages 368–376, White Plains, New York, 24–26 October 1988. IEEE Computer Society Press.
72. Luca Paolini. Call-by-value separability and computability. In Antonio Restivo, Simona Ronchi Della Rocca, and Luca Roversi, editors, *Theoretical Computer Science, 7th Italian Conference, ICTCS 2001, Torino, Italy, October 4-6, 2001*, volume 2202 of *Lecture Notes in Computer Science*, pages 74–89. Springer-Verlag, 2001.
73. Luca Paolini and Simona Ronchi Della Rocca. Call by value solvability. *Theoretical Informatics and Applications*, 33(6):507–534, nov 1999.
74. Luca Paolini and Simona Ronchi Della Rocca. The parametric parameter passing λ-calculus. *Information and Computation*, 189(1):87–106, feb 2004.
75. D. M. R. Park. The Y-combinator in scott's lambda-calculus models. Research Report CS-RR-013, Department of Computer Science, University of Warwick, Coventry, UK, June 1976.
76. Adolfo Piperno. An algebraic view of the Böhm-out technique. *Theoretical Computer Science*, 212(1–2):233–246, February 1999.
77. Andrew M. Pitts. Operational semantics and program equivalence. In G. Barthe, P. Dybjer, and J. Saraiva, editors, *Applied Semantics*, volume 2395 of *Lecture Notes in Computer Science*, pages 378–412. Springer-Verlag, 2002. (Revised version of lectures at the International Summer School On Applied Semantics, APPSEM 2000, Caminha, Minho, Portugal, 9-15 September 2000.).
78. Gordon D. Plotkin. Call-by-name, call-by-value and the λ-calculus. *Theoretical Computer Science*, 1:125–159, 1975.
79. Gordon D. Plotkin. LCF considerd as a programming language. *Theoretical Computer Science*, 5:223–225, 1977.
80. Gordon D. Plotkin. A structural approach to operational semantics. DAIMI FN-19, Aarhus University, Aarhus, Denmark, September 1981.
81. Gordon D. Plotkin. Domains. Dept. of Computer Science, University of Edinburgh, 1983.
82. Alberto Pravato, Simona Ronchi Della Rocca, and Luca Roversi. The call by value λ-calculus: a semantic investigation. *Mathematical Structures in Computer Science*, 9(5):617–650, 1999.
83. G. E. Revesz. *Lambda-Calculus Combinators and Functional Programming*, volume 4 of *Cambridge Tracts in Theoretical Computer Science*. Cambridge University Press, Cambridge, 1988.
84. Simona Ronchi Della Rocca. Discriminability of infinite sets of terms in the D_∞ -models of the λ-calculus. In Egidio Astesiano and Corrado Böhm, editors, *Proceedings of the 6th Colloquium on Trees in Algebra and Programming (CAAP'81)*, volume 112 of *Lecture Notes in Computer Science*, pages 350–364, Genova, Italy, March 1981. Springer-Verlag.
85. Simona Ronchi Della Rocca. Operational semantics and extensionality. In *Proceedings of the 2nd Imternational ACM SIGPLAN Conference on Principles and Practice of Declarative Programming (PPDP-00)*, pages 24–31, Montréal, September 20–23 2000. ACM Press.

86. Antonino Salibra. Topological incompleteness and order incompleteness of the lambda calculus. *ACM Transactions on Computational Logic*, 4(3):379–401, July 2003.
87. David A. Schmidt. *Denotational Semantics: A Methodology for Language Development.* Allyn and Bacon, Boston, 1986.
88. Dana S. Scott. Continuous lattices. In F. William Lawvere, editor, *Toposes, Algebraic Geometry, and Logic*, volume 274 of *Lecture Notes in Mathematics*, pages 97–136. Springer-Verlag, Berlin, Heidelberg, and New York, 1972.
89. Dana S. Scott. Data types as lattices. *SIAM Journal of Computing*, 5:522–587, September 1976.
90. Dana S. Scott. Relating theories of the λ-calculus. In J. P. Seldin and J. R. Hindley, editors, *To H. B. Curry: Essays on Combinatory Logic, Lambda Calculus and Formalism*, pages 403–450. Academic Press, 1980.
91. Joseph E. Stoy. *Denotational Semantics of Programming Languages: The Scott-Strachey Approach to Programming Language Theory.* The MIT Press, Cambridge, USA, 1977.
92. Christopher Strachey. Fundamental concepts in programming languages. *Higher-Order and Symbolic Computation*, 13(1–2):11–49, April 2000. Notes for the International Summer School in Computer Programming, Copenhagen, 1967.
93. Masako Takahashi. Parallel reductions in lambda-calculus. *Information and Computation*, 118(1):120–127, April 1995.
94. Daniele Turi and Gordon Plotkin. Towards a mathematical operational semantics. In *Proceedings, Twelth Annual IEEE Symposium on Logic in Computer Science LICS'97*, pages 280–291, Warsaw, Poland, 29 June–2 July 1997. IEEE Computer Society Press.
95. Alan M. Turing. The P-functions in λ-K-conversion. *The Journal of Symbolic Logic*, 2:164, 1937.
96. Philip Wadler. Call-by-value is dual to call-by-name. In Cindy Norris and Jr. James B. Fenwick, editors, *Proceedings of the Eighth ACM SIGPLAN International Conference on Functional Programming (ICFP-03)*, volume 38, 9 of *ACM SIGPLAN Notices*, pages 189–201, New York, August 25–29 2003. ACM Press.
97. Christopher P. Wadsworth. The relation between computational and denotational properties for scott's D_∞-models of the lambda-calculus. *SIAM Journal of Computing*, 5(3):488–521, September 1976.
98. Glynn Winskel. *The Formal Semantics of Programming Languages: An Introduction.* Foundations of Computing Series. The MIT Press, February 1993.
99. Glynn Winskel. Stable bistructure models of PCF. In Igor Prívara, Branislav Rovan, and Peter Ruzicka, editors, *Mathematical Foundations of Computer Science 1994, 19th International Symposium*, volume 841 of *Lecture Notes in Computer Science*, pages 177–197, Kosice, Slovakia, 22–26 August 1994. Springer.

Index

$\to_{\Delta \mathsf{O}\eta}$, 97
$(.)^*$, 212
$(.)^+$, 212
$=_{\Delta}$, 7
$=_{\mathsf{O}\eta}$, 97
$=_{\Delta \mathsf{O}\eta}$, 97
$=_{\Delta\eta}$, 23
$App_{\mathcal{H}}$, 133
$App_{\mathcal{L}}$, 168
$App_{\mathcal{N}}$, 154
$App_{\mathcal{V}}$, 195
$B \cup B'$, 133, 168, 195
B, 189
B^n, 29
$C_{\bowtie}$, 145
$C_{\angle}$, 163
C_{∞}, 120
$C_{\surd}$, 182
$Comp_{\mathcal{H}}$, 133
$Comp_{\mathcal{L}}$, 168
$Comp_{\mathcal{N}}$, 154
$Comp_{\mathcal{V}}$, 195
$Cond$, 227
E_{∞}, 128, 139
$I(C)$, 109
$I(C_{\bowtie})$, 145
$I(C_{\angle})$, 163
$I(C_{\infty})$, 120
$I(C_{\surd})$, 182
O^n, 29
$T(C)$, 109
$T(C_{\bowtie})$, 145
$T(C_{\angle})$, 163
$T(C_{\infty})$, 120
$T(C_{\surd})$, 182
U, 171
U^i_n, 29
Y, 27
Y_{Γ}, 238
Y_{Λ}, 234
$[f]$, 205
nf_{Λ}, 29
$\Downarrow_{\mathsf{O}}$, 67
Γ, 7
Γ-NF, 35
Γ-hnf, 39
Γ-lbnf, 66
Γ-nf, 35, 66
Γ_{Ω}, 184
$\bowtie$, 145
L_0, 85, 170
L_1, 85, 170
Λ, 3
$\Lambda\Omega$, 123
$\Lambda\eta$, 26
Λ-hnf, 25, 66
Λ-lhnf, 25, 66
Λ-nf, 25, 66
Λ-pnf, 144
$\Lambda\Omega$-nf, 124
Λ_I, 7
$\mathbf{N}$, 77
Ω, 123
$\Omega^{\mathcal{H}}$, 129
$\Phi^{\mathcal{H}}$, 130
Pred, 229
Ψ, 36
$\Psi\ell$, 36
$\Psi\ell$-nf, 36
$\Rrightarrow^{c}_{D}$, 138
$\Rrightarrow^{c}_{N}$, 159, 160
$\Rrightarrow_{\Gamma}$, 52
$\Rrightarrow_{\Lambda}$, 28
Succ, 229
T, 229
Test, 229
Θ, 65
$\Uparrow_{\mathsf{O}}$, 68
V_0, 93, 198
V_1, 93, 198
Var, 3
Θ, 65

$\rightarrow_{\Delta}$, 6
$\overset{\circ}{\rightarrow}_{\Delta}$, 9
$\searrow$, 39
$\Xi\ell$, 36
$\Xi\ell\Omega$-nf, 185
Ξ, 36, 66
Zero, 229
$\angle$, 163
$\approx_{\mathbf{N}}$, 78
$\approx_{\mathbf{H}}$, 73
$\approx_{\mathbf{V}}$, 90
$\approx_{\mathbf{L}}$, 82
$\approx_{\mathbf{O}}$, 68
args, 29
$\mathbf{H}$, 67, 73
$\mathbf{L}$, 67
$\mathbf{N}$, 67
$\mathbf{O}$, 66
$\mathbf{V}$, 67
$\overset{\circ}{\rightarrow}_{\Xi\ell}$, 42
$\circ$, 105
$\circ_{\nabla}$, 114
$\sqsubseteq_{\mathbb{L}}$, 207
$\mathbf{c}$, 210
$[\![.]\!]$, 105
$\sim_{\mathcal{LL}}$, 176
$\sim_{\mathcal{VV}}$, 206
$\equiv$, 3
$\mathcal{E}(\Delta, \Theta)$, 67
$\mathfrak{N}$, 229
$\frown$, 96
$\Downarrow_{\mathbf{H}}$, 73
∞, 120
λ-term, 3
$\langle.\rangle$, 44
$\Downarrow_{\mathbf{L}}$, 82
$\leq_{\bowtie}$, 145
$\leq_{\angle}$, 163
$\leq_{\infty}$, 120
$\leq_{\surd}$, 182
$\leq_{\nabla_{\mathbb{L}}}$, 212
$\ll$, 125
$\overset{\circ}{\ll}_c$, 139
$\ll_c$, 136
$\boldsymbol{\lambda}$, 208
∇, 109
$\Downarrow_{\mathbf{N}}$, 77
ϕ, 120
$\preceq_{\mathbf{V}}$, 90
$\preceq_{\mathbf{N}}$, 78
$\preceq_{\mathbf{H}}$, 73
$\preceq_{\mathbf{L}}$, 82
$\preceq_{\mathbf{O}}$, 68
ψ, 145
$\rightarrow_{\Omega}$, 123
$\overset{*}{\rightarrow}_{\mathbf{O}\eta}$, 97
$\overset{*}{\rightarrow}_{\Delta\mathbf{O}\eta}$, 97
$\overset{*}{\rightarrow}_{\Delta}$, 7
$\rightarrow_{\Gamma\Omega}$, 184
$\rightarrow_{\surd\Omega}$, 184
$\rightarrow_{\Lambda\angle\Omega}$, 165
$\rightarrow_{\Lambda\Omega}$, 123
$\rightarrow_{\angle\Omega}$, 165
$\rightarrow_{\mathbf{O}\eta}$, 97
$\rightarrow_{\Psi\ell}$, 36
$\rightarrow_{\Xi\ell\Omega}$, 185
$\trianglelefteq$, 202
$\trianglelefteq_{\sigma}$, 202
$\Downarrow_{\mathbf{V}}$, 89
$\simeq_c$, 29
$\smile$, 96
$\smile_{\mathbf{H}}$, 97
$\smile_{\mathbf{L}}$, 98
$\smile_{\mathbf{N}}$, 97
$\smile_{\mathbf{V}}$, 98
$\sqcap$, 207
$\sqsubseteq_k$, 128
$\sqsubseteq_{\mathcal{F}}$, 116
$\sqsubseteq_{\mathcal{H}}$, 125
$\sqsubseteq_{\mathcal{LL}}$, 176
$\sqsubseteq_{\mathcal{N}}$, 146
$\sqsubseteq_{\mathcal{VV}}$, 206
$\surd$, 182
$\uparrow$, 114
$\vdash_{\angle}$, 164
$\vdash_{\infty}$, 120
$\vdash_{\surd}$, 183
$\vdash_{\bowtie}$, 146
$\vec{M}$, 3
$\mathcal{A}$, 124
$\mathcal{A}^{\checkmark}$, 185
$\mathcal{A}^{\angle}$, 165
$\mathcal{E}$, 218
$\mathcal{F}^0(\surd)$, 204
$\mathcal{H}$, 121
$\mathcal{I}$, 215
$\mathcal{J}$, 217
$\mathcal{LL}$, 176
$\mathcal{L}$, 163
$\mathcal{N}$, 145
$\mathcal{U}$, 219
$\mathcal{U}^{\checkmark}$, 185
$\mathcal{VV}$, 205
$\mathcal{V}$, 183
$\mathbb{L}$, 207
$\mathbb{L}_{\infty}$, 210

abstraction, 3
active, 9
– $\Xi\ell$, 42
algorithm
– Γ-separability, 52
– Λ-separability, 32
– $\mathcal{N}$-Semiseparability, 160
– semiseparability, 138
application, 3
approximant, 124
– $\sqrt{\Omega}$, 185
– $\angle\Omega$, 165
– defined, 159
– maximal along a path, 139
– upper, 185
arguments, 8

basis, 109
– agree, 114
– finite, 133, 154, 168, 195
binder, 4
block
– head, 8
body, 8

calculus
– $\lambda\Lambda$, 25
– $\lambda\Gamma$, 35
– $\lambda\Delta$, 6
– complete, 69
– correct, 69
call-by-name, 8, 25
call-by-value, 8, 35
closure
– Var, 7
– reduction, 7
– substitution, 7
compact, 207
compact-as-types procedure, 212
complete development, 15
confluence, 8
– $\Xi\ell$, 41
congruence, 21
context, 6
– **H**-relevant, 75
– **L**-relevant, 83
– **N**-relevant, 79
– **V**-relevant, 91
– Δ-valuable, 12
– discriminating, 69
– head, 12
– relevant, 69
continuous, 207

contractum
– $\Psi\ell$, 36
– Δ, 6
correctness
– **H**, 75
– **L**, 83
– **N**, 79
– **V**, 90

D, 4
defined, 159
degree
– Γ, 39
– Λ, 26
derivation, 67, 110
– size, 67
discriminability
– $\mathcal{H}$, 127
– $\mathcal{N}$, 150
– head, 99
domain, 207
– constructor, 209

E, 4
embedding, 209
environment, 105
equivalence
– operational, 68
evaluation relation, 66
– deterministic, 68
– nondeterministic, 68
– uniform, 99
– universal, 71

filter, 113
fixed point
– Γ, 40
– Λ, 27
– operator, 27, 41
function
– Z, 226
– **O**-representable, 233
– composition, 226
– continuous, 207
– embedding, 209
– minimalization, 226
– monotone, 207
– partial recursive, 226
– primitive recursive, 226
– projection, 209, 226
– step, 208
– strict, 207
– successor, 226

greatest lower bound, 207

head
– block, 8
– context, 6
– redex, 8
– variable, 8
head normal form
– Γ, 39
– Λ, 25, 66
– Λ-lazy, 25, 66
– $\Psi\ell$, 36
hole, 6

I, 4
incompleteness
– **H**, 76
– **L**, 84
– **N**, 80
– **V**, 92
inhabited, 173
input value, 7
– approximated, 184
– standard set, 10
interpretation, 105, 114
intersection
– relation, 109
– – $\bowtie$, 145
– – $\angle$, 163
– – ∞, 120
– – $\surd$, 182
– type assignment system, 108
– types, 109
inverse limit, 210
isomorphic, 208

K, 4

language, 3
– $\Lambda\Omega$, 123
lattice
– ω-algebraic, 207
– complete, 207
lazy
– Ψ-contractum, 36
– Ψ-redex, 36
– Ψ-reduction, 36
lazy blocked nf
– Γ, 66
least upper bound, 207
legality, 112
– $\bowtie$, 145
– $\angle$, 163
– ∞, 121
– $\surd$, 183
lifting, 209

machine
– reduction, 68
minimalization, 226
model
– $\lambda\Delta$, 105
– $\mathcal{H}$, 121
– $\mathcal{LL}$, 176
– $\mathcal{L}$, 163
– $\mathcal{N}$, 145
– $\mathcal{VV}$, 205
– $\mathcal{V}$, 183
– complete, 106
– correct, 106
– filter, 116
– – complete, 117
– – correct, 117
– – fully abstract, 117
– fully abstract, 107
monotone, 207

normal form, 7
– Γ, 35, 66
– $\Gamma\Omega$, 184
– Λ, 25, 66
– $\Lambda\Omega$, 124
– $\angle\Omega$, 165
– $\Psi\ell$, 36
– Δ, 7
– $\Xi\ell\Omega$, 185
– persistent, 144
normalizing
– $\Psi\ell$, 36
– Δ, 7
numeral system, 228

O, 4
occur, 4
operational semantics
– **L**, 82
– **N**, 77
– **V**, 89
– **H**, 73
order
– Γ, 39
– Λ, 26
output value, 65

pair, 227
partial recursive, 226
Peano's natural numbers, 228

pointwise, 208
primitive recursive, 226
principality condition, 65
principle
– operational extensionality, 96
– operational functionality, 96
projection, 209

recursion
– call-by-value, 41
redex
– Γ, 35
– $\Psi\ell$, 36
– Δ, 6
– $\Xi\ell$-degree, 42
– $\Xi\ell$-principal, 42
– degree, 9
– head, 8
– principal, 9
reduction, 6
– Γ_Ω, 184
– $\sqrt{\Omega}$, 184
– $\Lambda\angle\Omega$, 165
– $\Lambda\Omega$, 123
– $\angle\Omega$, 165
– $\mathbf{O}\eta$, 97
– Ω, 123
– Ψ-lazy, 36
– $\hookrightarrow_\Delta$, 14
– Δ, 6
– $\Rightarrow_\Delta$, 14
– $\Rightarrow^{\circ}_\Delta$, 16
– $\Rightarrow^{i}_\Delta$, 19
– $\rightarrow^{i}_\Delta$, 19
– $\rightarrow^{p}_\Delta$, 9
– $\Xi\ell\Omega$, 185
– Ξ-lazy, 36
– α, 5
– η, 23
– parallel deterministic, 14
– parallel nondeterministic, 14
– principal, 9
– standard, 9
replacement, 4
– simultaneous, 5
retraction, 209
retraction sequence, 210

semantics
– denotational, 106
– operational, 68
semiseparability
– algorithm
– – $\mathcal{H}$, 137
– – $\mathcal{N}$, 159
separability, 22
– Γ, 40
– Λ, 26
separable, 22
sequentialization, 9
set of input types, 109
size, 67
solvability
– Γ, 39
– Λ, 26
SOS, 65, 101
standardization, 9
– $\Xi\ell$, 43
– theorem, 10
strict, 207
strongly normalizing
– $\Psi\ell$, 36
– Δ, 7
subdomain, 209
substitution, 5
subterm, 4

term
– Γ-solvable, 38
– Γ-valuable, 35
– $\mathbf{O}$-comparable, 96
– Δ-solvable, 12
– Δ-unsolvable, 12
– $\mathcal{H}$-computable, 133
– $\mathcal{L}$-computable, 168
– $\mathcal{N}$-computable, 154
– $\mathcal{V}$-computable, 195
– closed, 4
– occurrence, 4
– open, 4
– openly Λ-solvable, 120
– potentially Γ-valuable, 35
term model, 107
theory
– $\Delta\mathbf{O}\eta$, 97
– Δ, 21
– consistent, 22
– full extensional, 23
– inconsistent, 22
– input consistent, 22
– input inconsistent, 22
– maximal, 22
– semisensible, 22
– sensible, 22
type
– assignment system, 109
– constant, 109

– input, 109
– system, 109
–– $\bowtie$, 145
–– $\angle$, 163
–– ∞, 120
–– $\surd$, 182
–– legal, 112
– theory, 109

typing, 110

variable
– bound, 4
– free, 4
– head, 8

weight, 43

Monographs in Theoretical Computer Science · An EATCS Series

K. Jensen
Coloured Petri Nets
Basic Concepts, Analysis Methods and Practical Use, Vol. 1
2nd ed.

K. Jensen
Coloured Petri Nets
Basic Concepts, *Analysis Methods* and Practical Use, Vol. 2

K. Jensen
Coloured Petri Nets
Basic Concepts, Analysis Methods and *Practical Use,* Vol. 3

A. Nait Abdallah
The Logic of Partial Information

Z. Fülöp, H. Vogler
Syntax-Directed Semantics
Formal Models Based on Tree Transducers

A. de Luca, S. Varricchio
Finiteness and Regularity in Semigroups and Formal Languages

E. Best, R. Devillers, M. Koutny
Petri Net Algebra

S.P. Demri, E. S. Orłowska
Incomplete Information: Structure, Inference, Complexity

J.C.M. Baeten, C.A. Middelburg
Process Algebra with Timing

L. A. Hemaspaandra, L.Torenvliet
Theory of Semi-Feasible Algorithms

E. Fink, D. Wood
Restricted-Orientation Convexity

Zhou Chaochen, M. R. Hansen
Duration Calculus
A Formal Approach to Real-Time Systems

M. Große-Rhode
Semantic Integration of Heterogeneous Software Specifications

Texts in Theoretical Computer Science · An EATCS Series

J. L. Balcázar, J. Díaz, J. Gabarró
Structural Complexity I

M. Garzon
Models of Massive Parallelism
Analysis of Cellular Automata
and Neural Networks

J. Hromkovič
Communication Complexity
and Parallel Computing

A. Leitsch
The Resolution Calculus

G. Păun, G. Rozenberg, A. Salomaa
DNA Computing
New Computing Paradigms

A. Salomaa
Public-Key Cryptography
2nd ed.

K. Sikkel
Parsing Schemata
A Framework for Specification
and Analysis of Parsing Algorithms

H. Vollmer
Introduction to Circuit Complexity
A Uniform Approach

W. Fokkink
Introduction to Process Algebra

K. Weihrauch
Computable Analysis
An Introduction

J. Hromkovič
Algorithmics for Hard Problems
Introduction to Combinatorial
Optimization, Randomization,
Approximation, and Heuristics
2nd ed.

S. Jukna
Extremal Combinatorics
With Applications
in Computer Science

P. Clote, E. Kranakis
Boolean Functions
and Computation Models

L. A. Hemaspaandra, M. Ogihara
The Complexity Theory Companion

C.S. Calude
Information and Randomness.
An Algorithmic Perspective
2nd ed.

J. Hromkovič
Theoretical Computer Science
Introduction to Automata,
Computability, Complexity,
Algorithmics, Randomization,
Communication and Cryptography

A. Schneider
Verification of Reactive Systems
Formal Methods and Algorithms

S. Ronchi Della Rocca, L. Paolini
The Parametric Lambda Calculus
A Metamodel for Computation

Y. Bertot, P. Castéran
Interactive Theorem Proving
and Program Development
Coq'Art: The Calculus
of Inductive Constructions

L. Libkin
Elements of Finite Model Theory